Endocrine and Biochemical Development of the Fetus and Neonate

REPRODUCTIVE BIOLOGY

Series Editor: Sheldon J. Segal

The Rockefeller Foundation
New York, New York

THE ANTIPROGESTIN STEROID RU 486 AND HUMAN FERTILITY CONTROL
Edited by Etienne-Emile Baulieu and Sheldon J. Segal

AUTOCRINE AND PARACRINE MECHANISMS IN REPRODUCTIVE ENDOCRINOLOGY
Edited by Lewis C. Krey, Bela J. Gulyas, and John A. McCracken

CONTRACEPTIVE STEROIDS: Pharmacology and Safety
Edited by A. T. Gregoire and Richard P. Blye

DEMOGRAPHIC AND PROGRAMMATIC CONSEQUENCES OF CONTRACEPTIVE INNOVATIONS
Edited by Sheldon J. Segal, Amy O. Tsui, and Susan M. Rogers

ENDOCRINE AND BIOCHEMICAL DEVELOPMENT OF THE FETUS AND NEONATE
Edited by José M. Cuezva, Ana M. Pascual-Leone, and Mulchand S. Patel

GENETIC MARKERS OF SEX DIFFERENTIATION
Edited by Florence P. Haseltine, Michael E. McClure, and Ellen H. Goldberg

GOSSYPOL: A Potential Contraceptive for Men
Edited by Sheldon J. Segal

IMMUNOLOGICAL APPROACHES TO CONTRACEPTION AND PROMOTION OF FERTILITY
Edited by G. P. Talwar

A Continuation Order Plan is available for this series. A continuation order will bring delivery of each new volume immediately upon publication. Volumes are billed only upon actual shipment. For further information please contact the publisher.

Endocrine and Biochemical Development of the Fetus and Neonate

Edited by

José M. Cuezva

Universidad Autonoma de Madrid
Madrid, Spain

Ana M. Pascual-Leone

C.S.I.C. Universidad Complutense de Madrid
Madrid, Spain

and

Mulchand S. Patel

Case Western Reserve University
Cleveland, Ohio

SPRINGER SCIENCE+BUSINESS MEDIA, LLC

Library of Congress Cataloging-in-Publication Data

Spanish Biochemical Society. Perinatal Biochemical Group. Meeting
(1988 : Madrid, Spain)
  Endocrine and biochemical development of the fetus and neonate /
edited by José M. Cuezva, Ana M. Pascual-Leone, and Mulchand S.
Patel.
      p.   cm. -- (Reproductive biology)
  "Proceedings of the Annual Meeting of the Perinatal Biochemical
Group of the Spanish Biochemical Society, on biochemical development
of the fetus and neonate, held December 15-16, 1988, in Madrid,
Spain"--T.p. verso.
  Includes bibliographical references.
  Includes index.
  ISBN 978-1-4615-9569-4   ISBN 978-1-4615-9567-0 (eBook)
  DOI 10.1007/ 978-1-4615-9567-0

  1. Fetus--Physiology--Congresses. 2. Fetus--Metabolism-
-Congresses. 3. Fetus--Growth--Congresses. 4. Obstetrical
endoctrinology--Congresses.   I. Cuezva, José M. II. Pascual-Leone,
Ana M. III. Patel, Mulchand S. IV. Title. V. Series.
  [DNLM: 1. Endocrine Glands--physiology--congresses. 2. Fetal
Development--congresses. 3. Fetus--physiology--congresses.
4. Infant, Newborn--congresses.   WQ 210.5 S735e 1988]
RG616.S63 1988
612.6'47--dc20
DNLM/DLC
for Library of Congress   90-7827
                                                        CIP

Proceedings of the annual meeting of the Perinatal Biochemical Group
of the Spanish Biochemical Society, on Biochemical Development of
the Fetus and Neonate, held December 15–16, 1988, in Madrid, Spain

© 1990 Springer Science+Business Media New York
Originally published by Plenum Press, New York in 1990
Softcover reprint of the hardcover 1st edition

PREFACE

    The annual meeting of the Perinatal Biochemical Group of the Spanish
Biochemical Society held in Madrid on December 15-16, 1989, provided an
excellent opportunity to bring together a group of distinguished
investigators both from Spain and from abroad with a common interest in
developmental endocrinology and biochemistry. The aim of the symposium was
to present and discuss the most recent developments in the areas of
endocrine and biochemical processes critical to normal growth and
maturation of the newborn. To achieve a high degree of interaction among
the participants, subject reviews as well as short communications were
include in the program. The reviews provided in-depth information on
selected important topics. The purpose of short communications was two
fold: (i) to provide a forum to discuss on-going investigations on related
areas; and (ii) to present opportunities for active participation by young
investigators. This format proved very successful in generating fruitful
discussions among the participants. Taken together the review chapters and
the short communications have resulted in a coherent and unified subject
presentation.

    Advances in biochemistry, molecular biology and cell biology have
provided not only new and exciting experimental approaches but also have
opened up new directions in the investigation of differentiation and
developmental processes at cellular, molecular and biochemical levels
during the early stages of growth and maturation. In recent years a wealth
of information in these areas of development has emerged at a rapid rate.
To review recent developments in the study of endocrine systems and
metabolic processes during the perinatal period, three comprehensive themes
were developed. They were: (i) endocrine and functional aspects of
development; (ii) molecular and functional aspects of mitochondrial
maturation; and (iii) metabolic aspects of development. These themes are
interrelated and therefore distribution of papers in the assigned
categories merely represented convenience for presentation. The conference
theme gradually developed from cellular differentiation to whole body
metabolism in the newborn. The endocrine section addressed cellular
differentiation as well as development of hormonal transduction systems.
Mitochondrial maturation processes as influenced by hormones were then
considered, employing molecular biology techniques. The development and
regulation of major metabolic pathways in the fetus and newborn and their
relationship to maternal metabolism was brought into focus by several
investigators.

    We would like to thank the participants for presenting stimulating
talks, for penetrating scientific discussions, and for prompt submission
of manuscripts. We thank Professor F. Mayor for his invaluable help and
suggestions during the early stages of development of this symposium. The
organizers of the symposium and the editors of the present volume want to
specially acknowledge the Fundación Ramón Areces (Spain) for generously

supporting both the symposium and the edition of this volume. Financial contributions from Universidad Autónoma de Madrid, Consejo Superior de Investigaciones Científicas and Dirección General de Investigación Científica y Técnica for the organization of the symposium are also gratefully acknowledged. This support exemplified the longstanding interest of these organizations in research activities in Spain.

We hope that the contributed ideas and stimulating propositions will lead to further investigations by fellow researchers and a future development of the field. If so, this will be the most appreciated reward for our editorial efforts.

José M. Cuezva (Madrid, Spain)
Ana M. Pascual-Leone (Madrid, Spain)
Mulchand S. Patel (Cleveland, USA)

CONTENTS

ENDOCRINE AND FUNCTIONAL ASPECTS OF DEVELOPMENT

MOLECULAR AND FUNCTIONAL ASPECTS OF MITOCHONDRIAL DEVELOPMENT

METABOLIC ASPECTS OF DEVELOPMENT

ENDOCRINE AND FUNCTIONAL ASPECTS OF DEVELOPMENT

ENDOCRINE CYTODIFFERENTIATION OF

THE RAT FETAL TESTIS

Alfred Jost and Solange Magre

Collège de France
75231 Paris Cedex 05, France

INTRODUCTION

Growth of the fetus involves not only increase in the number of cells or in weight of the body, it is actually accompanied with functional differentiation of new cell types at definite developmental periods. This is especially obvious when endocrine glands are considered. These glands undergo first a local embryogenesis, before the day they enter physiologically into play and release appreciable amounts of hormones. Their cells acquire then characteristics of endocrine activity. The mechanisms responsible for the initial differentiation and sudden functional activation are still unknown.

For a few years our current work has been aimed at studying the factors involved in the differentiation of the endocrine cells of the rat fetal testis, namely the Sertoli cells which produce the Müllerian inhibitor (or anti-Müllerian Hormone, AMH, or Müllerian Inhibiting Substance, MIS) and the Leydig cells secreting androgens. We first observed that in *in vitro* cultures of testicular primordia we could experimentally dissociate testicular morphogenesis from endocrine cytodifferentiation of the Sertoli and Leydig cells. This result was obtained in experiments showing that if serum was added to the synthetic culture medium in which undifferentiated gonadal primordia were grown, it prevented the morphogenesis of the seminiferous cords (1,2) but not the endocrine differentiation of the testicular cells (3,4).

This finding opened the way to the current experiments. It will be briefly summarized in the first part of this presentation, the second part reporting the more recent observations.

TECHNIQUES

It is unnecessary to review in this paper all the techniques which were previously described. Suffice it to mention a few points. The age of the fetus was reckoned from the assumed hour of fertilization, 1-2 o'clock the night of cohabitation. Explantation of the gonads on their mesonephroi, cultures *in vitro* and sexing of the young by the sex chromatin test in amniotic cells were performed as reported (2,5). For histochemical detection of 3β-Hydroxysteroid-Dehydrogenase-Isomerase activity (3β-HSD)

Endocrine and Biochemical Development of the Fetus and Neonate
Edited by J. M. Cuezva *et al.*
Plenum Press, New York, 1990

and testosterone determinations, see 4 and 5, and for immunohistochemical study of laminin and fibronectin expression, see 6 and 7.

DISSOCIATION BETWEEN TESTICULAR ORGANOGENESIS AND ENDOCRINE CYTODIFFERENTIATION

In control rat fetuses, the initial Sertoli cells appear and progressively aggregate into seminiferous cords, which encompass germ cells, between 13 and 14 days after fertilization of the ova (8,9). The Leydig cells appear between the seminiferous cords somewhat later, on day 15 (9,10,11). Differentiation of the seminiferous cords, followed by that of the Leydig cells, can be obtained *in vitro*, in explants of the gonadal primordia inserted on their mesonephroi, taken before or at the very onset of Sertoli cell emergence and cultured in a synthetic medium (CMRL 1066) for 3 or 4 days (2). Even if the process is somewhat delayed *in vitro* as compared with the *in vivo* chronology, seminiferous cords formation and Leydig cell differentiation occur at fixed stages (4); the whole process is over in explants taken at 13 days following fertilization and cultured for 4 days.

Serum added to the medium (15% or less fetal calf or human serum) (12) prevented the differentiation of the seminiferous cords, though large cells resembling Sertoli cells scattered throughout the imperfect gonad had appeared (2). Endocrine studies were conducted as follows. The production of the Müllerian inhibitor was demonstrated using the *in vitro* test developed by Picon (13) (coculture of sex ducts from a 14.5 day old rat fetus with the gonad to be investigated). The cordless gonads produced the Müllerian duct inhibitor (3), this suggested that the large cells seen in histological sections were Sertoli cells, which for one reason or another did not aggregate.

Leydig cell differentiation and function were explored using both a histochemical technique for 3β-Hydroxysteroid Dehydrogenase activity (using 3β-etiocholanolone = 5β andro-stan-3βol-17 one as substrate) and testosterone determinations in the culture medium (4). Histochemically detected Leydig cells and testosterone release into the medium began simultaneously in control gonads (cultured in the synthetic medium) displaying seminiferous cords, and in cordless gonads cultured in the serum added medium. Although the testosterone release by the cordless gonads decreased rapidly after its initial spurt, it was clear that the onset of differentiation and functioning of the Leydig cells took place in these gonads in a way similar to that of controls.

The functional cytodifferentiation of the two types of testicular endocrine cells did not require the most characteristic aspects of testicular organogenesis, the morphogenesis of seminiferous cords (3,4). As to the mode of action of serum in these experiments, several hypotheses were considered. Serum could interfere with normal differentiation of the extracellular matrix or perhaps of the cytoskeleton and thus indirectly affect the morphogenetic process. The following experiments were undertaken with these possibilities in mind, and the extracellular matrix was studied first.

EXPERIMENTAL CONTROL OF THE DIFFERENTIATION OF THE LEYDIG CELLS

Several agents known to interfere with either collagen synthesis or secretion (αα dipyridyl, 10 μg/ml; or tunicamycin, 150 ng/ml) or with cross linking (β amino-propionitrile, 50 to 400 μg/ml) were added to the medium in which male gonadal primordia were cultured. Only the latter disturbed

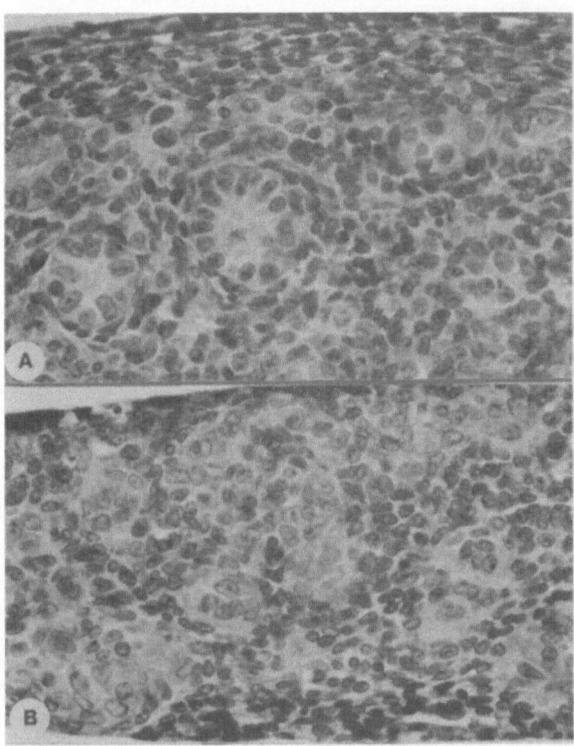

Fig. 1. Sections through gonads taken from 13 day 9h
fetuses and cultured for 4 days either: (A)
in a control medium,or (B) in a LACA added
medium, (100 µg/ml). In the control medium
well defined seminiferous cords have formed.
Under the influence of LACA no cord developed
but many large clear cells appeared. (x 460).

testicular morphogenesis but this was an inconstant effect under the
conditions which were used. On the contrary the proline competitor
L-azetidine-2 carboxylic acid (LACA) known for interfering with collagen
synthesis, constantly prevented the formation of seminiferous cords (6),
on condition that the concentration was at least 100 µg/ml (lower
concentrations, 12.5 - 25 - 50 µg/ml, had no or inconstant effects).
Therefore LACA at a concentration of 100 µg/ml was used in the series of
experiments to be reported below.

1) Effects of LACA on Testicular Morphogenesis

As mentioned before, in gonads from male fetuses explanted at the
stage of 13 day 9h., i.e. just before or at incipient appearance of initial
Sertoli cells in the developing testis, well defined seminiferous cords
differentiate during the following three days *in vitro* in the synthetic
medium (CMRL 1066). After four days these cords appear as shown in Figure
1A. The roundish section of the cord is surrounded by a definite basal
membrane. The peripheral cells are polarized against that membrane, with
their nuclei placed superficially and most of their abundant cytoplasm
lying centrally. Figure 1B shows a comparable level of a gonad from a male
fetus, after 4 days of culture in a medium containing LACA. The difference
is obvious. Irregular clusters of cytoplasm rich cells are seen between
undifferentiated mesenchymal cells : no typical epithelio-mesenchymal
contact has differentiated. This was studied in more detail using immuno-

-histochemical techniques for laminin and fibronectin expression (6). In control cultures the two components were expressed as a definite edging around the seminiferous cords and also in the tissue lying between the cords (Figure 2). On the contrary in the LACA treated gonads both components failed to being expressed, although they were seen in the neighbouring mesonephric structures (Figure 2). The gonads respond to LACA in their own way.

2) Production of the Müllerian Inhibitor

To verify whether the cordless male gonads developed under the influence of LACA had the capacity to produce the Müllerian inhibitor, the technique was the same as that previously used for cordless gonads resulting from the action of serum (3). The gonads were first cultured for two days in the LACA containing medium, and then cocultured, in the same medium, in contact with sex ducts from 14.5 day old female rat fetuses for three more days. Control assays showed that LACA by itself does not inhibit the Müllerian ducts. In the cocultures inhibition of the Müllerian ducts was obvious (5), which shows that the gonadal disorders resulting from the

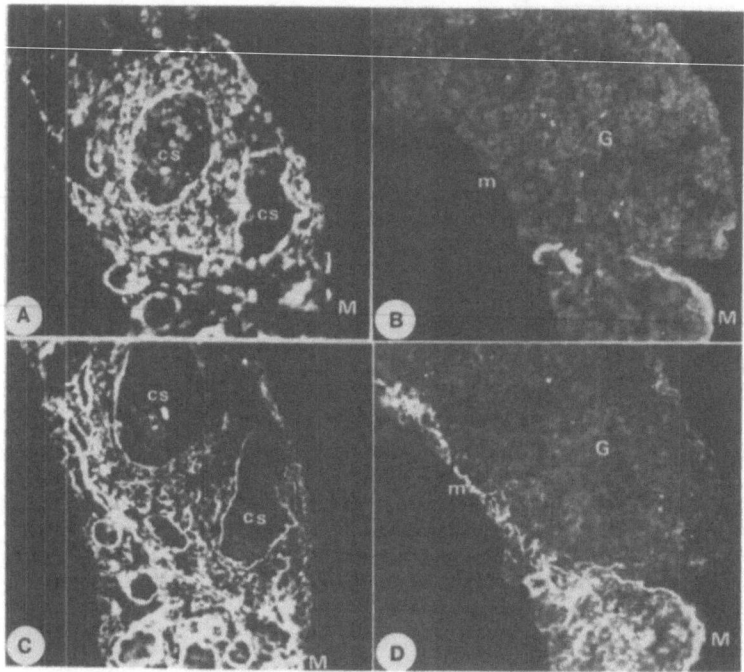

Fig. 2. Sections of male gonads cultured *in vitro* in the same way as those in Figure 1, and submitted to immuno-histochemical techniques for the expression of laminin in A and B, or of fibronectin in C and D. In the control medium (A and C) the gonads contain seminiferous cords (CS) surrounded by laminin (A) and fibronectin (C) deposits. In the LACA containing medium (B and D), the two components are not expressed in the gonads (G) although they are seen in the mesonephros (M). (A and C) and (B and D), respectively, are sections from the same gonads (from ref. 6). (x 130).

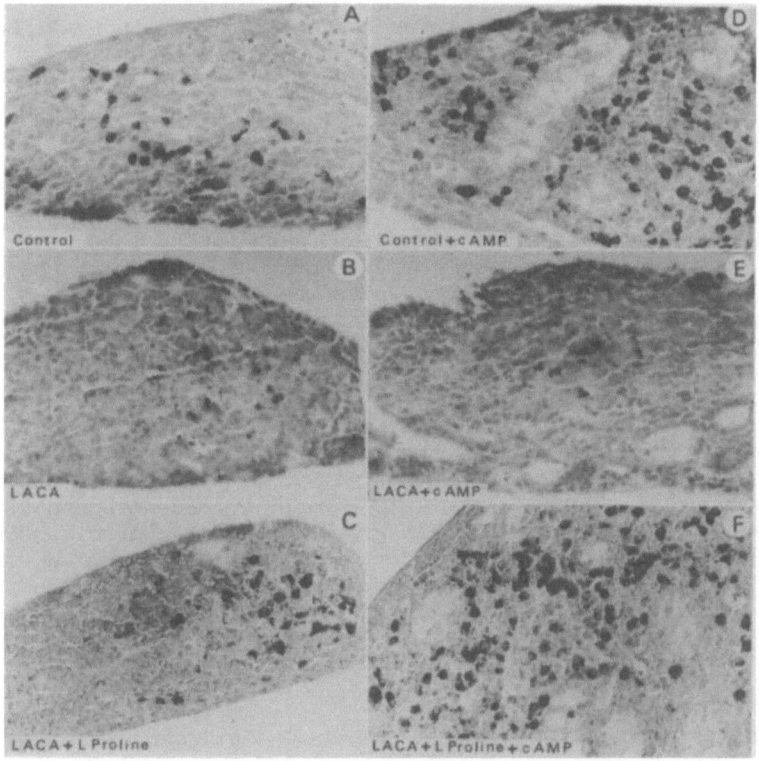

Fig. 3. Sections through male gonads explanted from 13 day
9 hours old fetuses and cultured for 4 days *in
vitro*. Histochemical reaction for 3β-HSD activity.
(A) Control medium; (B) medium containing LACA (100
μg/ml); (C) medium containing LACA (100 μg/ml) plus
L-Proline (250 μg/ml). D-F, the same media,
respectively, as A-C but supplemented with cAMP
dibutyrate (1 mM) during the final 24 hours (from
ref. 5).

action of LACA did not suppress the production of the Müllerian inhibitor.
The large clear cells seen in Figure 1B can be considered Sertoli cells.

3) Function of the Leydig Cells

The physiological activity of the Leydig cells in gonads developed
under the influence of LACA was investigated with the two tests mentioned
above : 1) histochemical detection of 3β-HSD; 2) release of testosterone
into the medium during the last 24h *in vitro*, cAMP not being or being added
to the medium for the last 24h. Since LACA produced striking effects and
since LACA is a competitor of proline, the inquiry was complemented with
assays where L-proline was added in excess in an attempt to counteract the
LACA competing effect. The results were very clear.

LACA almost completely prevented 3β-HSD activity in male gonads
(Figure 3) and testosterone release (Figure 4). It also impaired cAMP
effect on both tests.

L-Proline added in excess (250 μg/ml) suppressed the effects of LACA
(100 μg/ml) and permitted both normal or cAMP stimulated activities to be

observed. Thus proline seems to be an essential factor in fetal Leydig cell differentiation and function.

4) Compared 3β-HSD Activity in Testicular Leydig Cells and Adrenocortical Cells

The gonads differentiate in close vicinity of the adrenocortical glands. Therefore when the pieces of tissue containing the gonads on the mesonephroi are dissected out, care being taken not to injure the gonads, it happens that adrenal cortices are explanted together with the gonado-mesonephric primordia. In some cases this was done on purpose. By collecting explants from female fetuses it could be verified that the presence of adreno-cortical tissue did not result in testosterone production.

After 4 days in the control medium, 3β-HSD activity was expressed in the adrenals even more strongly than in the testes. Very interestingly, in explants submitted to LACA, 3β-HSD activity had developed in the adrenals while it was almost absent in the male gonads. We did not quantitatively compare the activity in the adrenals grown either in control or in LACA added medium, but the difference, if any, was not striking. We verified that no 3β-HSD activity could be detected with our technique neither in the adrenal cortex nor in the testis on day 13, when the tissues were

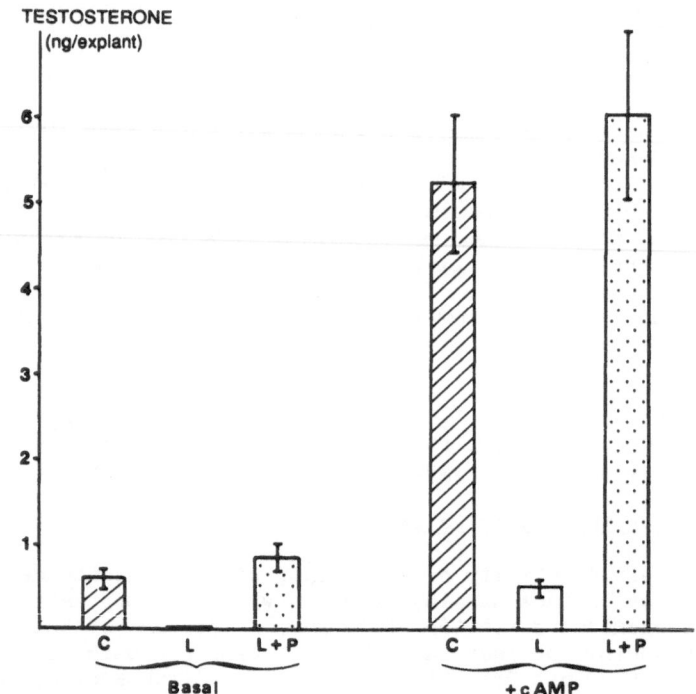

Fig. 4. Explants from male fetuses, containing both gonads, and cultured for 4 days (the medium was renewed after 3 days): Testosterone contents of the culture medium used for the last 24h. C = control medium; L = LACA containing medium; L + P = medium containing LACA + proline. On the left side (Basal): no other addition to the medium. On the right side (+ cAMP): cAMP dibutyrate (1 mM) was added to the medium during the final 24 h (mean ± SEM).

explanted. Moreover, it was observed that the enzyme activity becomes evident on day 14 in the adrenal whereas it is seen only on day 15 in the testes.

The difference in the timing of 3β-HSD activity appearance in the testis and in the adrenals, and the extremely pronounced dissimilarity in the response to LACA, suggest either that the control mechanisms of the same enzyme(s) are different in the two endocrine glands, or that the enzymatic systems are not identical in two glands. Existence of two 3β-HSD isoenzymes has been suggested in boys suffering from a 3β-HSD defect affecting adrenal but not testicular steroidogenesis (14, see also 15).

It is clear that 3β-HSD activity in the early fetal testis is dependant on proline, whatever the exact testicular process requiring proline; the adrenals seem not to depend on proline to the same extent.

CONCLUSION

Our experiments concerning Leydig cells open new ways since it has become possible to control the initial differentiation of one type of endocrine cell, the Leydig cell, by controlling proline availability. The key role played by proline still is unexplained. So far it is still impossible to determine which testicular cells and which synthetic process is so severely affected by the proline competitor as to stop the endocrine differentiation of the Leydig cells precursors. We previously considered the question that intratesticular paracrine relations between Sertoli and Leydig cells could be involved (5), but no definite evidence is available. One point of great interest in the present results concerns the difference in expression on 3β-HSD activity in the testes and adrenals. It deserves further studies.

REFERENCES

1. S. Magre, R. Agelopoulou and A. Jost, Action du serum de foetus de veau sur la différenciation in vitro ou le maintien des cordons séminifères du testicule du foetus de rat, C.R.Acad.Sc.Paris, Serie III, 292:85 (1981).
2. R. Agelopoulou, S. Magre, E. Patsavoudi and A. Jost, Initial phases of the rat testis differentiation in vitro, J.Embryol.Exp.Morph., 83:15 (1984).
3. S. Magre and A. Jost, Dissociation between testicular organogenesis and endocrine cytodifferentiation of Sertoli cells, Proc.Natl.Acad.Sci.USA, 81:7831 (1984).
4. E. Patsavoudi, S. Magre, M. Castanier, R. Scholler and A. Jost, Dissociation between testicular morphogenesis and functional differentiation of Leydig cells, J.Endocr., 105:235 (1985).
5. A. Jost, S. Perlman, O. Valentino, M. Castanier, R. Scholler and S. Magre, Experimental control of the differentiation of Leydig cells in the rat fetal testis, Proc.Natl.Acad.Sci.USA, 85:8094 (1988).
6. A. Jost, O. Valentino, R. Agelopoulou and S. Magre, Action d'un analogue de la proline (acide L-azétidine-2- carboxylique) sur la différenciation in vitro du testicule foetal de rat, C.R.Acad.Sc.Paris, Serie III, 301:225 (1985).
7. R. Agelopoulou and S. Magre, Expression of fibronectin and laminin in fetal male gonads in vivo and in vitro with and without testicular morphogenesis, Cell Differentiation, 21:31 (1987).
8. A. Jost, Données préliminaires sur les stades initiaux de la différenciation du testicule chez le rat, Arch.Anat.Microsc.Morphol.Expér., 61:415 (1972).

9. S. Magre and A. Jost, The initial phases of testicular organogenesis in the rat. An electron microscopy study, <u>Arch. Anat. Microsc. Morphol. Expr.</u>, 69:297 (1980).

10. E. C. Roosen-Runge and D. Anderson, The development of the interstitial cells in the testis of the albino rat, <u>Acta Anat.</u>, 37:125 (1959).

11. R. Narbaitz and R. Adler, Submicroscopical aspects in the differentiation of rat fetal Leydig cells, <u>Acta Physiol. Latino Americana</u>, 17:286 (1967).

12. I. Chartrain, S. Magre, M. Maingourd and A. Jost, Effect of serum on organogenesis of the rat testis *in vitro*, <u>In Vitro</u>, 20:912 (1984).

13. R. Picon, Action du testicule foetal sur le développement *in vitro* des canaux de Müller chez le Rat, <u>Arch. Anat. Microsc. Morphol. Expr.</u>, 58:1 (1969).

14. M. D. C. Cravioto, A. Ulloa-Aguirre, J. A. Bermudez, J. Herrera, R. Lisker, J. P. Mendez and G. Perez-Palacios, A new inherited variant of the 3β-Hydroxysteroid Dehydrogenase-Isomerase deficiency syndrome: evidence for the existence of two isoenzymes, <u>J. Clin. Endocrinol. Metabol.</u>, 62:360 (1986).

15. L. M. Miller, Molecular biology of steroid hormone synthesis, <u>Endocrine Reviews</u>, 9:295 (1988).

TRANSFER OF THYROID HORMONES

FROM THE MOTHER TO THE FETUS

Gabriella Morreale de Escobar,
Maria J. Obregón and Francisco Escobar del Rey

Instituto de Investigaciones Biomédicas, C.S.I.C.
Facultad de Medicina
28029 Madrid, Spain

INTRODUCTION

Impairment of thyroid function early in life results in developmental defects. The prevention of the CNS damage caused by congenital hypothyroidism is the aim which is being successfully achieved by neonatal thyroid screening programs. But CNS damage may also result from maternal hypothyroxinemia alone. Especially severe CNS damage is found in the endemic cretin of the neurological type born to iodine deficient mothers: these women are hypothyroxinemic throughout pregnancy, and their fetus has an impaired thyroid function caused by the low iodine supply. The CNS damage increases in severity from that caused by maternal hypothyroxinemia alone, congenital hypothyroidism alone, or the combined maternal and fetal impairment of thyroid function found in areas of marked iodine deficiency. But whereas the mental retardation caused by congenital hypothyroidism may still be prevented by early treatment of the newborn, the CNS damage caused by maternal hypothyroxinemia is prevented by adequate treatment during pregnancy, and the birth of neurological cretins requires adequate iodization of the mothers before conception, or very early in pregnancy. These associations between thyroid failure and CNS damage, and the experimental models used to study them, have been previously reviewed by us (1-4). These reviews contain pertinent bibliography of the abundant work on the subject carried out by many investigators.

The above considerations have centered our attention on the problem of maternal-fetal thyroid hormone relationships.

MATERNAL TO FETAL TRANSFER OF THYROID HORMONES BEFORE ONSET OF FETAL THYROID FUNCTION

Taking the day of appearance of the vaginal plug as gestational day (dg) =0, uterine implantation occurs at 6 dg, and birth occurs 22.7±0.2 (SEM) dg. By injecting radioiodide into pregnant rats, which were thyroidectomized (T) a few hours earlier, Nataf and Sfez (5), and Geloso (6) showed that the fetal thyroid starts secreting labeled T4 after 17.5-18 dg. Weiss and Noback (7) described changes in bone development of fetuses obtained at 16 dg from goitrogen-treated rats, that is, before onset of fetal thyroid secretion.

Endocrine and Biochemical Development of the Fetus and Neonate
Edited by J. M. Cuezva *et al.*
Plenum Press, New York, 1990

Sweney and Shapiro (8) showed labeled T4 and T3 in amniotic fluid, liver and palatal areas of 13 and 14-dg-old embryos from rats injected with labeled T4. Results from both studies clearly suggested the possibility that maternal thyroid hormones are available to the rat embryo before onset of fetal thyroid function, despite opinions to the contrary (9-13).

Having developed specific and sensitive RIAs (14), and extensive extraction and purification procedures which permitted the determination of very small concentrations of T4 and T3 in tissues (14-16), we decided to re-investigate this possibility by measuring steady-state levels of both iodothyronines in embryonic and fetal tissues obtained before onset of fetal thyroid function. Different types of samples were obtained from the embryonic compartment, at different stages of development, and after perfusion of the dams with phosphate buffered saline. As summarized in Figure 1 A, all samples obtained from the fetal compartment before onset of fetal thyroid function contained both T4 and T3 (15). They were found in molar ratios which were quite different from those of maternal plasma, thus excluding contamination with maternal blood. An independent study by Woods et al. (17) prompted by theoretical considerations (18), showed transfer of labeled iodothyronines from rat dams to their embryos very early in gestation. The results obtained in rats appear to be pertinent to early phases of human development. Bernal & Pekonen (19) have shown the presence of T3 in the brain of 10-weeks old human embryos (Figure 1 B). T4 was later also found (personal communication by J. Bernal) in a whole 7 week-old human embryo; the amount of tissue was too small to attempt the detection of T3. Secretion of T4 by the human fetal thyroid is very low until 16-18 weeks of gestation, and that of T3 until the last trimester (11).

Studies by Perez-Castillo et al. (20) have shown the presence of nuclear receptors for T3 in rat embryos as early as 13 dg, in brain by 14 dg, and in liver, heart and lung from 16 dg onwards. Data for the rat brain are shown in Figure 2 A. The nuclear T3 receptor was also found in the brain of 10 week-old human fetuses (19) (Figure 2 B), and in a 7 week-old whole embryo. Therefore, both the hormone and its receptor are present before onset of fetal thyroid function, and it does not appear far-fetched

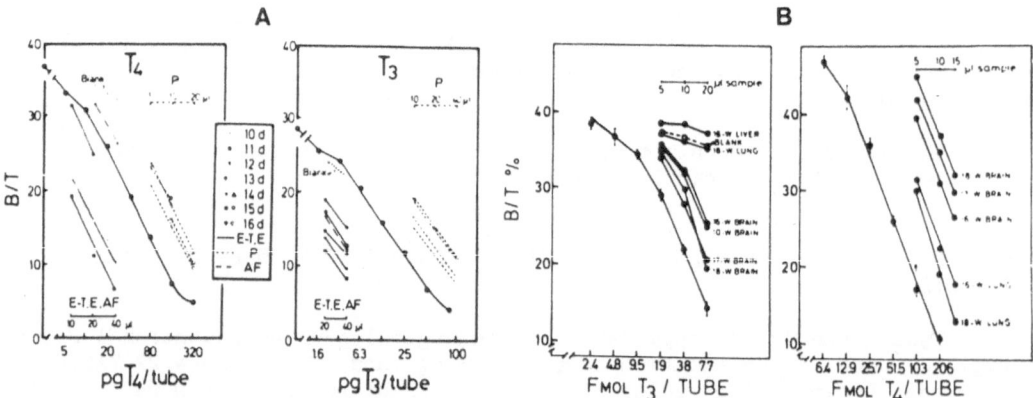

Fig. 1. 1A compares the T3 and T4 standard curves with serial dilutions of eluates obtained from embryo-trophoblasts (E-T), embryonic (E) placental (P) and amniotic fluid (AF) samples obtained in the rat at different gestacional ages, shown in the inset. 1B shows similar curves with extracts from human fetal brains, between 10 and 16 weeks of gestation. From Obregon et al. (15) and Bernal and Pekonen (19).

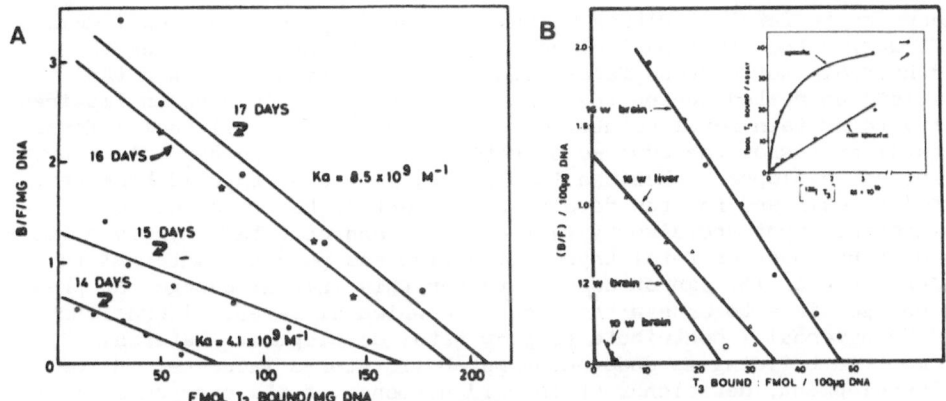

Fig. 2. Scatchard plots for nuclear T3 binding obtained with
rat fetal brain (panel A) at different gestational
ages, and with human fetal brain and liver (panel B)
at 10, 12 and 16 weeks for brain and 16 weeks for
liver. From Perez-Castillo et al. (20) and Bernal
and Pekonen (19).

to suggest that a biological effect might ensue, though this possibility
has not yet been confirmed nor the possible effect defined.

To confirm the maternal origin of the iodothyronines found in early
embryonic samples in the rat, we determined their concentrations in fetal
tissues obtained from normal (C) mothers and from dams which had been
thyroidectomized (T) a few months before mating (16). Data obtained from
C dams extended and confirmed the previous ones regarding the presence of
both T4 and T3 in all embryonic samples obtained well before the onset of
fetal thyroid function. In contrast, when the embryonic samples had been
obtained from T mothers, concentrations of both T4 and T3 were below the
limits of detection in all tissues, placentas included, at least up to 17
dg (Figure 3).

Starting at 18 dg, the thyroid was dissected out of the fetus before
homogenization for the determination of T4 and T3. The extrathyroidal
concentrations thus obtained are shown in Figure 3 on a logarithmic scale.
There is a marked increase in fetal T4 and T3 concentrations between 17
and 18 dg, in agreement with the timing previously reported (5,6) for onset
of fetal thyroid secretion in the rat. T4 and T3 found in fetuses before
and up to 17 dg are obviously of maternal origin.

Therefore, fetuses from T mothers develop in a thyroid hormone-
deficient condition, at least up to 17.5-18 dg. Their number is markedly
reduced (from 11.1±0.4 to 6.8±0.3), and the individual weight of fetuses
is smaller. Before onset of fetal thyroid function, fetal weight is reduced
by maternal T to 59 % of C values (16). After fetal thyroid function has
started, there appears to be a catch-up in growth, but even at 21 dg (near
term) the body weight of fetuses from T dams is only 80 % of normal
(16,21). Bonet and Herrera (22) have recently described that maternal T
impairs the physiological increase of maternal body weight (BW) which takes
place during gestation. The BW of the 21- dg-old fetuses was also reduced.
Treatment of the T dams with 1.8 µg T4/100 g BW/day throughout gestation
prevented these alterations. Interestingly, treatment with the same T4 dose
only during the first 12 days of pregnancy was enough to prevent these
changes, whereas treatment from day 12 to 21 did not. The fetal pituitary
growth hormone (GH) content was reduced to 31 % of that of fetuses from C
dams. As indicated for the increment in maternal BW, and for the fetal BW,

13

the decrease in fetal pituitary GH was prevented by treating the T dams with T4 during the first half of gestation; treatment only during the second half did not restore fetal pituitary GH. They also showed (23) that T dams left untreated during the first half of pregnancy have an impaired capacity to sustain fetal metabolic demands when food is withheld. These results clearly show the adverse effects of maternal hypothyroidism during a period of development when she is the only source of thyroid hormones, both for herself and for the developing conceptus. Later correction of both maternal hypothyroidism by T4 treatment, and of fetal thyroid hormone deficiency by onset of fetal thyroid function and possible maternal to fetal transfer of T4, cannot compensate for this initial damage. At present it is not possible to assess the degree to which different alterations caused by maternal T contribute to poor fetal development: maternal impairment in building up endogenous metabolic stores later needed for fetal development, deficiency of thyroid hormones in the embryonic tissues themselves, or both. Whichever the mechanism (s) involved, the adult progeny of T rats suffer from permanent behavioural defects (24). These are not avoided by the treatment of the dams near term with GH, despite the fact that such a treatment reverses many of the metabolic alterations found in fetuses from T dams near term (25,26).

There are similarities between the results obtained in rats and those described in man. We have already indicated that both the hormone and the nuclear T3 receptor are found in the human fetal brain several weeks before onset of active thyroid secretion, and during a phase of forebrain development when active neuroblast division is starting in the forebrain (for reviews, see 1-4). After 10 weeks of gestation, and coinciding with the active phase of neuroblast division, there was a rapid increase in cerebral DNA; receptor concentration increased ten-fold by 16 weeks of gestation (19). Present information thus shows that there is maternal transfer of thyroid hormone in man before the beginning of the second trimester of pregnancy. Therefore, maternal hypothyroidism could result in decreased availability of thyroid hormone during the initial phases of development and during the spurt in forebrain neuroblast division. It is known that maternal hypothyroxinemia results in lowered I.Q. of the progeny, even in areas without iodine deficiency (27,28). Although an

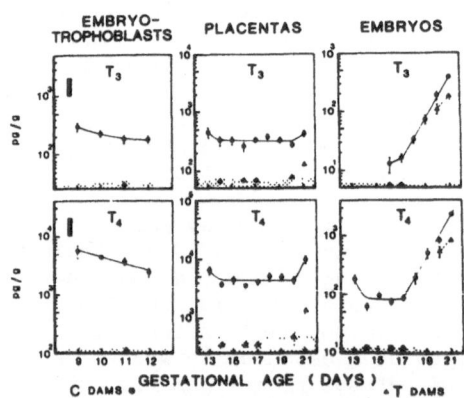

Fig. 3. T4 and T3 concentrations in embryo-trophoblasts (9-12 dg), placentas and fetuses (13-21 dg) obtained from C and T dams. Shaded areas correspond to the limit of detection. All differences between samples from C and T dams were statistically significant. From Morreale de Escobar et al. (16).

incapacity of the hypothyroxinemic woman to maintain an adequate metabolic milieau for the fetus might play an important role, defects in their progeny can no longer be attributed only to this (29), and the possibility should also be considered that a deficiency of thyroid hormones in the embryo itself might play and adverse role. This has received very little attention, because it was believed that maternal thyroid hormones were not available to the developing embryo before midgestation.

MATERNAL TO FETAL TRANSFER OF THYROID HORMONES AFTER ONSET OF FETAL THYROID FUNCTION

We have summarized evidence that maternal thyroid hormones cross the placenta into the fetal compartment early in gestation. Once the fetal thyroid starts functioning, this transfer might decrease and eventually stop, or it might continue until term. It is experimentally difficult to investigate these possibilities. Labeled-iodine containing hormones found in the fetal compartment after their injection into the mother do not necessarily indicate transfer, as the fetal thyroid might have synthesized them from labeled iodine deiodinated from the hormone administered to the dam. Determinations of T4 and T3 in fetal tissues by RIA do not permit to distinguish between maternal and fetal contributions. Either the maternal or the fetal thyroid function has to be blocked, or both. But conclusions may then only be drawn for maternal-fetal relationships where either the mother, the fetus, or both, are hipothyroid, and cannot be extrapolated to normal conditions.

Normal Maternal and Fetal Thyroid Status

As reviewed up to 1967 by Geloso (6), if labeled T4 is injected into pregnant females of different species (rat, mouse, guinea pig, rabbit, sheep, pig, monkey) labeled iodothyronines are always found in the fetal plasma, in varying concentrations depending on the species and on the gestational phase, except in sheep and pig. There was less information regarding passage of labeled T3 from mother to fetus, which appeared to be important in the rat, and minimal in the guinea pig. As pointed out (6), some of the experiments were carried out with perfused guinea pigs placenta left in situ, and in monkeys where the specific activity of the thyroxine in fetal plasma and fetal thyroid were measured. Results from such experiments strongly suggested that the labeled iodothyronines found in the fetal compartment were mostly derived from transplacental transfer, not from fetal thyroid synthesis.

More recent experiments carried out in the rat (30,31) have approached the problem by determining T4 and T3 kinetics in pregnant rats at the end of gestation, using labeled iodothyronines and giving potassium iodine to block radioiodide recycling. Placental transfer rates were calculated from labeled iodothyronine concentrations in maternal and fetal plasma. Placental transfer of the hormones were detected, but considered to represent a minimal part of total fetal production rates. As the authors themselves pointed out, the calculations were based on several assumptions regarding fetal T4 and T3 kinetics which could not be tested experimentally. The extensive investigations carried out in the sheep model have been reviewed by others (10-12,32,33), and indicate virtual placental impermeability to T4 and T3 in the maternal to fetal direction.

It might be pertinent to note here that Artiodactyla (among which sheep and swine) have epitheliochorial placentas, quite different from those of Anthropoidea (among which man and monkey) and most Rodentia (among which rat, mouse and guinea pig) which are of the same type, namely hemochorial. The first type of placenta contains maternal endothelium, reticulum and epithelium and fetal trophoblast and endothelium, whereas the latter type does not contain the maternal tissues.

As regards to transfer in normal women at the end of uncomplicated pregnancies, both Grumbach and Werner (34) and Kearns and Hutson (35) found that significant amounts of labeled T4 and T3 were transferred from the mother to the newborn, although at a slow rate. In both studies T3 was more easily transferred than T4. Recycling by the fetal thyroid of radioiodine deiodinated from the labeled iodothyronine given to the mother was avoided by the administration of potassium iodide, and the lack of fetal thyroidal uptake of the label was confirmed at birth.

The use of labeled iodothyronines was later avoided, and the problem was investigated by giving normal women near term relatively large doses of either T4 (36) or T3 (37,38). The end-points measured to assess maternal to fetal transfer were usually related to cord-blood T4, T3 or suppression of fetal TSH. The use of high doses of T4 or T3 might have altered the maternal-fetal transport system, either by enhancing transfer or by activating placental deiodinations (13), thus actually decreasing the proportion of hormone transferred. Despite these possibilities, results obtained in uncomplicated pregnancies suggest that transfer of both iodothyronines occurs near term, though neither T4 nor T3 cross the placenta barrier freely. Fisher et al. (36) found much smaller increases in cord-blood BEI (butanol extractable iodine) than in maternal plasma BEI after administration of a single very large dose of T4 (0.5-8 mg) 3 to 52 hr before delivery, and concluded that there was virtually no transfer of T4 in normal women near term. From the data obtained, Fisher later (39) calculated that as much as 36 µg T4 would be transferred to the fetus under normal conditions, and considered that this might be important for the protection of an athyreotic fetus. As measured by increased cord-blood T3 concentrations, transfer of T3 was observed in women given relatively low doses of T3 (50 µg/day for the last three weeks of pregnancy) (38), inadequate even to suppress maternal T4 levels (through suppression of maternal TSH) (37). However, up to 300 µg T3 per day were needed to observe a decrease in cord-blood T4 (37,38). This finding lead to the conclusion that appreciable transfer of T3 only occurs when high doses are administered to women. However, as we will later show for the rat, it is also possible that sensitivity of the fetal pituitary near term to T3 (but not to T4) is still very low as compared to adults. As reviewed by Fisher (32) there seems to be a progressive maturation of T3 negative feed-back control of pituitary TSH release with advancing gestational age in fetal sheep and monkeys and in neonatal rats.

In summary, existing evidence strongly suggests that both T4 and T3 cross the human placenta near term in normal pregnancies, though not freely. It is impossible at present to assess how much maternal transfer contributes to total fetal thyroid hormone economy, although it probably represents a smaller proportion than the fetal contribution.

Maternal and or Fetal Thyroid Failure

1. Impairment of maternal thyroid function. If maternal transfer of thyroid hormones continues after onset of fetal thyroid function, some change would be expected to occurr in fetal thyroid hormone concentrations or TSH secretion if the usual maternal contribution were decreased. This was indeed the case when total extrathyroid T4 and T3 concentrations were determined in fetuses from T dams (see data at 20 and 21 dg, Figure 3). It was also found that not all tissues were equally affected: by 21 dg the brain was spared from T4 and T3 deficiency (20). We suggested that this might involve a compensatory increase in the secretion of hormone by the fetal thyroid.

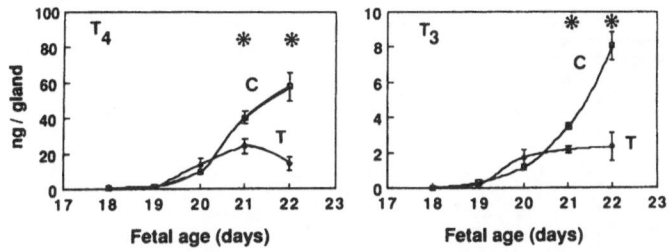

Fig. 4. Total T4 and T3 contents in proteolytic digests (Pronase)
of fetal thyroid glands obtained from C and T mothers
between 18 and 22 dg. Asterisks identify significant
differences between fetuses from C and T dams. From Ruíz
de Oña et al. (40).

Gray and Galton (41) found that thyroidal T4 content was decreased in
rat fetuses from T dams near term, and interpreted their data as indicating
increased fetal thyroidal secretion. We confirmed this also for the
thyroidal T3 content of 21-dg-old fetuses from T dams (21). Figure 4 shows
the changes in total T4 and T3 contents of the fetal thyroid from 18 until
22 dg, as measure by RIA in proteolytic digest. Total thyroidal content of
T4 increased from 0.66±0.07 (SEM) ng/gland at 18 dg to 57.7±8.2 at 22 dg;
data for T3 were 0.021±0.005 ng/gland and 8.07±0.78. The increases were
about a hundred-fold for T4, and four hundred-fold for T3. The T4 to T3
ratio was about 30: 1 at 18 dg and 7: 1 at 22 dg. When the mother had been
T, however, the fetal T4 and T3 contents were the same as those of fetuses
from C dams up to 20 dg, but then stopped increasing. By 22 dg the
thyroidal T4 and T3 contents in fetuses from T dams were 24 % and 28 %,
respectively, of those of fetuses from C dams. This finding confirms the
previous observations at 21 dg (21) and support the suggestion that the
thyroid of the fetuses from T mothers is stimulated to secrete increasing
amounts of hormones to compensate for the absence of the maternal
contribution. Both the data and their interpretation agree with those of
Gray and Galton (41), and are further supported by the finding of increased
plasma TSH levels in fetuses from T dams (22).

It is likely that the fetal thyroid would not be able to compensate
completely for the lack of maternal contribution if it were faced with
conditions impairing its function. This is the case when the iodine intake
is low (42), as summarized in Figure 5. In such a situation normal T4 and
T3 levels might not be attained in any of the fetal tissues, the brain
included.

In summary, data from experiments in rats near term show that the
fetus senses maternal thyroid failure, and the fetal thyroid reacts to this
situation with compensatory mechanism which attempt to maintain normal
thyroid hormone concentrations. A pausible explanation is that the stimulus
for activation of these mechanism is directly related to the decreased
maternal contribution of thyroid hormones to fetal economy, though it
cannot be excluded that the signal is related to other metabolic
alterations of the T dam.

2. Fetal hypothyroidism. Models of fetal hypothyroidism alone are
difficult to achieve in small laboratory animals. Geloso and Bernard (43)
decapitated fetuses at 17.5 dg and compared their BEI (isotope
equilibration method) at 21 dg with that of age-paired normal fetuses, and
that of decapitated fetuses from T mothers. Data are summarized in Table 1.

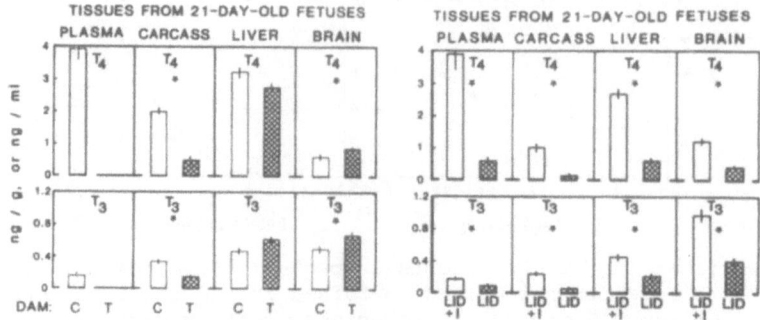

Fig 5. Concentrations of T4 and T3 in plasma and tissues
of fetuses from C and T dams, and from rats on a
low iodine diet (LID) or the same diet supplemented with KI (LID+I). Asteriscs identify statistically significant differences. Data are from
Morreale de Escobar et al.(16) and from Escobar
del Rey et al. (42). Reproduced from Morreale de
Escobar et al., Frontiers in Thyroidology, Plenum
Press (1986).

These results indicate that at the end of gestation the rat fetal
thyroid is the main source of the fetal plasma T4, as decapitation of the
fetus results in a 60 % decrease of the fetal plasma BEI. But it is not the
only source, as maternal T further decreases the fetal plasma T4 levels.
The authors concluded that a certain amount of maternal thyroid hormones
cross the placenta, but that the transfer is too limited to compensate for
the absence of the fetal thyroid (43).

Apart from the technical difficulties involved, this approach does
not permit to study the effects of possible maternal transfer of thyroid
hormones on the fetal hypothyroid brain, and fetal TSH secretion. We
therefore recurred, as many others (44-46), to the use of goitrogens which
cross the placenta and block the production of fetal thyroid hormones. This
treatment obviously also blocks maternal thyroid function so that the
experimental design includes groups of MMI-treated dams on a replacement
dose of thyroid hormone, given by constant infusion to avoid supraphysio-
logical peaks of thyroid hormone levels.

Table 1. Fetal plasma butanol extractable iodine (BEI),
by isotopic equilibrium (means±95% confidence
limits) (43)

Mothers	Fetuses	ng BEI/ml	ng T4[$]/ml
Normal	Normal	2.6 ± 0.6	3.9 ± 0.9
Normal	Decapitated	0.9 ± 0.2	1.4 ± 0.3
Thyroidectomized	Normal	2.0 ± 0.7	3.0 ± 1.1
Thyroidectomized	Decapitated	0.3 ± 0.1	0.5 ± 0.2

[$]Assuming the BEI was composed mainly of T4: BEI x
1.5. The corresponding BEI values for the normal and
thyroidectomized mothers were 6.2 ± 1.5 and 0.7 ± 0.5
ng/ml; T4, 9.3 ± 2.3 and 1.1 ± 0.8 ng/ml.

Table 2. Experimental design for experiments A, B and C

Exper.	Group of dams :	MMI[#] (0.02%)	T4 infused[$] (µg/100 g)	T3 infused[$] (µg/100 g)
A	C (or T)	–	–	–
	C (or T) + MMI	+	–	–
	C (or T) + MMI + T4	+	+(1.8)	–
B	C (or T)	–	–	–
	C (or T) + MMI	+	–	–
	C (or T) + MMI + T4	+	+(1.8)	–
	C (or T) + MMI + T3	+	–	+(0.45)
C	C	–	–	–
	C + MMI	+	–	–
	C + MMI + T4 (3 doses)	+	+(1.8; 2.4; 3.6)	–
	C + MMI + T4 (3 doses)	+	–	+(0.5; 1.5; 4.5)

[#]Given as drinking water, from 16-21 dg (Exp. A) or from 14-21 dg (Exp. B and C).
[$]In µg T4 or T3/100 g BW of dam/day, from 17-21 dg (Exp. A) or from 15-21 dg (Exp. B and C).

Table 2 outlines the experimental design which we used. Methimazole (MMI) was used to block maternal and fetal thyroid function. Some of the MMI-treated dams received a replacement dose of T4 (1.8 µg/100 g BW/day) by constant infusion, as described in detail (21) elsewhere. The main findings are summarized in Figure 6, which shows the T4 and T3 concentrations in fetal thyroid digest carcass and brain. The concentrations of T4 and T3 in the fetal thyroid were markedly decreased by MMI treatment, irrespective of the infusion of T4 into the MMI-treated dams. Hormone levels in tissues were also decreased by the MMI treatment. Comparison of the hormone levels in fetuses from C + MMI + T4 (or T + MMI + T4) dams with those from C + MMI (or T + MMI) mothers shows that infusion of T4 into the mothers not only ameliorates fetal deficiency of T4, but also of T3. This occurs without an increase of fetal plasma T3 (not shown) (21), indicating that the fetal brain derived its T3 from local deiodination of T4. Our conclusion that some maternal T4 crosses the placenta near term, at least if the fetal thyroid is impaired, is in agreement with the data and conclusions by others (22,41), who injected physiological T4 doses.

When the rat fetal thyroid is impaired (as would be the case in human congenital hypothyroidism) maternal T4 has a protective effect on the fetal brain and on other fetal tissues. This might mitigate adverse effects of fetal thyroid failure on the developing brain. It appeared important to determine whether, or not, T3 is also transferred from the mother to the fetus near term, and also mitigates the effects of fetal hypothyroidism. Experiment B was carried out (Table 2) similar to the one summarized above, but with several modifications. MMI was started two days earlier, at 14 dg, and the infusion of hormone into the dams was started at 15 dg. Several MMI-treated dams were again infused with the same dose of T4 used for Experiment A, and other MMI-treated dams were infused with T3 at a dose (0.45 µg/100 g BW/day) equivalent for adult rats to that of T4. Data for fetuses from dams infused with T4 confirmed those of the previous experiment, illutrated in Figure 6 as regards the transfer of T4 and its effects on the T4 and T3 deficiency of fetal tissues, the brain included. The new finding (47) was that infusion of T3 into the MMI-treated dams increased the concentration of T3 in fetal plasma, and in all fetal tissues studied, except the brain (Figure 7). MMI treatment had increased both the maternal and fetal plasma TSH levels (not shown). Infusion of both T4 and T3 decreased plasma TSH to normal levels in the mothers. Infusion of T4 decreased fetal TSH, although it remained elevated as compared to normal fetuses. In contrast, the infusion of T3 had no effect on fetal plasma TSH. Data shown in Figure 7 correspond to the groups from C dams only; results from the four groups of fetuses from T dams were very similar.

Previous data summarized here suggested an important role of local generation of T3 from T4 in providing the developing brain with increasing amounts of T3. The results of this last experiment show that the brain of rat fetuses near term is entirely dependent on T4 availability, as T3 does not enter the brain, at least at physiological plasma T3 levels. 5' D-II activities were measured in the cerebral cortex of the fetuses corresponding to Figure 7. Activity increased with MMI treatment, in response to thyroid hormone deficiency. Infusion of T4 into the mothers decreased enzyme activity markedly, although not to control levels, whereas infusion of T3 did not, showing that the changes in thyroid hormone level were accompanied by a biological effect.

Preliminary results have been obtained in fetuses from Experiment C (Table 2). The intact (C) dams were infused with saline; the MMI-treated dams were infused with 3 different T4 doses (1.8, 2.4 and 3.6 µg/100 g BW/day) and 3 different T3 doses (0.5, 1.5 and 4.5 µg/100 g BW/day), chosen on the basis of the results described by Knobil and Yasimovich (44). The lowest of the T4 and T3 doses already reversed maternal plasma TSH to C values. The 1.8 ug T4 dose was insufficient to raise the maternal hepatic T3 level and alfa-glycerophosphate dehydrogenase (α-GPD) activity of the MMI-treated dams to normal. A dose of 2.4 µg T4 reversed both variables to C values. The 3.6 µgT4 dose resulted in liver T3 levels and α-GPD activities which were higher than C values, although not higher than values for non-pregnant females. The lowest of the T3 doses already reversed liver T3 and α-GPD activity to C values, the higher doses resulting in dose-related increases of hepatic T3 and α-GPD activity.

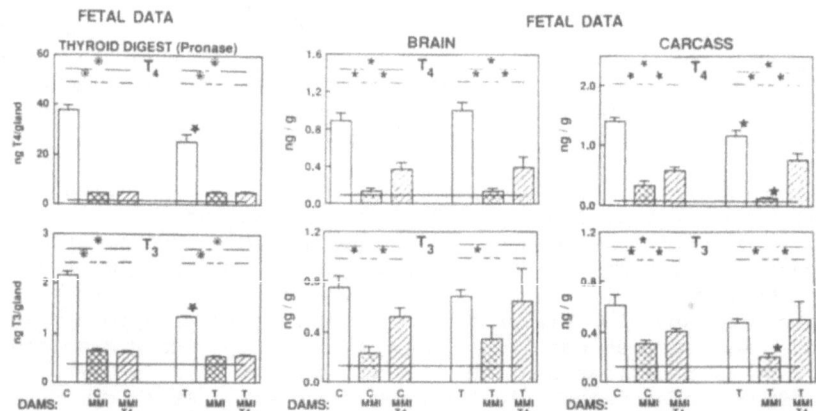

Fig. 6. Concentrations of T4 and T3 in fetal thyroid digests, and in the brain and carcass of fetuses from the six groups of Experiment A (Table 2). Asteriscs between data bars identify statistically significant differences. Horizontal lines indicate limits of detection. From Morreale de Escobar et al. (21) as illustrated in (4).

20

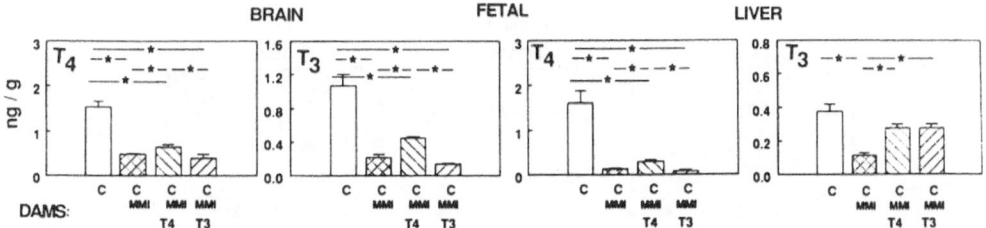

Fig. 7. Concentrations of T4 and T3 in the brain and lung of fetuses
from four of the groups of Experiment B (Table 2), corresponding to the intact (C) dams. Asteriscs placed mid-way between
groups identify statistically significant differences. From
Morreale de Escobar et al.(47).

The doses indicated from this point onwards are the amount infused/
100 g BW/day, the BW being that of the dams at 15 dg.

Findings in fetal tissues depended on the iodothyronine used for
infusion, on the dose, on the end-point measured in the fetus (T4, T3,
TSH), and on the tissue being considered. An example is shown in Figure 8,
for the fetal lung and brain T3, and plasma TSH. The results obtained in
fetal lung (and, although not shown, in fetal plasma and fetal liver)
demostrate that when T3 is infused, there is a dose-related transfer of T3,
accompanied by a dose-related increase of T3 in several fetal tissues. But,
as already found at a single T3 dose in Experiment B, fetal plasma TSH was
scarcely affected by the infusion of this iodothyronine. This occured even
at the highest T3 dose. Again, there was no increase in cerebral T3 above
the level found in MMI-treated fetuses with any of the T3 doses used, the
highest being 9 times the replacement dose for the adult rat. In contrast,
the maternal infusion with 2.4 μg T4 raised cerebral T3 levels to C values.
It also depressed plasma TSH. We wish to point out that the dose of T4
which normalized cerebral T3 levels in the fetus was lower than the dose
needed to normalize fetal plasma TSH (>3.6 μg T4, Figure 8) and fetal
plasma T4 (not shown), and was also the dose normalizing maternal liver T3
and α-GPD.

Recent studies on the role of the choroid plexus in the transport of
T4 and T3 into the brain show that after injection of labeled thyroid
hormones, the choroid plexus and the brain avidly accumulate T4, but not T3
(49). Such a mechanism might afford an explanation for the lack of effect
of T3 on fetal brain T3 concentrations.

In summary, results obtained with this experimental design show that
maternal thyroid hormones also cross the placenta late in gestation, at
least when the fetal thyroid is impaired, and despite active deiodinations
by the placenta (13). Maternal T4 has a more important role, as it
mitigates T3 deficiency of the brain, whereas maternal T3 does not. Neither
does maternal T3 decrease fetal TSH secretion. We believe it is quite
interesting that the fetal brain already posesses mechanism(s) which
achieve maintainance of normal cerebral T3 concentrations at levels of
maternal T4 which are within the physiological range for the mother, and
which are not sufficient to increase fetal plasma T4 (or to decrease fetal
TSH) to normal values, nor to normalize T3 concentrations in other fetal
tissues. If the mother is hypothyroxinemic, the fetal brain is not
protected, even if maternal T3 levels are normal.

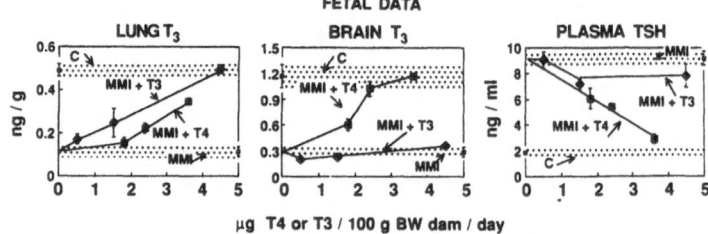

Fig. 8. "Dose-response" curves between fetal concentrations of T3
in lung and brain, or fetal plasma TSH, and the dose of
T4 or T3 infused into the mothers. Groups are those of
Experiment C (Table 2). Dotted areas correspond to means
± SEM for C or MMI-treated groups. From Calvo et al.(48).

We wish to point out that the conclusion that the human placenta at
term is virtually impermeable to the iodothyronines partly based on studies
(39,40) where normal women were given T3 near term, and some parameter
related to fetal TSH secretion was measured. The data obtained in the rat
suggest that this may be a misleading end-point to assess transfer of T3.
Moreover, the results should not be extrapolated to T4.

If experimental data obtained in the rat are pertinent to human babies
with congenital hypothyroidism, they strongly support the use of T4 for
early treatment, or in those cases where the condition is diagnosed in
utero. They also indicate that T3 might actually be harmful, if given in
doses which depress maternal T4 levels. On the other hand, normal maternal
T3 levels might protect tissues other than the brain from T3 deficiency and
thus avoid the appearance of overt signs of hypothyroidism at birth, even
if their plasma TSH is high. The possible relevance of these findings to
human endemic cretinism has been discussed elsewhere (4).

But are any of these findings pertinent to human congenital
hypothyroidism?. We have already discussed more extensively the possibility
that they are (21). Autonomy of the pituitary-thyroid feedback system in
human fetuses during the third trimester of pregnancy is well established,
but the finding that fetal thyroid failure is accompanied by high TSH
levels at birth does not exclude that TSH would have been higher (and
cord-blood T4 lower) if there had been no transfer of hormones from the
mother to the hypothyroid fetus. Transfer of maternal thyroid hormones to
athyreotic fetuses would explain that in some of them T4 levels are clearly
detectable in cord-blood, and then decrease rapidly (50). It would also
explain the variability with which athyreotic babies present signs of
hypothyroidism at birth, as this might be related to the maternal thyroid
hormone levels and the adequacy of placental transfer, which might vary for
different pregnancies (39). These same causes for variability might account
for differences in permanent brain damage presented by athyreotic babies.
It has been established that the babies with the lowest plasma T4 levels at
birth present more frequently signs of having suffered from hypothyroidism
in utero, as assessed from delayed maturation of prenatal ossification
centers; these babies tend to have the lowest IQ (although within the
normal range) even if treated promptly after birth (51).

FINAL COMMENTS

A comprehensive program with the aim of preventing CNS damage caused
by thyroid failure should include: i) neonatal thyroid screening, treatment

as soon as possible after birth (or even before birth) being the principal aim; ii) iodine prophylaxis to avoid maternal hypothyroxinemia combined with fetal thyroid failure and iii) careful monitoring of plasma T4 and free T4 levels during pregnancy to detect and treat cases of maternal hypothyroxinemia, with or without clinical hypothyroidism.

REFERENCES

1. G. Morreale de Escobar, and F. Escobar del Rey, Brain damage and thyroid hormones, in: "Neonatal Thyroid Screening", G. N. Burrow, ed., Raven Press, New York (1980).
2. G. Morreale de Escobar, A. Ruiz-Marcos, and F. Escobar del Rey, Thyroid hormones and the developing brain, in: "Congenital Hypothyroidism", J. H. Dussault and P. Walker, eds., Marcel Dekker New York (1983).
3. G. Morreale de Escobar, M. J. Obregon, and F. Escobar del Rey, Fetal and maternal thyroid hormones, Hom. Res 26:12 (1987).
4. G. Morreale de Escobar, C. Ruiz de Oña, M. J. Obregon, and F. Escobar del Rey, Models of fetal iodine deficiency , in: "Iodine in Neuro-development", G. R. DeLong and J. Robbins, eds., Plenum Press, New York, (in press).
5. B. Nataf and M. Sfez, Debut du fonctionnement de la thyroide foetale du rat, Compt. Ren. Soc. Biol. 156:1235 (1961).
6. J. P. Geloso, Fonctionnement de la thyroide et correlations thyreohypo-physaires chez le foetus de rat, Annales d'Endocrinologie 28:1 (1967).
7. R. M. Weiss and C. R. Noback, The effects of thyroxine and thiouracil on the time of appearance of ossification centers of rat fetuses, Endocrinology 45:389 (1949).
8. L. R. Sweney and B. L. Shapiro, Thyroxine and palatal development in the rat, Dev. Biol 42:19 (1975).
9. M. Hamburgh, The role of thyroid and growth hormone in neurogenesis, in: "Current Topics in Developmental Biology", A. A. Moscona and A. Monroy, eds., Academic Press, New York, vol 4:109 (1969).
10. D. A. Fisher, J. H. Dussault, J. Sack, and I. J. Chopra, Ontogenesis of hypothalamic-pituitary-thyroid function in man, sheep and rat, Rec. Progr. Horm. Res. 33:59 (1977).
11. D A. Fisher, and A. K. Klein, Thyroid development and disorders of thyroid function in newborn, New Engl. J. Med. 304:702 (1981).
12. D A. Fisher The unique endocrine milieu of the fetus, J. Clin. Invest. 78:603 (1986).
13. E. Roti, A. Gnudi, and L. E. Braverman, The placental transport, synthesis and metabolism of hormones and drugs which affect thyroid function, Endocr. Rev. 4:131 (1983).
14. M. J. Obregon, G. Morreale de Escobar, and F. Escobar del Rey, Concentrations of triiodothyronine in the plasma and tissues of normal rats as determined by radioimmunoassay: Comparison with results obtained by an isotopic equilibration technique, Endocrinology 103:2145 (1978).
15. M. J. Obregon, J. Mallol, R. Pastor, G. Morreale de Escobar, and F. Escobar del Rey, Thyroxine and triiodothyronine in rat embryos before onset of fetal thyroid function, Endocrinology 114:305 (1984).
16. G. Morreale de Escobar, R. Pastor, M. J. Obregon, and F. Escobar del Rey, Effects of maternal hypothyroidism on the weight and thyroid hormone content of rat embryonic tissues, before and after onset of fetal thyroid function, Endocrinology 117:1890 (1985).
17. R. J. Woods, A. K. Sinha, and R. Ekins, Uptake and metabolism of thyroid hormones by the rat fetus in early pregnancy, Clin. Sci. 67: 359 (1984).

18. R Ekins, Roles of thyroxine-binding proteins and maternal thyroid hormones in development, Lancet 1:1129 (1985).
19. J.Bernal and F.Pekonen, Ontogenesis of the nuclear 3, 5, 3'-triiodothyronine receptor in the human fetal brain. Endocrinology 114:677 (1984).
20. A.Perez-Castillo, J.Bernal, B.Ferreiro, and T.Pans, The early ontogenesis of thyroid hormone receptor in the rat fetus. Endocrinology 117:2457 (1985).
21. G.Morreale de Escobar, M.J.Obregon, C.Ruiz de Oña, and F.Escobar del Rey, Transfer of thyroxine from the mother to the rat fetus near term: Effects on brain 3, 5, 3'-triiodothyronine deficiency, Endocrinology 122:1521 (1988).
22. B.Bonet and E.Herrera, Different responses to maternal hypothyroidism during the first and second half of gestation in the rat, Endocrinology 122:450 (1988).
23. E.Herrera and B.Bonet, Effect of maternal hypothyroidism during two phases of gestation on fetal growth and metabolic response to starvation in the rat, Annales d'Endocrinologie 44:17A, abstract (1983).
24. C.E.Hendrich, W.J.Jackson, and S.P.Porterfield, Behavioral testing of progenies of Tx (hypothyroid) and growth hormone-treated Tx rats: an animal model for mental retardation, Neuroendocrinology 38:429 (1984).
25. S.P.Porterfield and C.E..Hendrich, The effects of growth hormone treatment of thyroid-deficient pregnant rats on maternal and fetal carbohydrate metabolism, Endocrinology 99:786 (1976).
26. C.E.Hendrich, S.P.Porterfield, and V.A.Galton, Pituitary-thyroid function of fetuses of hypothyroid and growth hormone treated hypothyroid rats, Horm.Met.Res 11:362 (1979).
27. G.W.Greenman, M.O.Gabrielson, J.Howard-Flanders, and M.A.Wessel, Thyroid function in pregnancy. Fetal loss and follow-up evaluation of surviving infants, New Engl.J.Med. 267:426 (1962).
28. E.B.Man and S.A.Serunian, Thyroid function in human pregnancy. IX. Development or retardation of 7-year-old progeny of hypothyroxinemic women, Am.J.Obst.Gynecol. 125:949 (1976).
29. A.J.McMichael, J.D.Potter, and B.S.Hetzel, Iodine deficiency, thyroid function and reproductive failure, in: "Endemic Goiter and Endemic Cretinism", J.B. Stanbury and B.S.Hetzel, eds., John Wiley & Sons, New York, (1980).
30. J.D.Dubois, A.Cloutier, P.Walker, and J.H.Dussault, Absence of placental transfer of L-triiodothyronine in the rat, Pediat.Res 11:116 (1977).
31. J.H.Dussault and P.Coulombe, Minimal placental transfer of L-thyroxine in the rat, Pediat.Res. 14:228 (1980).
32. D.A.Fisher, Ontogenesis of hypothalamic-pituitary-thyroid function in the human fetus, in: "Pediatric Endocrinology", F.Delange, D.A. Fisher, and P.Malvaux, eds., Karger, Basel, (1985).
33. D.A.Fisher, Development of fetal thyroid function and its control, in: "Iodine in Neurodevelopment", G.R.DeLongand J.Robbins, eds., Plenum Press, New York, (in press).
34. M.M.Grumbach and S.H.Werner, Transfer of thyroid hormone across the human placenta at term, J.Clin.Endocrinol. 16:1392 (1956).
35. J.E.Kearns and W.Hutson, Tagged isomers and analogues of thyroxine (Their transmission across the human placenta and other studies), J. Nuclear Med. 4:453 (1963).
36. D.A.Fisher, H.Lehman, and C.Lackey, Placental transfer of thyroxine, J.Clin.Endocrinol.Metab. 24:393 (1964).
37. S.Raiti, G.B.Holzman, R.L.Scott, and R.M.Blizzard, Evidence for the placental transfer of tri-iodothyronine in human beings, New Engl.J.Med. 277:456 (1967).

38. J.H.Dussault, V.V.Row, G.Lickrish, and R.Volpe, Studies of serum
    triiodothyronine concentration in maternal and cord blood: Transfer
    of triiodothyronine across human placenta, J.Clin.Endocrinol. 29:595
    (1969).
39. D.A.Fisher, Panel discussion on hyperthyroidism in the pregnant woman
    and neonate, J.Clin.Endocrinol.Metab. 27:1639 (1967).
40. C.Ruiz de Oña, M.J.Obregon, G.Morreale de Escobar, and F.Escobar del
    Rey, Thyroid hormone economy in fetal rats after onset of fetal
    thyroid function, 17th Meeting of the European Thyroid Association,
    Montpellier, (1988).
41. B.Gray and V.A.Galton, The transplacental passage of thyroxine and
    foetal thyroid function in the rat, Acta Endocrinol. (Kbhvn) 75:725
    (1974).
42. F.Escobar del Rey, R.Pastor, J.Mallol, and G.Morreale de Escobar,
    Effects of maternal iodine deficiency on the L-Thyroxine and 3, 5,
    3'-triiodo-thyronine contents of rat embryonic tissues before and
    after onset of fetal thyroid function, Endocrinology 118:1259 (1986).
43 J.P.Geloso and G.Bernard, Effects de l'ablation de la thyroide mater-
    nelle ou foetale sur le taux des hormones circulantes chez le foetus
    de rat, Acta Endocrinol. (Kbhvn) 56:561 (1967).
44. E.Knobil and J.B.Josimovich, Placental transfer of thyrotropic
    hormones, thyroxine, triiodothyronine and insulin in the rat,
    Ann.N.Y.Acad.Sci. 75:895 (1959).
45. S.K.Roy and Y.Kobayashi, Placental transfer of L-thyroxine and tri-
    iodo-L-thyronine in rats during late pregnancy, Proc.Soc.Exper.
    Biol.Med. 110:699 (1962).
46. F.Comite, G.N.Burrow, and E.C.Jorgensen, Thyroid hormone analogs and
    fetal goiter, Endocrinology 102:1670 (1978).
47. G.Morreale de Escobar, M.J.Obregon, C.Ruiz de Oña, Comparison of
    maternal to fetal transfer of T3 versus T4 in rats, as assessed by T3
    levels in fetal tissues, Annales d'Endocrinologie 48:178A, abstract
    (1987).
48. R.Calvo, C.Ruiz de Oña, M.J.Obregon, F.Escobar del Rey, and G.Morreale
    de Escobar, Differential effects of maternal T4 and T3 on hypothyroid
    rat fetal tissues, 17th Meeting of the European Thyroid Association,
    Montpellier, (1988).
49. P.W.Dickson, A.R.Aldred, J.G.T.Menting, P.D.Marley, W.H.Sawyer, and
    G.Schreiber, Thyroxine transport in choroid plexus, J.Biol.Chem.
    262:13907 (1987).
50. D.A.Price, R.M.Erlich, and P.G.Walfish, Congenital hypothyroidism.
    Clinical and laboratory characteristics in infants detected by
    neonatal screening, Arch.Dis.Child. 56:845 (1981).
51. J.Letarte and S.La Franchi, Clinical features of congenital hypo-
    thyroidism, in: "Congenital Hypothyroidism", J.H.Dussault and
    P.Walker, eds., Marcel Dekker, New York, (1983).

NEONATAL DEVELOPMENT OF THE ANGIOTENSIN

CONVERTING ENZYME. EFFECT OF HYPOTHYROIDISM

Eugenio Jimenez, Maximiliano Ruiz and Mercedes Montiel

Departamento de Bioquímica y Biología Molecular
Universidad de Málaga
Málaga, Spain

INTRODUCTION

The renin-angiotensin system consist of a short sequence of enzyme reactions which produces angiotensin II and III, whose role is intricately involved in the homeostasis of peripheral vascular resistance, as well as volume and electrolyte composition of corporal fluid.

It is well known that in normotensive pregnant women complex interrelated alterations develop in the cardiovascular system and in the volume and composition of the extracelular fluid. Simultaneously to these alterations, changes of the renin-angiotensin system, not only quantitative but also qualitative, occur during pregnancy. An increase in renin and renin substrate concentrations, as well as alterations in enzyme and/or substrate affinity, have been reported (1), possibly as results of maternal adaptation to this situation. Since these components can cross the placental barrier (2), a close maternal-fetal relation could exist in the renin-angiotensin system. This could be interrupted after birth then the renin-angiotensin system of the newborn would have to fit to extra-uterine environment.

The angiotensin converting enzyme (ACE), a peptidyl-dipeptide hydrolase (E.C. 3.4.15.1) principally found on the surface of pulmonary endothelial cells, is one of the key enzymes in this cascade since it is directly involved in the angiotensin II formation by removing the carboxyterminal histidyl-leucine residue from angiotensin I. In the present report the effect of the thyroid hormones deficiency upon the lung and serum ACE activities, and serum aldosterone concentration, has been studied in adult and newborn Wistar rats during the two first weeks of life, a critical period in the differentiation and functional maturation of several organs and tissues.

RESULTS AND DISCUSSION

ACE has been identified in a wide variety of organs such as lung, brain, liver and kidney, although angiotensin II is mainly generated during the transit of angiotensin I through pulmonary circulation, since lung enzyme represent more than 90% of total corporal ACE. For this reason, the changes in pulmonary ACE content seem to be a better representative index

Endocrine and Biochemical Development of the Fetus and Neonate
Edited by J. M. Cuezva *et al.*
Plenum Press, New York, 1990

27

of physiological enzyme activity than its peripheral levels, being frequently used as a marker of lung injury. The serum ACE activity, insufficient to account for generation of angiotensin II *in vivo*, can represent a balance between its tissular release and its metabolic clearance.

In the present study ACE activities were determined in serum and homogenized lung by measuring the hippuric acid release from hippuryl-L--histidyl-L-leucine, a synthetic substrate for ACE. The neonatal developmental patterns of pulmonary ACE activities in normal and hypothyroid animals are shown in Figure 1A. According to a previous study (3), lung ACE activity is low at birth and increases with age. An increase in lung ACE activity was observed between the 5th and 15th day of life in normal animals, which could be considered as a consequence of enzyme induction, as occurs with other enzyme systems involved in the intermediary metabolism for adaptation of the newborn to extra-uterine life. An increase of serum ACE activity was also observed only during the first 7 days of life (Figure 1B). A decline in serum activity is observed in adult animals. These changes could be due to either the increase in the ACE hepatic metabolic clearance or the decrease in its pulmonary vascular secretion, as evident from lung/serum activities ratios, which do not change during the first 10 days of life.

Among the numerous factors which control aldosterone synthesis and secretion, the renin-angiotensin system plays a priority role in the rat, acting the angiotensin II directly on adrenal gland provoking its release to blood. The serum aldosterone concentration during the postnatal development in normal and hypothyroid animals is shown in Figure 2. A decrease in serum aldosterone concentration up to 7 days of life in normal animals is observed, gradually increasing to reach adults levels later. These results suggest that there is no relationship between the ACE activities, or angiotensin II levels, and serum aldosterone concentrations in the first 2 weeks of life, due to immaturity of angiotensin II receptor

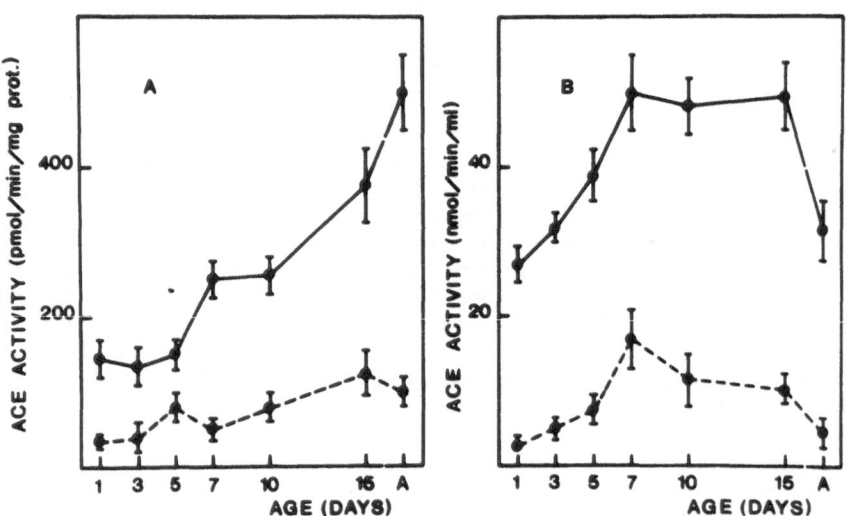

Fig. 1. Lung (A) and serum (B) angiotensin converting enzyme (ACE) activities during postnatal development (in days) and adulthood (A) in normal (•-•) and hypothyroid (•--•) rats. Mean values ±SEM are given for 5 determinations. Student's t test was applied for statistical evaluation.

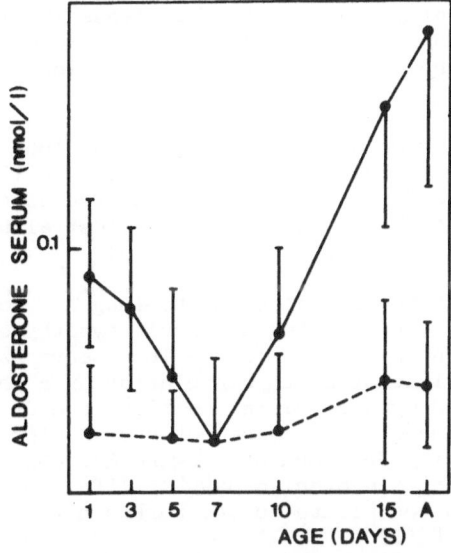

Fig. 2. Serum aldosterone concentrations during postnatal development (in days) and adulthood (A) in normal (●-●) and hypothyroid (●--●) rats. Mean values ± SEM are given for 5 determinations. Student's t test was applied for statistical evaluation. (8).

in the adrenal cortex (4). These data indicate that angiotensin II is not a predominant regulating factor controlling aldosterone at this age.

Previous studies have shown that ACE activity is regulated by hormones such as glucocorticoids, and particularly by the thyroid hormones (5). Although in recent investigations serum ACE activity is apparently not directly associated with serum thyroid hormone concentrations in acutely starved patients, possibly due to differences in the nutritional, metabolic and circulatory changes (6), most groups have demonstrated that thyroid hormones may affect the synthesis and release of ACE, without modifying endothelial cell number or protein content (5). According to these observations increasing serum ACE activity in hyperthyroid patients has been described (7). In addition, a decrease in serum and hepatic ACE activities without a change in lung and renal ACE activities has been found in hypothyroid status induced by propylthiouracil administration in adult animals (8).

In neonatal hypothyroidism, induced during fetal life by administration of methylmercaptoimidazole in the drinking water of pregnant rats from day 14 of gestation till term, a decrease in the lung ACE activity has been observed during postnatal life (Figure 1A), in contrast to the results previously described in adult hypothyroidism (8). This finding is another example of the importance of age, in which hypothyroidism is induced, for the normal morpho-functional development. As it is well known, thyroid hormones regulate the growth and maturation of various tissues and organs, such as brain, lung and kidney. A deficit of thyroid hormones retards cellular differentiation and can produce severe morphological and functional damages, which are not generally observed when the hypothyroidism appears, or it is induced, in later stages of the life. Thus, there is a critical period for each organ during which the thyroid hormones are necessary for maintaining the normal development. Different alterations in the renal renin releasing mechanism during hypothyroid status in an early stage of life and adult have been reported

(9). While in adult hypothyroid animals a decrease in plasma renin concentration has been observed, an increase in this enzyme activity occurs when the deficit of thyroid hormones is induced prior to maturation of kidney function.

The neonatal hypothyroidism could affect the lung ACE synthesis since the thyroid hormones exert their effects in various cells and tissues by stimulating the accumulation of mRNAs, and a marked decrease in the transcription of nuclear RNA is one of the earliest alterations produced by neonatal deprivation of thyroid hormones (10).

Concomitantly with the decrease in the lung ACE activity a reduced peripheral activity in both suckling and adult hypothyroid animals is observed (Figure 1B). In these animals the lung/serum ratio of activities suggest a lower enzyme liberation to circulation as a possible mechanism to counteract its decreased synthesis. In addition, the developmental pattern in serum aldosterone concentration does not show a significant change in hypothyroid animals during development (Figure 2), observing significant differences to normal animals over 15 days of life. This reduction in aldosterone level could contribute to the marked natriuresis found during thyroid hormones deficit (9).

On the other hand, the results of this study show that the serum ACE activities were very similar between 1 day-old animals and adults. This finding suggests that maternal ACE, as other component of the renin-angiotensin system, could cross the placental barrier in order to contribute to homeostasis of blood pressure.

In summary, during postnatal development of the newborn rat, an increase in lung ACE occurs with age. This increase is only found during the first days of life at the peripheral level. Neonatal hypothyroidism induces a marked decrease in both lung and serum ACE activity. The absence of correlation between ACE activity and serum aldosterone concentration suggests that angiotensin II is not a predominant factor in the control of aldosterone secretion in the rat at this age. This finding can contribute to renal disturbance of water and electrolyte handling in situations of thyroid hormones deficiency.

ACKNOWLEDGMENTS

We wish to acknowledge the skilled technical assistance of Mr. J.C. Montilla and Mr. J.L. Dieguez.

REFERENCES

1.   C.R.Daniels, V.Eisen, J.D.H.Slater, The renin-angiotensinogen reaction during pregnancy and oral contraception: estimation of kinetic parameters by an autologous plasma renin assay, J.Endocr. 112:465 (1987).
2.   L.D.Longo, The role of the placenta in the development of the embryo and fetus, in: "The physiological development of the fetus and newborn", C.T.Jones and P.W.Nathanielsz, eds. pp.1, Academic Press, New York, (1985).
3.   K.B.Wallace, M.D.Bailie and J.B.Hook, Angiotensin converting enzyme in developing lung and kidney, Am.J.Physiol. 234:R141 (1978).
4.   N.E.Sirett, J.I.Hubbard and J.J.Bray, Brain angiotensin II receptors. in: "The renin-angiotensisn system in the brain", D.Ganten, M. Printz, M.I.Phillips and B.A.Scholkens, Eds. pp. 233, Springer-Verlag, Berlin (1982).

5. A.H.Krulewitz, W.E.Baur and B.L.Fanburg, Hormonal influence on endothelial cell angiotensin converting enzyme activity, <u>Am.J. Physiol.</u> 247:163 (1984).

6. G.Komaki, H.Tamai, T.Mori, T.Nakagawa and S.Mori, Changes in serum angiotensin converting enzyme in acutely starved non-obese patients. A possible dissociation between angiotensin converting enzyme and thyroid state, <u>Acta Endocrinologica</u> (Kbh) 118:45 (1988).

7. Y.Nakamura, T.Takeda, M.Ishii, K.Nishiyama, M.Yamakada, Y.Hirata, K. Kimura and S.Murao, Elevation of serum angiotensin converting enzyme activity in patients with hyperthyroidism, <u>J.Clin. Endocrinol.Metab.</u>55:931 (1982).

8. M.Montiel, M.Ruiz, E.Jimenez and M.Morell, Angiotensin converting enzyme in hyper- and hypothyroid rats, <u>Horm.Metab.Res</u> 19:90 (1987).

9. E.Jimenez, M.Montiel, J.A.Narvaez and M.Morell, Renin-angiotensin system in hypothyroid rats: effects of potassium iodide and triiodo-thyronine, <u>Acta Endocrinologica</u> (Kbh) 105:505 (1984).

10. L.N.Berti, C.Sato, C.J.Gomez and L.Krawiec, Thyroid hormones effects on RNA synthesis in brain and liver of neonatal hypothyroid rats, <u>Horm.Metab.Res</u> 13:691 (1981).

DEVELOPMENT OF THE PANCREATIC B-CELLS:

GROWTH PATTERN AND FUNCTIONAL MATURATION

Bernard Portha

Laboratoire de Physiologie du Développement-GR
Régulations Métaboliques et Diabète-CNRS UA-307
Université Paris 7, Tour 33
75252 Paris Cedex 05, France

INTRODUCTION

The pancreatic B-cell may be regarded as a biological sensor which
plays a key role in controlling the disposal of glucose. It is conceivable
that the ontogenic development of such a system is of decisive importance
for its proper function in postnatal life. While both the growth and
functional differentiation of the fetal B-cell have been the subject of
numerous studies, there are still large gaps in our knowledge as to the
molecular mechanism of these events. It is the purpose of this review to
summarize recent observations on the growth, differentiation and functional
maturation of the fetal B-cell.

DEVELOPMENT OF ISLET CELLS

Basically, the process of differentiation of the islet cells is a
reflection of the activation of genes coding for the various islet
hormones. With regard to the B-cell, the nucleotide sequence of insulin
gene has been known for several years (1,2). Recent studies have aimed at
clarifying the mechanism that governs the expression of this gene and the
control of its phenotypic activity. In several experiments using mice, a
DNA fragment including the human insulin gene was microinjected into
fertilized mouse eggs. These experiments resulted in transgenic progeny
containing the human insulin gene in its genomic DNA (3,4). Transcripts of
the human insulin gene were found in the pancreas but not in other tissues,
and production of human insulin was revealed by the detection of human
C-peptide in the plasma and urine of the transgenic mice.
Immunocytochemical studies clearly showed that human insulin, identified
by an anti-human C-peptide antibody, was present only in pancreatic islet
cells (4). It was also found that the serum levels of human insulin were
properly regulated by the glucose, amino acids, and tolbutamide. These
combined observations indicate that the human DNA fragments present in the
genomic DNA of the transgenic mice carried, in addition to the insulin
gene, the sequences necessary for the tissue-specific expression of this
gene and that these regulatory sequences reacted to homologous signals in
the mouse.

Further studies of the molecular biology of the insulin gene have
indicated a tight regulation of its cellular transcription. Evidence has

Endocrine and Biochemical Development of the Fetus and Neonate
Edited by J. M. Cuezva *et al.*
Plenum Press, New York, 1990

been presented for the existence, in the 5' flanking DNA of the rat insulin I gene, of sequences controlling its cell-specific expression (5).

Analyses of this region suggested that a transcriptional enhancer might be located in the distal position of the 5' flanking DNA. They also presented evidence for the existence of another controlling element located more proximally to the transcription start site and for the view that the activity of both these controlling sequences is restricted to pancreatic B-cells. Recent studies have also demonstrated the existence of nuclear factors derived from an insulin-producing tumor cell line which interact with three distinct regions within the insulin enhancer (6). Further data suggest that the insulin enhancer is controlled by both positive and negative regulators and that the negative regulators may repress the enhancer in non-islet cells (7). Taken together, these studies suggest that the cell-specific regulation of insulin gene transcription involves a complicated pattern of molecules which can act upon a relatively small DNA domain.

The genes for both glucagon and somatostatin have been cloned and sequenced, but to date virtually nothing is known of their regulation (8,9). Therefore, the molecular biology of the differentiation of the A-, D-, and PP-cells is still virgin territory.

Origin of Islet Cells

The embryological origin of pancreatic islet cells has been widely debated. Although it is generally accepted that the endoderm is the source of the exocrine pancreas, it is unclear which germ layer gives rise to the islet of Langerhans. Two theories have been put forth to explain the ontogeny of islets: the neuroectodermal hypothesis and the endodermal hypothesis. The original evidence for a neuroectodermal origin of the islets came from the work of Pearse (10), who showed that the islet cells share with a subset of neuronal cells the characteristic of amine precursor uptake and decarboxylation (APUD). The "APUD" cells, according to Pearse, can take up and decarboxylate amine precursors, and can synthesize and store peptides. The expression of tyrosine hydroxylase (TH), the first enzyme of the catecholamine biosynthetic pathway, in endocrine cells of the pancreas also supports the possibility that the islet cells bear a relationship to neural tissue (11). Other neural markers have been found in the endocrine pancreas, including PGP9.5, neuron-specific enolase, synaptophysin, dopa decarboxylase, A2B5, and phenylethanolamine N-methyl-transferase.

Two sets of experiments argue against Pearse's hypothesis of a neuroectodermal origin for the islet cells. Transplants of vagal quail neural crest into chick embryos produce chimeras that lack quail cells in the pancreatic islets, although the quail cells do populate other sites such as the myenteric plexus of the gut (12). Additionally, when the ectoderm was removed form rat embryos at day 9, prior to migration of the neural crest, the remaining endomesoderm could still generate the endocrine pancreas *in vitro* (13). Since neither of these experiments eliminates the possibility of an early migration of ectodermal cells into the enteropancreatic region, the origin of the pancreatic islets still remains an open question.

The cell lineage relationships among the different populations of pancreatic cells are also unknown. There are several indications that progenitor cells for the islet of Langerhans may be identified by their ability to coexpress markers which normally characterize distinct cell types. First, the possibility of a common precursor to the A and B-cells has been suggested by experiments documenting the presence of cells that transiently contain TH, the first enzyme of the catecholamine biosynthetic pathway, and either glucagon or insulin (11) during development. In addition, cloned cell lines derived from islet cell tumors have been observed to express multiple pancreatic hormones, including glucagon, somatostatin and insulin (14,15). Additional markers are, however, needed to establish the ontogenic relationship between the different pancreatic cell types.

Recently the use of transgenic mice to study vertebrate development has turned out to be a valuable tool. Alpert, Hanahan and Teitelman (16) have analyzed pancreatic development in mice harboring a distinct transgene consisting of the regulatory information flanking the rat insulin II gene linked to the sequences encoding the simian virus 40 (SV-40) large T antigen (Tag). Tag is a nonsecreted nuclear antigen and therefore identifies the cells where it is being synthesized. Insulin appears in the developing mouse pancreas at embryonic day 12. Transgenic mice first express the transgene product two days earlier at day 10, in a few cells of the pancreatic bud. Throughout development and postnatal life, all of the insulin producing B-cells coexpress the hybrid insulin gene. In addition, islet cells containing glucagon, somatostatin, pancreatic polypeptide, and the neuronal enzyme tyrosine hydroxylase coexpress the transgene when they first arise. Similarly, coexpression of these normally distinct cell markers occurs during differentiation of the four endocrine types. The transgene product also appears transiently during embryogenesis in cells of the neural tube and in neural crest. These results suggest a common precursor for the endocrine cells of the pancreas. Moreover, they imply a relationship between neural and pancreatic endocrine tissue.

Ontogeny of Islet Cells

While it is only in the last few years or so that it has been possible to analyze the molecular biology of islet cell gene expression, the morphology of this process, as reflected in islet cell differentiation, has been a matter of study for decades. Earlier observations in this field have been extensively reviewed (17,18). The introduction of sensitive and highly immunocytochemical methods has greatly contributed to the precise definition of the developmental stage when the islet cells first appear.

In the rat pancreas, the first cell to be identified was the A-cell. It appeared in the embryonic pancreas on day 11 of gestation which is just before the fusion of the ventral and dorsal rudiments of the embryonic rat pancreas. At this stage glucagon-reactive cells were found in the epithelium of the foregut and in cap-like pancreatic primordia on both the ventral and the dorsal intestinal walls. The next endocrine cells to appear in the embryonic rat were the B-cells on day 12.5-13.5 and the PP-cells on day 14. These cells seemed to be located entirely within the pancreatic bud, in which islet-like structures could be seen as early as day 13. D-cells appeared 3 days after the B-cells and were located within the primitive islets. The typical adult arrangement of the islet cells in the rat, which consists of a B-cell core surrounded by the other cell types, was attained on gestational day 18.

In the human fetal endocrine pancreas, the first appearance of endocrine cells has been demonstrated as early as the 8th fetal week, when A-, B-, D- and PP cells were found either in primitive islets or as isolated cells adjacent to duct cells (19,20). No immunoreactive cells could however be traced at 7 weeks of intrauterine age (19). It is noteworthy that in the 8-week-old fetal pancreas there was a sizable population of endocrine cells reacting only with antiglicentin antiserum, while the A-cells reacting with both antiglucagon and antiglicentin antisera appeared at 12-13 weeks of gestational age (20).

The morphological observations on the differentiation of human islet cells indicate that the four cell types appear earlier than was hitherto believed. It also seems that the different human endocrine islet cells begin their intrauterine hormone production more simultaneously than is the case with the fetal rat, in which a distinct sequential appearance of islet cells seems to occur.

GROWTH OF THE ISLET CELL MASS IN FETAL AND POSNATAL LIFE

The growth pattern of B-, A- and D-cell populations in late fetal and postnatal life of the rat has been comprehensively studied by McEvoy (21,22). At gestational day 16, insulin-containing B-cells were present but only as scattered, single cells. At this fetal age, glucagon-containing A-cells were the most numerous endocrine cells, accounting for no less than 2% of the total pancreatic mass and over 96% of the endocrine cell mass. During the subsequent period up to birth, there was a rapid increase in the mass of all the endocrine cell types and particularly of the B-cells. B-cells were the predominant endocrine cell type at birth, comprising about 65% of the total islet mass. By comparison, the A-cells made up about 32% and the D-cells only 2.5%. During postnatal life, the percentage volume contribution of islet cells to the total pancreatic volume diminished from just over 4% on the 10th day after birth to about 1.5% at day 210. The total islet volume continued to expand throughout this period, though there were marked differences between the different islet cell types. The B-cell volume was relatively constant between, days 10 and 20 and subsequently increased 10- fold up to day 210. In contrast, the A-cell volume observed on day 50 was nor significantly different from that observed on day 210. The D-cell population had already reached its adult volume on day 35. The pancreatic contents of insulin and glucagon closely reflected the growth of the B- and A-cell volumes respectively. These data suggest that in the rat the total B-cell volume shows a continuous increase with age, while the total A- and total D-cell volumes reach a maximum value relatively early in life. The growth of the total B-cell volume during the adult life of the rat conforms to previous observations which indicate that the rat islet volume continues to increase up to 700 days of age (23,24).

Concerning the possible mechanisms which can account for the growth of the B-cell mass, two mechanisms have been identified as operative in order to increase the B-cell mass, at least in the rat: 1) the replication of pre-existing differentiated B-cells and 2) the differentiation of new B-cells from a pool of nondifferentiated precursor cells.

The main difficulty encountered in the search for these precursor cells committed to become B-cells, is the matter of identification. If such precursor cells could be selectively labeled, prospective studies would indicate their role in islet growth, but unfortunately such methods are not presently available. Nevertheless, alternative approaches have been developed which emphasize the importance of this pool of precursor cells. It has been recently shown that, in the rat fetus, the increase of the B-cell mass was more than twofold between the 2 last days of intrauterine life (25) (between days 20 and 22). However, calculations on the formation of new cells from pre-existing B-cells indicate that this process cannot account for more than 20% of total growth, and thus the remaining 80% must be accounted for by mechanisms other than B-cell division. Neoformation of B-cells from rapidly proliferating nondifferentiated precursors appears to be the most likely explanation. Also in the early postnatal period, observations from our group suggest that neogenesis of B-cells from precursor cells might contribute to the growth of the B-cell mass. We have shown (26) that, when newborn rats are injected with streptozotocin on the day of birth, they exhibit a transient hyperglycemia that is restored to normal levels after the first postnatal week. These changes were accompanied by a marked initial destruction and loss of B-cells, rapidly followed by signs of repair, as evidenced by the typical pattern of the insulin pancreatic stores which are rebuilt after 1 week. This spontaneous recovery process was characterized by the reappearance of insulin-positive cells, as observed in a study by Bonneir-Weir et al. (27) related to quantitative estimation of B-cells in the islets after neonatal streptozotocin treatment. By studying the mitotic rate of the B-cells in the streptozotocin-treated neonates, we have found that their mitotic index was similar to that in the controls (28), thus suggesting that replication of surviving B-cells was not of major importance in the regeneration of the B-cell population. By contrast, using immunocytochemistry, we have found numerous insulin-positive cells within the duct epithelium (29) and, using electron microscopy, we have obtained evidence that apparent budding of B-cells from ducts was a prominent and specific feature of the regeneration period (28). These observations are taken to indicate a rapid formation of B-cells primarily through multiplication and differentiation of undifferentiated precursor cells that would be located in the duct epithelium. To what extent such precursor cells contribute to the adult B-cell growth and whether they exist in species other than the rat, remains to be elucidated.

ROLE OF NUTRIENTS AND GROWTH FACTORS IN THE REGULATION OF B-CELL PROLIFERATION

The B-Cell Cycle

Knowledge of the characteristics of the cell cycle of the pancreatic B-cell is of considerable importance for the full understanding of the mechanism of B-cell proliferation. By using isolated fetal rat islets, in which the progress of the B-cells through the cell cycle had been synchronized *in vitro* with the aid of hydroxyurea, Swenne (30) recently made an extensive study of the lengths of the various phases of the cell cycle. As in other somatic cells, the time between two B-cell divisions can be subdivided into the G1, S and G2 phases. The period of DNA synthesis is confined to the S phase. After cell division, the daughter B-cells can progress through a new cell cycle or enter a resting state denoted G0. From here, B-cells may either be recruited back into the cell cycle or die. By monitoring *in vitro* the time course of the DNA synthesis in synchronized B-cells, Swenne (30) was able to calculate the duration of the generation time, that is, the duration of a full cell cycle: about 15h. The next necessary step was to estimate the proportion of B-cells which move through the cycle at a given moment; this is named the proliferative compartment. The results of Swenne's studies indicate that the size of the proliferative compartment probably never exceeds 10% of the islet cells. Such an observation suggests that only a minor proportion of all B-cells are able to divide but, that these cells, due to a relatively rapid cell cycle, may nevertheless contribute substantially to the growth of the islet organ. It may indeed be calculated that these few, but actively dividing B-cells are able to double the total B-cell number within a period of about 1 month, provided simultaneous loss is negligible.

The above observations were made on B-cells fetal or neonatal rats. In subsequent studies, Swenne (31) showed that aging does not affect the duration of the generation time, but that the amplitude of the DNA biosynthesis decreases as a function of age (from 3 weeks to 3 months). Thus, the proliferative capacity of the B-cells is restricted by a decreased number of cells capable of entering the cell cycle. For example, in 3-month-old rats, the proliferative compartment of the B-cells makes up only about 3% compared with 10% in the perinatal rats. This diminution with age of B-cell proliferation makes it conceivable that the capacity of the islets to respond with proliferation to a diabetogenic stimulus decreases with age.

Some Regulators of B Cell Proliferation

Adjustments of the B-cell mass probably represent a way by which the organism can meet more lasting changes in the demands for insulin production. This can be examplified by the physiological B-cell growth observed in pregnancy and in obesity. It also seems plausible that in diabetes, and especially in Type-2 diabetes, the capacity of the B-cells to respond with proliferation is disturbed. Therefore knowledge of the factors that control the growth of the B-cells is of decisive importance for the understanding of both normal and pathologic blood glucose homeostasis.

The most important physiologic regulator for B-cell growth is probably the glucose molecule. The stimulating effect of glucose on B- cell replication is now well documented. Such an effect is observed *in vivo* as for example in rats maintained hyperglycemic with a chronic glucose infusion (32). It is also observed *in vitro* in tissue culture of fetal rat islets (30). Using the *in vitro* approach, Swenne (30) showed that glucose directly stimulates tritiated-thymidine incorporation in the islets, whatever the methodology used to quantitate DNA replication (autoradiography or radioactivity incorporated in the TCA-precipitable DNA). Moreover, he obtained evidence that the B-cell cycle proceeds at the same rate irrespective of the glucose concentration in the culture medium, and that the glucose in fact acts by increasing the proportion of B-cells which, at a given moment, move through the cell cycle. To summarize, these important observations suggest that the effect of glucose is to increase the size of the proliferative compartment, this compartment representing 3% of the B-cell population at low glucose levels, and 7% of the B-cell population at high glucose levels.

At the moment the molecular mechanism by which glucose is able to initiate the mitotic events is less well understood. Among the other carbohydrates, only D-mannose (18) was reported to enhance DNA synthesis in the B-cell; L-glucose, D-fructose and 3-O-methyl-glucose (18) were without effect; mannoheptulose (18) was inhibitory. It is known that glucose and mannose are readily utilized as metabolic substrates by the B-cell, while mannoheptulose blocks this process by competitive inhibition of sugar phosphorylation. Therefore the effects of glucose and mannose on the replication process could be related to their role as nutrients to the B-cell.

Stimulation of B-cell growth has also been observed in the presence of other metabolic substrates, such as alpha-ketoisocaproic acid (18), which is the first intermediate in the catabolism of leucine and which also stimulates insulin biosynthesis and secretion (18). Thus it seems that in the B-cell, the control of cell replication is linked to a substrate site, but the nature of this site remains to be elucidated.

As indicated above, proliferating islet cells maintain a similar response to glucose during aging from late fetal to adult life. However, up to day 20 of gestation, the fetal rat islet cells do not respond to a high glucose concentration with an increased proliferation (33,34). In this respect, the regulation of islet cell replication forms a parallel with the development of insulin biosynthesis and secretion in that both attain adult characteristics during the perinatal period. The factors governing these processes are unknown. The possibility that hormones may influence the sensitivity of islet cell replication to glucose has been put to a test, but neither growth hormone(GH) nor thyroid hormone had any effect in this respect (34). While the rat fetal B-cell responds poorly to glucose until the end of gestation, a mixture of essential amino acids strongly stimulates islet cell proliferation prior to this period (33,34). Indeed, the amino-acid-enriched medium resembles the physiological condition for the rat fetus late in gestation when the serum level of glucose is low and that of amino acids high. It may be that during fetal development an ample supply of amino acids is neccesary both as a source of energy and as a source of building blocks for constructing peptides.

Although nutrients seem to be the most important regulators of B-cell proliferation, macromolecular growth factors may be necessary to modulate, or even maintain, the proliferative response to sugars and amino acids. This notion is supported by the stimulatory effect of serum on islet cell proliferation (35,36) and by the lack of a replicatory response to glucose in islets cultured in the absence of serum (37). However, fetal rat islets can be maintained in a serum-free, chemically defined medium provided it is supplemented with, among other substances: transferrin, insulin, prolactin, growth hormone, and epidermal growth factor (38). Among specific growth factors, growth hormone (GH), insulin-like growth factor 1 (IGF-1) and platelet derived growth factor (PDGF) have been shown to stimulate B-cell proliferation in islets obtained from rat fetuses at the end of gestation (37,39). The stimulatory effects were substrate dependent in that they were more pronounced at a high glucose concentration.

In order to better understand the intracellular signals required to initiate the DNA replication, fetal B-cells obtained from 22-day old rats fetuses were transfected with oncogene constructs containing the promoter sequences of the insulin gene (40). Transfection with v-src, c-myc or mutated c-Ha-ras oncogenes induced an intense stimulation of DNA synthesis (3-6 fold) of the fetal B-cells. Similar experiments, performed in B-cells from adult obese-hyperglycemic mice showed a weaker stimulation (40). In all experiments the DNA replication was more pronounced in the presence of the high glucose concentration. It was noteworthy that src, the gene product of which acts as a tyrosine kinase, was the most potent in this respect. Since also PDGF and IGF-I activate receptors with tyrosine kinase activity it appears that src somehow shortcuts the normal growth factor mediated pathway in the stimulation of B-cell replication.

FUNCTIONAL MATURATION OF THE B CELLS

The B-cells in the fetal rat pancreas appear relatively responsive to stimulation of insulin biosynthesis by glucose (41,42).

Concerning the secretion process a number of *in vitro* studies have found repeatedly glucose to be ineffective in eliciting insulin release from the pancreas of the fetal rat and human fetus (review in 43). Despite these negative reports, many papers have now been published showing that the fetal pancreas does exhibit responsiveness to glucose. This has been demonstrated in the fetus *in vivo*, in pancreas pieces and in isolated islets (review in 43). A slight but transiently significant "first phase" of secretion was detected at 17 days by Rhoten (44). No responses to glucose or amino-acids were detected on days 15 and 16. However, on day 17 when the response to glucose was absent (or minimal), clear responses to arginine and arginine plus leucine were seen. Thus the failure to respond to glucose lies at some point in the sequence of stimulus-secretion coupling and not in the exocytotic mechanism *per se*. From day 18 of gestation, a stimulated level of secretion in response to sustained glucose was detectable (43-45). However, given the data of the perifusion studies presently available (43), it is clear that there are qualitative differences between fetal and adult islets. In fetal islets 1) the release profile is not typically biphasic and 2) the sustained response that occurs at the time of a second phase does not have the characteristic of the adult second phase because it is not inhibited by Ca^{2+} channel blockers (43). In the neonatal islet (3 days of age) the response has the release profile of an adult islet and is dependent on increased Ca^{2+} entry through the voltage-dependent Ca^{2+} channels.

In attempting to understand the difference between the insulin release responses of fetal and adult pancreases, one can assume that the ability of the B-cell to release insulin is adequately functional and that the exocytotic machinery is not rate limiting. This appears clear from the facts that amino acids stimulate insulin secretion at an early stage in development, when glucose does not (46), and that theophylline or isobutylmethylxanthine both strongly potentiate the fetal pancreatic response to glucose (46-48). Current understanding of the mechanism by which glucose stimulates insulin release is still rudimentary. However, it is known that the second phase of insulin secretion is heavily dependent on increased uptake of extracellular Ca^{2+} through voltage- dependent Ca^{2+} channels. In addition, it is known that in the course of stimulus-secretion coupling with respect to nutrient secretagogues like glucose, metabolism generates signals, one of which leads to closure of K^+ channels in the plasma membrane, reduces K^+ efflux, and leads to membrane depolarization. It is this depolarization that causes the voltage-dependent Ca^{2+} channels to open and allows increased Ca^{2+} entry into the cell down its electro-chemical gradient. An immature response to glucose could result from a missing signal, channel, or sensor in this chain of events.

In the fetal islet, from the lack of inhibition of insulin release by verapamil it appeared probable that voltage-dependent Ca^{2+} channels were not present or not activated by high glucose. With their presence confirmed by insulin release in response to a high KCl concentration, which was blocked by verapamil, an inability of glucose to depolarize the membrane was probable (43). This in turn could be due to a lack of glucose-sensitive K^+ channels or an inability by the B-cell to produce the signal required to block the channel. The presence and effectiveness of the K^+ channels in the fetal islets were suggested with quinine (a blocker of the glucose-sensitive K^+ channels) which potentiated the fetal islet response to glucose and conferred sensitivity to verapamil on the response(43). This conclusion that the K^+ channel does not respond to glucose in the fetal islet is also supported by the experiments of Ammon et al. (47) indicating that ^{86}Rb efflux from fetal islets was unaffected by an increased glucose concentration (at variance with the decrease of the ^{86}Rb efflux rate that occurs in response to glucose in adult islets).

This hypothesis that the immaturity of the response reflects a deficient coupling between the metabolism of nutrient secretagogues and the transmembrane flux of K^+ in the fetal islet was recently put to a more direct test by investigating the ATP-regulated K^+-channels (G- channels) in 21 day-fetal rat B-cells with the patch-clamp technique (49) with corresponding measurements of voltage-activated Ca^{2+} currents, insulin release, glucose metabolism and ATP content. It was found that the G-channels of the fetal B-cells were insensitive to glucose but had otherwise similar characteristics to those of adult rat B-cells. The properties of the voltage-activated Ca^{2+} currents were also similar in the fetal and adult B-cells, whereas the glucose metabolism appeared impaired in the fetal islets. In particular, during stimulation with high glucose concentrations there was no increase in the cellular ATP content. These observations therefore suggest that the deficient insulin response to glucose in the fetal B-cell reflects an immature glucose oxidation leading to an impaired regulation of the otherwise normal G-channel.

In summary the immature B-cell in the fetal pancreas displays an impaired responsiveness to glucose despite relatively adequate insulin reserves and insulin biosynthetic capacity: such poor recognition of the secretory signal seems to be specific of the glucose molecule since a variety of other secretagogues were able to elicit insulin secretion. Such a pattern of decreased or desensitized secretory state in the fetal B-cell in many ways resembles that recognized in the B-cells of human non-insulin-dependent diabetes and in the early stages of human insulin-dependent diabetes. Therefore new insights in the maturation of stimulus-secretion coupling in fetal islets may have ramifications for patho-physiology and therapeutic strategies which extend beyond fetal life.

REFERENCES

1. B.Cordell, G.I.Bell, E.Tischer, F.M.Denoto, A.Ullrich, R.Pictet, W.J.Rutter and H.M.Goodman, Isolation and characterization of a cloned rat insulin gene, Cell, 18:533 (1979).
2. P.Lomedico, N.Rosenthal, A.Efstratiadis, W.Gilbert, R.Kolodner and R.Tizard, The structure and evolution of the two nonallelic rat preproinsulin genes, Cell, 18:545 (1979)
3. D.Bucchini, M-A Ripoche, M.G Stinnakre, P.Desbois, D.Lore, E. Montehioux, J.Absil, J-A Lepesant, R.Pictet and J.Jami, Pancreatic expression of human insulin gene in trangenic mice, Proc.Natl.Acad. Sci.USA 83:2511 (1986)
4. R.F.Seldon, M.J.Skoskiewicz, K.B.Howie, P.S.Russel and H.M.Goodman, Regulation of human insulin gene expression in transgenic mice, Nature 321:525 (1986)
5. T.Edlund, M.D.Walker, P.J.Barr and W.J.Rutter, Cell-specific expression of the rat insulin gene: evidence for role of two distinct 5'flanking elements, Science 230:912 (1985)
6. H.Olssonand and T.Edlund, Sequence-specific interactions of nuclear factors with the insulin gene enhancer, Cell 45:35 (1986)
7. U.Nir, M.D.Walker, P.J.Barr and W.J.Rutter, Regulation of rat insulin I gene expression: Evidence for negative regulation in non-pancreatic cells, Proc.Natl.Acad.Sci.USA 83:3180 (1986)
8. G.Bell, R.F.Santerreand and G.T.Mullenbach, Hamster preproglucagon contains the sequence of glucagon and two related peptides, Nature 302:716 (1983)
9. L.P.Shenand and W.J.Rutter, Sequence of human somatostatin I gene, Science 224:168 (1984)
10. A.G.Pearse, The APUD concept and its implications: related endocrine peptides in brain, intestine, pituitary, placenta, and anuran cutaneous glands, Med.Biol. 55:115 (1977)
11. G.Teitelmanand and J.K.Lee, Cell lineage analysis of pancreatic islet cell development: glucagon and insulin cells arise from catechol-aminergic precursors present in the pancreatic duct, Dev.Biol. 121: 454 (1987).
12. N.Le Douarin, The embryological origin of the endocrine cells associated with the digestive tract: experimental analysis based on the use of a stable cell marking technique, in: "Gut Hormones", S.R. Bloom, ed. Churchill Livingstone, London. pp. 49 (1978)
13. R.Pictet, L.B.Rall, P.Phelpsand and W.J.Rutter, The neural crest and origin of the insulin-producing and other gastrointestinal hormone-producing cells, Science 191:191 (1976)
14. J.Philippe, W.L.Chickand and J.F.Habener, Multipotential phenotypic expression of genes encoding peptide hormones in rat insulinoma cell lines, J.Clin.Invest. 79:351 (1987).

15. O.D.Madsen, L.I.Larsson, J.F.Rehfeld, T.W.Schwartz, A.Lernmark, A.D. Labrecque and D.F.Steiner, Cloned cell lines from a transplantable islet cell tumor are heterogeneous and express cholecystokinin in addition to islet hormones, _J.Cell.Biol_. 103:2025 (1986).

16. S.Alpert, D.Hanaban and G.Teitelman, Hybrid insulin genes reveal a developmental lineage for pancreatic endocrine cells and imply a relationship with neurons, _Cell_ 53:295 (1988).

17. R.Pictet and W.J.Rutter, Development of the embryonic endocrine pancreas, _in_: "Handbook of physiology", sect 7, vol 1. D.F.Steiner and N.Freinkel eds. American Physiological Society, Washington DC pp. 25 (1972)

18. C.Hellerstrom and I.Swenne, Growth pattern of pancreatic islets in animals, _in_: "The diabetic pancreas", 2nd edn. B.W.Volk and E.R. Aequilla eds. Plenum Press, New York, pp. 53 (1985).

19. A.Clark and A.M.Grant, Quantitative morphology of endocrine cells in human fetal pancreas, _Diabetologia_ 25:31 (1983).

20. Y.Stefan, S.Grasso, A.Perrelet and L.Orci, Quantitative immuno-fluorescent study of the endocrine cell populaion in the developing human pancreas, _Diabetes_ 32:293 (1983).

21. R.McEvoy and K.Madson, Pancreatic insulin-, glucagon- and somatostatin-positive islet cell populations during the perinatal development of the rat, _Biol.Neonate_ 38:248 (1980).

22. R.McEvoy, Changes in the A-, B- and C-cell populations in the pancreatic islets during the postnatal development of the rat, _Diabetes_ 30:813 (1981).

23. B.Hellman, The total volume of the pancreatic islet tissue at different ages of the rat, _Acta Pathol Microbiol Scand_. 47:35 (1959).

24. C.Remacle, N.Hauser, M.Jeanjean and A.Gommers, Morphometric analysis of endocrine pancreas in old rats, _Exp.Gerontol._ 12:207 (1977).

25. C.Hellerstrom, The life story of the pancreatic B-cell, _Diabetologia_ 26:393 (1984).

26. B.Portha, C.Levacher, L.Picon and G.Rosselin, Diabetogenic effect of streptozotocin in the rat during the perinatal period, _Diabetes_ 23: 889 (1974).

27. S.Bonner-Weir, D.F.Trent, R.N.Honey and G.C.Weir, Responses of neonatal rat islets to streptozotocin. Limited B-cell regeneration and hyerglycemia, _Diabetes_ 30:64 (1981).

28. M.C.Dutrillaux, B.Portha, C.Roze and E.Hollande, Ultrastructural study of pancreatic B-cell regeneration in newborn rats after dewstruction by streptozotocin, _Virchows Arch (Cell Pathol)_ 39:173 (1982).

29. D.Cantenys, B.Portha, M.C.Dutrillaux, E.Hollande, C.Roze and L.Picon, Histogenesis of the endocrine pancreas in newborn rats after destruction by streptozotocin, _Virchows Arch (Cell Pathol)_ 35:109 (1981).

30. I.Swenne, Role of glucose in the _in vitro_ regulation of cell cycle kinetics and proliferation in fetal pancreatic B-cells, _Diabetes_ 31: 754 (1982).

31. I.Swenne, Effects of ageing on the regenerative capacity of the pancreatic B-cell of the rat, _Diabetes_ 32:14 (1983).

32. J.Logothetopoulos, N.Valiquette and D.Cvet, Glucose stimulation of beta-cell DNA replication in the intact rat and in pancreatic islets in suspension culture. Effects of alpha-keto-isocaproic acid, dibutyryl cyclic AMP, and 3-isobutyl-1-methyl-xanthine in the _in vitro_ system, _Diabetes_ 32:1172 (1983).

33. M.De Gasparo, G.R.Milner, P.D.Norris and R.D.G.Milner, Effects of glucose and amino acids on foetal rat pancreatic growth and insulin secretion _in vitro_, _J.Endocrinol_. 77:241 (1978).

34. I.Swenne, Glucose-stimulated DNA replication of the islets during development of the rat fetus: Effects of nutrients, growth hormone and triiodothyronine, _Diabetes_ 34:803 (1985).

35. A. Rabinovitch, C. Quigley and M. M. Rechler, Growth hormone stimulates islet B-cell replication in neonatal rat pancreatic monolayers, Diabetes 32:307 (1983).

36. I. Swenne, A. J. Bone, S. L. Howell and C. Hellerstrom, Effects of glucose and amino acids on the biosynthesis of DNA and insulin in fetal rat islets maintained in tissue culture, Diabetes 29:686 (1980).

37. I. Swenne, D. J Hill, A. J. Strain and R. D. G. Milner, Growth hormone regulation of somatomedin C/insulin-like growth factor I production and DNA replication in fetal rat islets in tissue culture, Diabetes 36:288 (1987).

38. R. C. Mc Evoy and P. E. Leung, Tissue culture of fetal rat islets: comparison of serum-supplemented and serum free, defined medium on the maintenance, growth, and differentiation of A, B and D-cells, Endocrinology 111:1568 (1982).

39. I. Swenne, C. Heldin, D. J. Hill and C. Hellerstrom, Effects of platelet-derived growth factor and IGF I on the deoxyribonucleic acid replication of fetal rat islets of Langerhans in tissue culture, Endocrinology 122:214 (1988).

40. M. Welsh, N. Welsh, T. Nilsson, P. Arkhammar, R. Blake-Pepinsky, O. Steiner and P. O. Berggren, Stimulation of pancreatic islet B-cell replication by oncogenes, Proc. Natl. Acad. Sci 85:116 (1988).

41. A. Asplund, A. Andersson, C. Jarrousse and C. Hellerstrom, Function of the fetal endocrine pancreas, Israel J. Med. Sci. 11:581 (1975).

42. E. Heinze, H. Schatz, C. Nicole and E. F. Pfeiffer, Insulin biosynthesis in isolated pancreatic islets of fetal and newborn rats, Diabetes 24:373 (1975).

43. R. L. Hole, C. May, M. Pian-Smith and G. W. G. Sharp, Development of the biphasic response to glucose in fetal and neonatal rat pancreas, Am. J. Physiol. 254:E167 (1988).

44. W. B. Rhoten, Insulin secretory dynamics during development of rat pancreas, Am. J. Physiol. 239:E57 (1980)

45. A. Kervran and J. Randon, Development of insulin release by fetal rat pancreas in vitro: effect of glucose, amino acids and theophylline, Diabetes 29:673 (1980).

46. E. Heinze and J. Steinke, Insulin secretion during development: response of isolated pancreatic islets of fetal, newborn and adult rats to theophylline and arginine, Horm. Metab. Res. 4:234 (1972).

47. H. P. T. Ammon, A. Fahmy, M. Mark, W. Strolin and M. A. Wahl, Failure of glucose to affect Rb efflux and ^{45}Ca uptake of fetal rat pancreatic islets, J. Physiol. Lond. 358:365 (1985).

48. V. Grill, K. Asplund, C. Hellerstrom and E. Cerasi, Decreased cyclic AMP and insulin response to glucose in isolated islets of neonatal rats, Diabetes 24:746 (1975).

49. P. Rorsman, K. Bokvist, P. Arkhammar, P. O. Berggren, C. Hellerstrom, T. Nilsson, M. Welsh and N. Welsh, Defective regulation of glucose-and ATP-regulated K^{+} channels results in disturbed glucose sensitivity in fetal B-cells, Diabetologia 31:537A (1988).

INSULIN RECEPTOR BINDING ACTIVITY IN

SKELETAL MUSCLE DURING PREGNANCY

Marta Camps, Manuel Palacín, Xavier Testar
and Antonio Zorzano

Departamento de Bioquímica y/o Fisiología
Facultad de Biología
Universidad de Barcelona
08028 Barcelona, Spain

INTRODUCTION

During pregnancy, the growth of fetal structures represents an important energetic effort for the mother. In that regard, glucose is particularly critical since it represents the major fetal oxidative substrate and as a result the glucose turnover is accelerated during late pregnancy. In parallel, and as an attempt to spare glucose for sustained fetal needs, insulin resistance develops during pregnancy both in women (1,2) and in experimental animals (3-5). In the pregnant rat, insulin resistance has been traced in skeletal muscle (6), adipose tissue (7) and liver (4). However, a defect in insulin action has been only clearly substantiated in adipose tissue and liver when the euglycemic-hyper-insulinemic clamp technique was employed.

Special attention should be paid to skeletal muscle since it is a major target tissue for insulin and it plays quantitatively a predominant role in the *in vivo* insulin-stimulated glucose utilization. During pregnancy, modification of both insulin sensitivity and responsiveness has been reported in muscle(4,6), however the mechanisms involved have not been delimited.

The initial step in the cellular actions of insulin is the binding of the hormone to its receptor located on the cell surface. Insulin receptor from muscle has been isolated and characterized, and consists of two α-subunits (135,000 Da) that bind insulin and two β-subunits (95,000 Da) that exhibit ligand-stimulated tyrosine kinase activity, although it shows some differences compared to receptors isolated from other tissues (8).

In the present study we have attempted to explore possible mechanisms by which insulin action is decreased in muscle during pregnancy. To that end, we have examined the binding properties as well as the structural characteristics of the insulin receptors partially purified from muscle of control (non-pregnant) and pregnant (late gestation) rats. Our data suggest that the insulin resistance found in muscle during late pregnancy is not consequence of an alteration at the level of the insulin receptor binding.

Endocrine and Biochemical Development of the Fetus and Neonate
Edited by J. M. Cuezva *et al.*
Plenum Press, New York, 1990

RESULTS AND DISCUSSION

Insulin Binding in a Lectin-Purified Extract of Skeletal Muscle

In order to obtain evidence for a possible role of the insulin receptor in the development of the insulin resistance during pregnancy, insulin binding was initially measured in wheat germ agglutinin (WGA) purified receptor preparations from skeletal muscles of control and late pregnant rats. The yield of glycoprotein eluted from the WGA chromatography did not differ between control (1.02 ± 0.07 mg/g muscle) and pregnant (1.04 ±0.09 mg/g muscle) groups. Scatchard analysis of ^{125}I-insulin binding yielded identical curvilinear plots for both groups (Figure 1). The high affinity component was characterized by performing a linear regression on

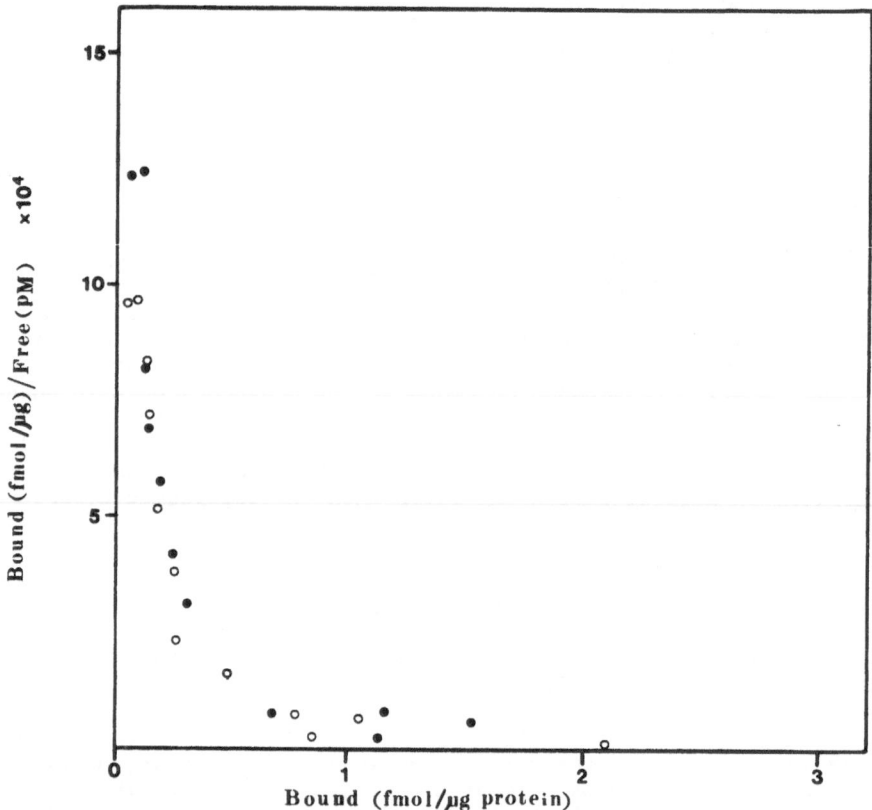

Fig. 1. Insulin binding to solubilized insulin receptors from
pregnant and control rat muscles. Insulin receptors were
partially purified from skeletal muscle of virgin control
(o) and late pregnant rats muscles (●), following
solubilization in Triton-X-100 and ultracentrifugation,
using lectin affinity chromatography. The wheat germ
agglutinin eluate (20µl) was incubated for 1 h at 22°C in
a buffer containing 25 mM Hepes, 0.1 mg/ml bovine serum
albumin, 100 units/ml bacitracin, ^{125}I-insulin, and
varying concentrations of unlabeled insulin in a total
volume of 200µl. Receptors were precipitated with 0.5 ml
of bovine γ-globulin (1 mg/ml) and 0.5 ml of polyethylene
glycol (25% w/v). The data shown are from a representative Scatchard plot (n=5). Each point is the mean of
triplicate determinations.

the initial six binding data points. The affinity of insulin for its receptor was identical in virgin and pregnant rats (Kd=0.25 nM). Also no differences were found in the number of high affinity binding sites in control (0.62 ± 0.11 fmol/μg protein) and pregnant (0.50 ± 0.10 fmol/μg protein) rats.

Additional experiments were performed to examine whether a possible change in receptor phosphorylation state compromising insulin binding in muscle from pregnant rats was missed due to the action of cellular phosphatases during the receptor purification procedure. Thus, receptors were partially purified from muscle of control and pregnant rats in the presence of phosphatase inhibitors NaF (10 mM), Na_3VO_4 (100 μM) and sodium pyrophosphate (10 mM). Using a similar procedure it has been shown that the insulin receptor activated *in vivo* maintains its enhanced phosphorylation state and tyrosine kinase activity after purification (9). In our conditions, the presence of phosphatase inhibitors did not modify the binding characteristics of the insulin receptor in the control or pregnant group (data not shown).

In a separate series of experiments, insulin receptors were purified using WGA chromatography from red and white portions of skeletal muscle in control and pregnant rats. As before, glycoprotein yield and Scatchard plots were superimposable when the control and pregnant groups were compared (data not shown).

Thus, after partial purification of muscle insulin receptors from red and white muscle, no differences in binding properties (high affinity binding sites or dissociation constants) were detected in control and pregnant rats. Overall, unless a differential distribution of insulin receptors is substantiated in muscle during pregnancy, insulin resistance cannot be explained on the basis of insulin receptor number or affinity.

Very recently, an inhibitor of insulin receptor binding activity has been found in skeletal muscle (10). This inhibitor which is removed from a skeletal muscle sarcolemmal fraction after solubilization and wheat germ agglutinin affinity chromatography, has a molecular weight in between 75,000 and 160,000 Da and is heat-resistant. Therefore, it might be postulated that a) this inhibitor affects insulin binding properties *in vivo* or, b) presents a greater inhibitory activity during late pregnancy. These possibilities deserve further study.

Structural Analysis of Insulin Receptors

The molecular mass of the insulin receptors species present in the partially purified receptor preparation in control and pregnant groups, was determined on SDS-polyacrylamide gel electrophoresis (PAGE) following affinity cross-linking of ^{125}I-insulin to the receptor using disuccinimidyl suberate (Figure 2). In keeping with previous reports (11), under non-reducing conditions, a high molecular weight protein band was specifically labeled with ^{125}I-insulin in muscle from control and pregnant rats (Figure 2A). This band corresponds to the holoreceptor ($\alpha2\beta2$), indicating the absence of proteolysis of the insulin receptor during the purification procedure. In addition, no differences in migration were detected in the holoreceptor band from control and pregnant muscle preparations. Therefore, these studies suggest the existence of no major structural differences of insulin receptors in muscle during late pregnancy. Identical results were observed when insulin receptors were purified from red and white muscle in control and pregnant rats (data not shown).

When affinity-labeled muscle insulin receptors from control and pregnant rats were run on SDS-PAGE under reducing conditions (Figure 2B),

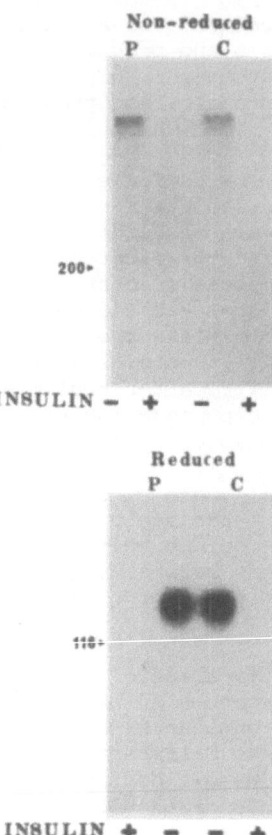

Fig. 2. Affinity cross-linking of ^{125}I-insulin to the insulin
receptor from pregnant and control rat muscles. Partially
purified insulin receptors were obtained from control and
pregnant rat skeletal muscles as described in legend to
Figure 1. Partially purified receptor (40μl) was
incubated at 22°C for 60 min in 30 mM Hepes buffer
containing 0.5nM [^{125}I-TyrB26] insulin in the absence or
presence of 5μM unlabeled insulin. After 5 min incubation
at 0°C, disuccinimidyl suberate was added at a final
concentration of 1 mM, and samples were incubated for a
further 15 min at 0°C. The reaction was stopped by the
addition of Laemmli sample buffer (without or with
dithiothreitol). Samples were subjected to electrophoresis in 5% (nonreducing conditions) or 7.5% (reducing
conditions) polyacrylamide gel. Autoradiograms of representative gels are shown.

a band (Mr=135,000 Da) was specifically labeled with ^{125}I-insulin, which
corresponds to the α-subunit of the insulin receptor. Again, no differences
in the pattern of migration of the α-subunit of the insulin receptor was
detected between control and pregnant group. Under both conditions,
unlabeled insulin (10^{-6} M) in excess completely inhibited the labeling of
the α subunit of the receptor.

Our findings indicate that the decreased insulin action in muscle
during late pregnancy does not involve a reduction in insulin binding.
This is, in contrast, to the situation in adipose tissue where an enhanced

number of surface insulin receptors has been reported (12). On the other hand, contradictory findings regarding insulin binding in hepatocytes or liver plasma membranes have been reported (13,14).

There are a number of other situations characterized by insulin resistance in skeletal muscle such as diabetes, obesity, denervation or uremia. In some of these situations, insulin resistance is associated with either increase in insulin binding (diabetes), decrease in insulin binding (obesity) or no modification of this parameter (denervation, uremia or pregnancy). In addition, the insulin receptor tyrosine kinase activity has been found to be decreased in diabetes or obesity, and unaltered in uremia and denervation. Therefore, all that information allows us to suggest that insulin resistance in muscle can be the result of a variety of molecular mechanisms, either at receptor or post-receptor level.

The data of the present study indicate that the loss of insulin response of skeletal muscle at late pregnancy is the result of ·a defect distal to hormone binding in the pathway of signal transduction of insulin action. Therefore, insulin resistance could be due to a post-receptor defect or alternatively, consequence of an alteration in the tyrosine kinase activity of the insulin receptor. These possibilities are currently under investigation.

ACKNOWLEDGMENTS

This work was supported in part by a grant from the D.G.I.C.T. (PB-573/86) and from F.I.S. (87/1718 and 89/0179), Spain. Pig insulin was kindly provided by Mr. M.L. Johnson (Eli Lilly, Indianapolis, IN, USA).

REFERENCES

1. W.N.Spellacy and F.C.Goetz, Plasma insulin in normal pregnancy, N.Engl.J.Med. 268:988 (1963).
2. E.A.Ryan, M.J.O'Sullivan and J.S.Skyler, Insulin action during pregnancy. Studies with the euglycemic clamp technique, Diabetes 34:380 (1985).
3. R.H.Knopp, H.J.Ruder, E.Herrera and N.Freinkel, Carbohydrate metabolism in pregnancy. VII. Insulin tolerance during late pregnancy in the fed and fasted rat, Acta Endocr. 65:352 (1970).
4. A.Leturque, A.F.Burnol, P.Ferré and J.Girard, Pregnancy-induced insulin resistance in the rat: assessment by glucose clamp technique, Am.J.Physiol. 246:E25 (1984).
5. A.Martin, A.Zorzano, I.Caruncho and E.Herrera, Glucose tolerance test and in vivo response to intravenous insulin in the unanaesthetized late pregnant rat and their consequences to the fetus, Diabet.Metab. 12:302 (1986).
6. R.J.Rushakoff and R.K.Kalkhoff, Effect of pregnancy and sex steroid administration in skeletal muscle metabolism in the rat, Diabetes 30: 545 (1981).
7. A.Leturque, P.Ferré, A.F.Burnol, J.Kande, P.Maulard and J.Girard, Glucose utilization rates and insulin sensitivity in vivo in tissues of virgin and pregnant rats, Diabetes 35:172 (1986).
8. C.F.Burant, M.K.Treutelaar, N.E.Block and M.G.Buse, Structural differences between liver- and muscle-derived insulin receptors in rats, J.Biol.Chem. 261:14361 (1986).
9. C.F.Burant, M.K.Treutelaar and M.G.Buse, In vitro and in vivo activation of the insulin receptor kinase in control and denervated skeletal muscle, J.Biol.Chem. 261:8985 (1986).

10. R. H. Whitson, G. K. Grimditch, E. Sternlicht, S. A. Kaplan, R. J. Barnard and K. Itakura, Characterization of rat skeletal muscle sarcolemmal insulin receptors and a sarcolemmal insulin binding inhibitor, J. Biol. Chem. 263:4789 (1988).

11. D. E. James, A. Zorzano, M. Boni-Schnetzler, R. A. Nemenoff, A. Powers, P. F. Pilch and N. B. Ruderman, Intrinsic differences of insulin receptor kinase activity in red and white muscle, J. Biol. Chem. 261:14939 (1986).

12. D. J. Flint, P. A. Sinnett-Smith, R. A. Clegg and R. G. Vernon, Role of insulin receptors in the changing metabolism of adipose tissue during pregnancy and lactation in the rat, Biochem. J. 182:421 (1979).

13. D. J. Flint, Changes in the number of insulin receptors of isolated rat adipocytes during pregnancy and lactation, Biochim. Biophys. Acta 628: 322 (1980).

14. G. Baumann, G. Puavilai, N. Freinkel, L. A. Domont, B. E. Metzger and H. B. Lerene, Hepatic insulin and glucagon receptors in pregnancy: their role in the enhanced catabolism during fasting, Endocrinology 108: 1979 (1981).

INSULIN BINDING AND RECEPTOR TYROSINE KINASE

ACTIVITY IN RAT LIVER AT TERM GESTATION

Carmen Martínez, Pilar Ruiz, Antonio
Andrés and José M. Carrascosa

Departamento de Biología Molecular
Centro de Biología Molecular C.S.I.C.
Universidad Autónoma de Madrid
28049 Madrid, Spain

INTRODUCTION

It is now well established that in humans and rats late pregnancy is associated with a state of insulin resistance, manifested by decreased sensitivity and responsiveness to the hormone of target tissues (1,2). The growth of the conceptus requires supply of substrates like glucose from the mother, and this is only possible if a mechanism develops that prevents the uptake of glucose by the otherwise consuming tissues. On the other hand, the glucose production by liver is also augmented in rats at term gestation (3,4) in order to maintain the glucose homeostasis and satisfy the increasing fuel needs of fetal tissues. Since glucose oxidation by peripheral tissues is stimulatable by insulin, it seems likely that insulin resistance in producing and consuming tissues could contribute to provide metabolic fuels necessary for the growth of the fetuses.

The first two events in insulin action are the binding of the hormone to the α subunit of its membrane receptor, and the insulin-dependent tyrosine autophosphorylation of the receptor β subunit (5). In the present work we have studied both functions of the rat liver insulin receptor in virgin and 22-day pregnant animals. Our result suggest that the insulin resistance in rat liver at term gestation could be due to a decreased responsiveness to insulin of the receptor tyrosine kinase activity.

RESULTS AND DISCUSSION

Insulin Binding to Solubilized Receptors

Insulin binding studies were performed using wheat germ agglutinin (WGA) purified insulin receptors from rat liver plasma membranes. Receptor preparations were incubated in presence of a tracer amount of (^{125}I-A14) Insulin (0.2 ng/ml) and increasing concentrations of unlabeled hormone (0.3-3000 ng/ml). Binding data were analyzed by the Scatchard method and the use of a computer program based in the two binding sites model (6) resulted in the K_D and Bmax values shown in Table 1. Solubilized insulin receptors from virgin and pregnant rats exhibit a similar K_D for the high and low affinity component of the binding. Regarding to the maximal binding

Endocrine and Biochemical Development of the Fetus and Neonate
Edited by J. M. Cuezva *et al.*
Plenum Press, New York, 1990

51

Table 1. Characteristics of insulin binding to WGA
purified rat liver receptors

Animals	High affinity		Low affinity	
	K_D	Bmax	K_D	Bmax
Virgin rats	4.5×10^{-10}	36.7	0.7×10^{-8}	73.6
Pregnant rats	3.9×10^{-10}	39.7	0.7×10^{-8}	59.6

Scatchard plots were constructed from binding data
obtained in the text. The best fit curve was obtained
using a computer program. K_D values are expressed in M,
and Bmax in pg of insulin/μg of protein.

capacity no differences were also found between virgin and pregnant rats.
These results indicate that insulin signal transduction is probably not
impaired during gestation at the level of hormone binding to the α subunit
of the receptor and the cause of insulin resistance should be attributed
to a postbinding deffect. It should be noted, however, that binding was
measured using partially purified insulin receptors and it might be argued
that insulin binding could be modulated in pregnant rats by an inhibitory
compund removed during the purification process. Such an inhibitor has been
recently found in skeletal muscle (7) and it seems intersting to pay more
attention to this putative regulation mechanism of insulin binding in rat
liver. Experiments are currently under progress in our laboratory to
elucidate this issue.

Kinase Activity of Rat Liver Insulin Receptor

After insulin binds to its receptor in the plasma membrane, autophosphorylation of the β subunit takes place leading to the stimulation of the
receptor intrinsic tyrosine kinase activity. In order to explore the
possibility insulin signal transduction to be negatively modulated at the
level of receptor kinase activity in pregnant rats we have partially
isolated rat liver insulin receptors by affinity chromatography in WGA-
Sepharose in presence of phosphatase inhibitors (100 mM NaF, 10 mM sodium
pyrophosphate) to preserve the original phosphorylation state of the
receptor molecule. Kinase activity of receptor preparations was tested
using casein as exogenous substrate. Equal amounts of insulin receptors
from virgin and 22-day-pregnant rats were preincubated overnight in
presence or absence of insulin (1 ng/ml) and phosphorylated for 15 min with
50 μM unlabelled ATP. Casein (final concentration 0.3 mg/ml) was added
together with 4 μCi of (γ-^{32}P) ATP, and the reaction continued for 10 min.
After addition of 2% SDS / 1% glycerol / 1% β-mercaptoethanol, samples were
boiled and proteins separated by SDS-PAGE. Radioactive proteins were
identified by autoradiography. Gels were alternatively treated with 1 N
KOH in order to remove serine and threonine-bound phosphate from proteins:
This method allows the quantification of the ^{32}P incorporated in tyrosine
residues exclusively (8).

Table 2 shows the results obtaines after densitometry of the
corresponding autoradiograms. Both basal and insulin stimulated phosphorylation of casein were decreased when receptors from pregnant rats were
used, and the same was observed after KOH treatment indicating that the
tyroisne kinase activity of rat liver insulin receptor is disminished at
term gestation. This fact is in good agreement with the observed decreased
^{32}P incorporation of the β subunit of insulin receptor as shown in Table 2.
Since receptors were preincubated with unlabelled ATP, the radioactivity
incorporated does not represent the total phosphorylation achieved by the

Table 2. Phosphorylation of casein by WGA purified liver
insulin receptor

| Animals | Before KOH treatment | | | After KOH treatment | | |
| | Insulin | | Insulin | Insulin | | Insulin |
	−	+	effect	−	+	effect
Virgin	33.1	71.4	2.15	1.9	16.7	8.8
Pregnant	17.6	28.4	1.6	0.75	0.9	1.2

Autophosphorylation of insulin receptor β subunit

Virgin	15.8	323.7	20	6.7	200.4	29.9
Pregnant	16.7	115.7	6.9	2.6	33.3	12.0

Incorporation of ^{32}P into casein and the β subunit of rat
liver insulin receptor is expressed in arbitrary units
obtained from densitometry of autoradiograms. One unit is
equivalent to 3 fmol of Pi incorporated/10 min/30 μg of
casein. The arbitrary units for the receptor phosphorylation do not correspond to the total ^{32}P incorporation
as discussed in the text.

receptor molecule but could be indicative of the auto-phosphorylation rate
of insulin receptor.

In conclusion, the data presented here suggest that the diminished
insulin response of rat liver at term gestation could be the result of a
decreased stimulation by insulin of the insulin receptor tyrosine kinase
activity. Our present knowledge does not allow us to determine if the
alteration of the kinase activity is consequence of a decreased sensitivity
or a reduced responsiveness to the hormone action (9), and if this is due
to a structural modification of the ATP-binding site or of the putative
phosphorylation sites. On the other hand, the interaction between insulin
and its receptor, at least in solubilized preparations, seems to be
unaffected during late pregnancy.

ACKNOWLEDGMENTS

This work was supported by a grant from the Universidad Autónoma de
Madrid. The Centro de Biología Molecular is recipent of institutional
grants from the Ramón Areces Foundation and the FIS, Spain. We thank
Alberto Martínez for the computer analysis of the binding data.

REFERENCES

1. A. Leturque, P. Ferré, P. Satabin, A. Kervran and J. Girard, *In vivo* insulin
    resistance during pregnancy in the rat, Diabetologia 19:521 (1980).
2. G. Puavilai, E. C. Drobny, L. A. Domont and G. Bauman, Insulin receptors and
    insulin resistance in human pregnancy: Evidence for a postreceptor
    defect in insulin action, J. Clin. Endocrinol. Metab. 54:247 (1982).
3. C. Valcarce, J. M. Cuezva and J. M. Medina, Increased gluconeogenesis in
    the rat at term gestation, Life Sci. 37:553 (1985).
4. J. M. Cuezva, C. Valcarce, M. Chamorro, A. Franco and F. Mayor, Alanine and
    lactate as gluconeogenic substrates during late gestation, FEBS
    Letters 194:219 (1986).
5. O. M. Rosen, After insulin binds, Science 237:1452-1458(1987).

6. P.J.Munson and P.Rodbard, LIGAND: A versatile computerized approach for characterization of ligand-binding systems, <u>Anal.Biochem.</u> 107: 220 (1980).

7. R.H.Whitson, G.K.Grimditch, E.Sternlicht, S.A.Kaplan, R.J.Barnard and K.Itakura, Characterization of rat skeletal muscle sarcolemmal insulin receptors and a sarcolemmal insulin binding inhibitor, <u>J.Biol.Chem.</u> 263:4789 (1988).

8. J.A.Cooper and T.Hunter, Changes in protein phosphorylation in rous sarcoma virus-transformed chicken embryo cells, <u>Mol.Cell.Biol.</u> 1: 165 (1981).

9. C.R.Kahn, The molecular mechanism of insulin action, <u>Ann.Rev.Med.</u> 36: 429 (1985).

GLUCOSE HOMEOSTASIS IN SUCKLING RATS

UNDERNOURISHED DURING INTRAUTERINE LIFE

Fernando Escrivá, Consuelo Rodríguez, Carmen Alvárez
and Ana M. Pascual-Leone

Instituto de Bioquímica
(C. Mixto C.S.I.C.-Universidad Complutense)
Facultad de Farmacia
28040 Madrid, Spain

INTRODUCTION

Nutrional status is an important factor with regard to glucose
tolerance and insulin release. Modifications observed in over and undernourished men and experimental animals have been the subject of numerous
studies. It is well known that a relatioship exists between obesity and
some types of diabetes, in fact, it was the first risk factor for diabetes
to be identified. Overnutrition causes glucose intolerance because of a
defect in stimulus-secretion coupling in the beta cell and insulin
resistance (1).

More recently, it has also been recognised that the opposite (that
is, dietary protein-calorie deficiency) causes perturbations of glucose
homeostasis that could lead to several forms of diabetes. Undernutrition
must be regarded as an important factor not only in the cause of diabetes
but also in clinical features of diabetic syndrome, as it has been recently
reviewed (2). The incidence of malnutrition-related diabetes is important
in populations of developing countries (3).

In the past years, a lot of work has been reported showing that
impaired glucose tolerance can be associated with chronic malnutrition in
man as well as in animals. In a minority of undernourished children,
nutritional rehabilitation does not lead to recovery in glucose
intolerance and, consequently, irreversible damage may have been caused
(4). Recently, two studies confirm that undernutrition in rats brings
about an impaired insulin secretory response which persists in decreased
state after refeeding (5,6).

Most experimental models concenrning this subject try to characterize
these aspects of glucose homeostasis in rats submitted to protein-calorie
malnutrition during a variable length of time, but ussually in the post-
weanning stages. Less attempt has been made to face this aspect of undernutrition in experimental models in which the dietary restriction was
impossed at the fetal stage. However, such situations are closer to
majority of malnutrition-diabetes related cases in impoverished humans
from countries of the third world (2). Maternal food restriction produces
a significant decrease in plasma insulin concentrations in both fetal and

Endocrine and Biochemical Development of the Fetus and Neonate
Edited by J. M. Cuezva *et al.*
Plenum Press. New York. 1990

55

suckling rats. This could be of importance in the light of the role of pancreatic hormones with respect to suitable disposal of nutrients during the perinatal period. During this period, metabolism is necessary to achieve normal development (7).

The aim of the present work is to study the possibility that food restriction in pregnant and nursing rats could lead to an alteration in glucose homeostasis in their pups. This has been evaluated by measuring the basal levels of circulating glucose and insulin and by glucose tolerance tests.

In our experimental model, undernutrition of Wistar rats was established beginning from the 16th day of pregnancy. Undernourished pregnant and lactating rats were given a restricted mass of the standard laboratory food (18% protein), roughly 50% of that consumed by the control. This amount is according to a protocol previously described (8); controls fed the same food *ad libitum*. An intraperitoneal glucose tolerance test (IPGTT) was performed at 8, 14 and 22 days of age in suckling rats. The pups were removed from their mothers and 70 minutes later they were injected i.p. with a 20% (w/v) glucose solution at a dose of 2 g/kg bw. At different times, as indicated in Figures, the rats were killed without anaesthesia by decapitation and arteriovenous blood was collected from the trunk. Glucose was measured enzymatically and insulin by radioimmunoassay using rat insulin as standard. Data are mean ± SEM for 8-10 rats; statistical comparison of the results were analysed with Student's "t" test.

RESULTS

Food restriction to pregnant rats prevented the 20% increase of body weight that the control group underwent from the 16th day of gestation until delivery. During lactation, the pups from restricted mothers gained body weights at a significantly lower rate than controls (results not shown).

Basal blood glucose concentrations were lower in undernourished rats at all stages studied (Figure 1). IPGTT caused an increase in blood glucose

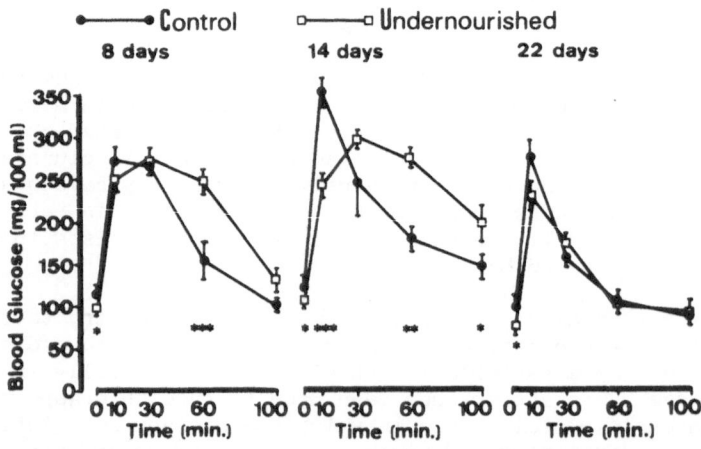

Fig. 1. Blood glucose concentrations during an IPGTT at different ages in suckling rats whose mothers were undernourished from 16th day of gestation. *p<0.05; **p<0.01; ***p<0.001.

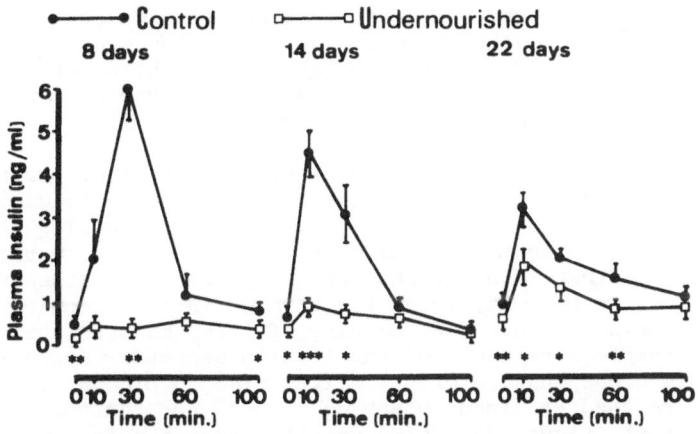

Fig. 2. Plasma insulin levels during an IPGTT at different
ages in suckling rats whose mothers were undernourished from 16th day of gestation. *, p<0.05;
**, p<0.01; ***, p<0.001.

level and the peak was reached at 10 or 30 minutes (Figure 1). In 8 day-
old undernourished rats blood glucose level was significantly greater than
in the controls at 60 minutes after i.p. injection. The glucose level was
also greater at 60 and 100 minutes in 14 day-old undernourished animals.
Integrated values for glucose concentrations during the test, calculated
as areas under the curves and above basal levels (6), were also greater in
undernourished rats at 8 and 14 day of life. In contrast, both 22 day-old
control and undernourished rats underwent a similar pattern of changes in
blood glucose levels all through the test. The integrated glucose responses
were very close.

Glucose load was effective in promoting a rapid insulin secretory
response in control rats (Figure 2). The highest peaks was obtained in the
8 day-old rats, the response being less marked in the succesive stages of
life. On the contrary, this secretory response was almost completely absent
in the 8 day-old malnourished animals and severely depressed at 14 and 22
days. Therefore, plasma insulin levels in these rats were significantly
lower than the control values during most of the test. Basal plasma insulin
concentrations were also affected by restriction on mother's diet, and they
remained lower than controls during suckling; these values, expresed as
ng/ml, were 0.75±0.30 for 8 day-old control rats; 0.79±0.27 for 14 day-old
control rats and 0.92±0.31 for 22 day-old control rats; and 0.51±0.15,
0.45±0.15 and 0.33±0.17 ng/ml for their undernourished counterparts.

DISCUSION

It has been observed that rats given a protein-deficient food
voluntarily reduce their daily intake (9) so that experimental animals fed
with a low-protein but isocaloric diets are actually submitted to protein-
calorie malnutrition. This was the dietary condition impossed to pregnant
and nursing rats in our study.

Restricted availability of food beginning from intrauterine life is
a common situation in children from undernourished populations.
Consequently, our model based on reduced daily intake from a gestational
period is closer to malnourished human beings than other studies in which
nutritional stress is imposed in post-weaning stages. Undernutrition of

pups is brought about inderectely since food restriction in mothers decreased milk production without major changes in the composition of the milk. The significant failure to gain weight proves the negative effect of maternal restriction on the offspring. This is a feature commonly found in human and animal protein-energy malnutrition.

In the model we used for undernutrition, the rats from restricted mothers presented a poor insulin response to i.p. glucose load during suckling. Despite the high blood glucose levels reached during the test (Figure 1) (the most important stimulus of insulin release) 8 and 14 day-old malnourished rats did not show the expected marked increase in plasma insulin concentrations, which is observed in their control. Rather this response remained blunted. At 22 days of life, malnourished animals underwent an incomplete recovery of the insulin secretory response so that exogenous glucose induced only a slight rise in plasma insulin. The attained level reached remained in any case below the control values throughout the test (Figure 2). These results suggest a functional impairment of the pancreatic beta cell because of the fact that circulating basal insulin levels were lower in undernourished rats than in controls. In adults or children after weaning, as in animals it has been reported in numerous studies that protein-energy malnutrition leads to a lack of insulin secretion in response to various stimuli. This contributed to glucose intolerance (1,2).

In view of the fact that pancreatic response to exogenous glucose is weak or absent, we could expect that glucose handling by peripheral tissues was clearly poorer in malnourished rats. An impaired glucose tolerance is shown in the 8 and 14 day-old undernourished animals. However, the differences in relation to controls during the test suggest that intolerance is relatively mild. At 22 days of life, the glucose tolerance had normalized and did not differ from that of the control animals (Figure 1).

The most surprising feature in our study is that the weak pancreatic response to glucose-load in malnourished rats is not paralleled by a high degree of glucose intolerance. Eight and 14 day-old undernourished rats have only a moderate impairment in glucose tolerance, regardless of low levels in plasma insulin during the test. At 22 days of life, malnourished animals become tolerant to glucose, while insulin secretory response is improved but not fully recovered. These results could be explained if impaired insulin secretion were associated with an increased sensitivity to insulin. Swenne et al.(5) and Okitolonda et al.(6) have shown that long-lasting protein-calorie malnutrition induces a persistent high tissue sensitivity to insulin. In that line, it has been recently shown that a diet containing 5% of protein chronically supplied to rats induces an increase in the number of insulin receptors in hepatocytes (10).

In order to verify if the described condition persists after weaning we have continued the food restriction to rats until two months of age. Preliminary results (not shown) are in the line of those discussed above: basal plasma insulin levels remain still lower in malnourished than in control adult animals. Glucose tolerance is not impaired although insulin released by a glucose-load is very reduced. The degree of the decline in blood glucose level after injection of exogenous insulin is greater in malnourished rats, and the return to initial values is slow and incomplete.

In conclusion, these results agree with recent studies showing an adaptation of glucose homeostasis in the rats submitted to long-term protein-calorie deficiency. Such an adaptation would depend on an increase in the sensitivity of the tissues to the insulin. Because of that, an exogenous glucose-load is tolerated by undernourished rats, in spite of low levels of plasma insulin secreted by these animals.

ACKNOWLEDGMENTS

Supported by DGICYT grant PM88-0021, Spain.

REFERENCES

1. L.S.Levine, P.G.Writght and F.Marcus, Failure to secrete immuno-reactive insulin by rats fed a low protein diet, Acta Endocrinol. 102:240 (1983).
2. R.Harsha-Rao, Diabetes in the undernourished: coincidence or consequence?, EndocrineRev. 9:67 (1988).
3. WHO Study Group, "Diabetes Mellitus", Tech. Rep. Ser. 727:20 World Health Org., Geneve (1985).
4. G.C.Cook, Glucose tolerance after kwashiorkor, Nature (Lond.) 215:1295 (1967).
5. I.Swenne, C.J.Crace and R.D.G.Milner, Persistent impairment of insulin-secretory response to glucose in adult rats after limited period of protein-calorie malnutrition early in life, Diabetes 36:454 (1987).
6. W.Okitolonda, S.M.Brichard and J.C.Henquin, Repercussions of chronic protein-calorie malnutrition on glucose homeostasis in the rat, Diabetol. 30:945 (1987).
7. E.Alvarez, S.Fernández and E.Blázquez, Effect of maternal food restriction on circulating insulin and glucagon levels and or liver insulin and glucagon binding sites of fetal and suckling rats, Diabetes and Metab. 12:337 (1986).
8. F.Escriva, C.Rodríguez and A.M.Pascual-Leone, Glycemia, ketonemia and brain enzymes of ketone body utilization in suckling and adult rats undernourished from intrauterine life, J.Neurochem. 44:1358 (1985).
9. L.S.Phillis, in: "Endocrine Control of Growth", W.M.Daughaday, ed., Elsevier, New York (1981).
10. U.S.Thakur, V.Thakur, O.Singh, A.Mittal, M.G.Karmarkar and M.M.S.Ahuja, Protein calorie malnutrition: is the basic anomaly in insulin secretion or tissue receptors?, Diabetol. 31:403 (1988).

GLUCOSE METABOLISM AND INSULIN SENSITIVITY

DURING SUCKLING PERIOD IN RATS

Tarik Issad, Marçal Pastor-Anglada[+], Christine Coupé,
Pascal Ferré and Jean Girard

Centre de Recherches sur la Nutrition
C.N.R.S.
92190 Meudon-Belleuve, France

[+]Departamento de Bioquímica y Fisiología
Universidad de Barcelona
08028 Barcelona, Spain

INTRODUCTION

In most mammals the mechanism contributing to the regulation of
glucose metabolism does not seem to be fully developed during suckling.
Glucose infusion to the suckling rat (1) is followed by a marked
hyperglycemia concomitant to the persistence of hepatic glucose production.
Moreover, glucose tolerance capacity during suckling, as determined by
intragastric tolerance test, is lower in suckling than in weaned rats (2).
Either hormonal or nutritional changes occurring during the suckling-
to-weaning transition are likely to contribute to the full development of
the metabolic capacities of the pups. In fact, in most mammals weaning
corresponds to a change from a low-carbohydrate, high-fat diet (milk) to a
high-carbohydrate, low-fat diet and the role of this nutritional transition
on the appearance of the adult characteristics is not well understood.

The goal of the present work was to study glucose metabolism before
and after weaning as well as to determine the role of insulin in regulating
glucose homeostasis during this period. Thus, we infused glucose to
anesthesized suckling and weaned rats in the postabsorptive state, we
measured glycemia as well rates of hepatic glucose production (Ra) and of
glucose disappearance (Rd) by means of isotopic tracers. To study insulin
sensitivity of glucose metabolism we used the euglycemic hyperinsulinemic
clamp coupled to the measurement of whole body glucose kinetics and glucose
utilization by individual tissues as described previously (3). The studies
were performed on both suckling and weaned rats, but to ascertain the role
of the nutritional transition in the development of glucose metabolism and
insulin sensitivity, suckling rats were weaned either on to a high-
carbohydrate (HC group) or a high-fat (HF group) diet.

RESULTS

Basal glycemia and glucose turnover rates were similar in both
suckling and HC-weaned rats. When animals were infused with glucose at

Endocrine and Biochemical Development of the Fetus and Neonate
Edited by J. M. Cuezva *et al.*
Plenum Press. New York. 1990

61

20 mg·min^{-1}·kg^{-1} (140% of basal turnover rate), HC-weaned rats showed a sustained hyperglycemia over 8.3mM, but in suckling rats glucose levels increased with time until reaching the value of 15.5 mM 90 min after the beginning of the glucose infusion. At this time plasma insulin levels were markedly higher in suckling than in HC-weaned rats (data not shown), thus indicating that hyperglycemia was not due to impaired insulin secretion. Furthermore, the insulin/glucagon ratio after 90 min of glucose infusion was two-fold higher in suckling than in HC-weaned rats. Rates of glucose utilization increased during glucose infusion and reached a plateau of over 110 μmol·min^{-1}·kg^{-1} in both experimental groups. Meanwhile, hepatic glucose production was fully abolished in HC-weaned rats but not in suckling animals, where it remained constant over 33 μmol.min^{-1}.kg^{-1} during all the infusion period. Thus, it seems clear that the persistence of the hepatic glucose production, which is likely to be due to gluconeogenesis (4), is actively contributing to the hyperglycemia of suckling rats. As an attempt to further study the role of the residual hepatic glucose production in the development of hyperglycemia, we infused glucose to suckling rats that had been previously treated with 3-mercaptopicolinate, an inhibitor of the gluconeogenic flux. In these conditions endogenous glucose production was abolished, but hyperglycemia did not plateau and remained higher than in the HC-weaned rats (data not shown), suggesting that the residual hepatic glucose production was not the only factor responsible for hyperglycemia.

The fact that suckling rats develop high hyperglycemia despite a large increase in plasma insulin concentrations suggests the presence of a not well developed insulin sensitivity of glucose metabolism. We studied this specific point by using the euglycemic hyperinsulinemic clamp. For the same insulin infusion rate (12 munits·min^{-1}·kg^{-1} after a priming dose of 0.33 units·kg^{-1}) suckling rats developed a higher insulinemia than that found in weaned animals (Table 1), suggesting a lower clearance rate of insulin in suckling rats. Basal glucose turnover rate of HF-weaned rats was similar to that of either suckling or HC-weaned animals, but glycemia was higher (Table 1). Under euglycemic clamp, glucose utilization rate increased in the three experimental groups, but to a lesser extent in suckling than in weaned animals (Table 1). No differences induced by diet were observed in weaned animals (Table 1), but since glycemia was clamped at basal value, which were higher in HF- than in HC-weaned rats, glucose metabolic clearance rate was significantly lower in HF- than in HC-weaned animals (28 to 35 ml·min^{-1}·kg^{-1}). Under euglycemic clamp, endogenous glucose production was abolished in HC-weaned rats but not in suckling and HF-weaned animals (Table 1).

The 2-deoxyglucose technique (3) allowed measurements of glucose utilization by peripheral insulin-dependent tissues in either basal or euglycemic hyperinsulinemic states. Since the glycemia of HF-weaned rats was higher than in other groups, we used indexes of glucose metabolic clearance rates instead of glucose utilization indexes. These results are shown in Table 2. No differences were found in basal glucose clearance indexes of the extensor digitorum longus in the three groups of animals. Under clamp, indexes increased to a similar extent in all groups. Basal glucose clearance index of soleus muscle was lower in HF-weaned rats than in other groups, but under clamp, indexes of soleus from HC-weaned animals were markedly higher than those of suckling and HF-weaned rats. Diaphragms from HC-weaned animals showed higher glucose clearance indexes than those from the other groups either in the basal or in the clamped state. White adipose tissue in the basal state showed a higher glucose clearance index in HC-weaned rats than in the other groups, but under clamp, the differences were still more marked, because white adipose tissue from HC-weaned animals was extremely sensitive to insulin, conversely to what was found for the other groups. Basal glucose clearance of brown adipose tissue was lower in HF-weaned rats than in suckling and HC-weaned rats. In

Table 1. Glucose kinetics, plasma insulin and plasma free-fatty acids in the basal state and during euglycemic hyperinsulinemic clamp

	Suckling		HC-weaned		HF-weaned	
	Basal	Clamp	Basal	Clamp	Basal	Clamp
Blood Glucose (mM)	6.0±0.1	5.6±0.2	5.7±0.2	5.6±0.2	6.9±0.2**	7.2±0.3**
Ru (μmol /min/kg)	79±4	111±5	78±5	196±6***	72±5	204±8***
RH (μmols /min/kg)	79±4	49±5	78±5	6±4***	72±5	50±5
IRI (μunits /ml)	18±3	833±108	44±5**	405±36**	31±6	50±10**
FFA (μM)	400±30	430±80	470±60	180±40***	430±50	175±30***

Ru: Glucose utilization rate; RH: Hepatic glucose production; IRI: Immunoreactive insulin; FFA: Free Fatty Acids. Results are mean ± S.E.M. of five to seven determinations. Statistical comparisons were done vs. suckling rats by using the Student's t test, **p<0.01, ***p<0.001.

the three groups there was a marked stimulation of glucose clearance index by insulin, but differences between them remained.

DISCUSSION

Our experiments either with glucose infusion or under euglycemic hyperinsulinemic clamp, both coupled to kinetic measurements, have shown that liver and extrahepatic tissues contribute to glucose intolerance during suckling, the former due to the inability to suppress glucose production when plasma insulin levels are much higher than those needed to abolish hepatic glucose production in HC-weaned rats and the latter due to their decreased sensitivity to insulin, which leads to a lower enhancement of their glucose utilization rates.

Insulin Effect on Hepatic Glucose Production

Liver of suckling rats shows an insulin resistance, which disappears only if the animals are weaned on to a HC-diet. Thus it is clear that the nutritional change that occurs during the suckling to weaning transition is actively contributing to the development of liver insulin sensitivity. An anti-insulin effect exerted by glucagon could be involved in this insulin resistance because plasma glucagon levels are higher in suckling and HF-weaned than in HC-weaned rats (data not shown). Since it has been reported that the binding of insulin to liver plasma membranes is not significantly different before and after weaning (5), it is likely that post-receptor events are involved in the insulin resistance. As confirmed by the 3-mercaptopicolinate experiments, glucose production in suckling

Table 2. Glucose clearance indexes in muscles and adipose tissues in the
basal state and during euglycemic hyperinsulinemic clamp

	Suckling		HC-weaned		HF-weaned	
	Basal	Clamp	Basal	Clamp	Basal	Clamp
Diaphragm	20±4	60±2	58±9**	123±5**	13±2	46±2
Soleus	9±1	15±2	12±2	34±1**	5±1**	19±4
EDL	9±1	28±3	12±1	32±1	10±2	38±5
WAT	3±1	3±1	6±1*	23±2***	4±1	6±1*
BAT	10±2	100±7	6±1	140±24	2±0.5	60±12*

EDL: Extensor Digitorum Longus; WAT: White Adipose tissue; BAT: Brown
Adipose Tissue. Clearance indexes are given as ml/min per kg. Results
are the mean ± S.E.M of five to seven determinations. Comparisons are
done vs. suckling rats under the same experimental state. Statistical
significance by means of the Student's t test, *p<0.05, **p<0.01,
***p<0.001.

rats results from gluconeogenesis; this pathway is less sensitive to
insulin that glycogenolysis (6), which is probably providing glucose in
HC-weaned animals. Furthermore, as pointed out by Terretaz et al. (7),
suppression of hepatic glucose production during euglycemic hyperinsulinemic clamp in adults rats fed a HC-diet is linked to an acceleration
of the glycolytic and the lipogenic fluxes. Both fluxes are decreased
during suckling and after weaning on a HF-diet because of the carbohydrate-
dependent development of key enzymes of these pathways, such as glucokinase
and acetyl-CoA-carboxylase (4).

Insulin Effect on Glucose Clearance Rates

    Glucose infusion experiments showing that for a higher hyperglycemia
in suckling than in HC-weaned rats glucose disappearance rates remained
similar in both groups, strongly suggested what thereafter was confirmed
by using the euglycemic hyperinsulinemic clamp: decreased sensitivity to
insulin of whole-body glucose utilization in suckling rats. The actual
magnitude of the different response of each group of rats might be even
underestimated if we consider that plasma insulin levels were higher in
suckling than in weaned rats. Weaning on to a HF-diet also induced a lower
sensitivity to insulin as shown in terms of glucose metabolic clearance
rates.

    Diaphragm and soleus from suckling rats showed insulin resistance,
although the possibility cannot be excluded that extensorum digitorum
longus could be also resistant, because, as indicated above, plasma insulin
levels during clamp were higher in suckling than in weaned rats. Very
similar results were found for muscles from HF-weaned animals; moreover,
the overall skeletal muscle contribution to the total body weight is
greater in weaned than in suckling animals (25% vs.16%), and this might
explain why the insulin resistance at the whole body level is milder in
HF-weaned rats. Mechanisms contributing to this insulin resistance are
still to be determined, but it is likely that either a decrease in insulin

binding or post-receptor events might be involved. Indeed, it has been shown that muscles isolated from HF-diet fed rats show a decreased effect of insulin and glucose transport and metabolism concomitant with a low insulin binding (8).

Furthermore, plasma non-sterified fatty acid levels are not decreased during clamp in suckling rats (Table 1) and this is likely to contribute to a low glucose clearance because fatty acid are used as alternative oxidative substrates (9).

Insulin resistance in white adipose tissue from suckling and HF-weaned rats is likely to be due to the very low activities of the key enzymes of lipogenesis (2). Preliminary studies from our laboratory suggest that neither a decreased insulin receptor number nor a lack of stimulation of glucose transport by insulin binding to isolated adipocytes are solely involved in insulin resistance in both suckling and HF-weaned animals (unpublished observations).

The lack of insulin resistance in brown adipose tissue of suckling and HF-weaned rats despite the low lipogenic capacity, might be due to the fact that HF-diets enhance its thermogenic activity and it is known that insulin potentiates glucose oxidation when the tissue is activated (10).

CONCLUSION

These results show that there is a marked insulin resistance during suckling which is mediated by a low insulin sensitivity of liver, muscles and white adipose tissue. The nutritional change that occurs during the suckling-weaning transition is a major factor contributing to the development of insulin sensitivity at weaning.

REFERENCES

1. P.Ferré, P.Turlan and J.Girard, Effects of medium chain triglyceride feeding or glucose infusion on glucose kinetics in the newborn rat, J.Develop.Physiol. 7:37 (1985).
2. M.Tsujikawa and S.Kimura, Effect of litter size on glucose metabolism in rat pup adipose tissue during the suckling period, J.Nutr.Sci. Vitaminol 27:361 (1981).
3. P.Ferré, A.Leturque, A.F.Burnol, L.Pénicaud and J.Girard, A method to quantify glucose utilization in vivo in skeletal muscle and white adipose tissue of the anaesthetized rat, Biochem.J. 228:103 (1985)
4. P.Ferré, J.F.Decaux, T.Issad and J.Girard, Changes in energy metabolism during the suckling and weaning period in the newborn, Reprod.Nutr.Develop. 26:619 (1986).
5. E.Blázquez, E.Rubalcava, R.Montesano, L.Orci and R.H.Unger, Development of insulin and glucagon binding and the adenylate cyclase response in liver membranes of the prenatal, postnatal and adult rat: Evidence of glucagon "resistance", Endocrinology 98:1014 (1976).
6. J.L.Chiasson, J.E.Liljenquist, F.E.Finger and W.W.Lacy, Differential sensitivity of glycogenolysis and gluconeogenesis to insulin infusion in dogs, Diabetes 25:283 (1976).
7. J.Terretaz, F.Assimacopoulos-Jeannet and B.Jeanrenaud, Inhibition of hepatic glucose production by insulin in vivo in rats: Contribution of glycolysis, Am.J.Physiol.250:E346 (1985).
8. M.L.Grundleger and S.W.Thenen, Decreased insulin binding, glucose transport and glucose metabolism in soleus muscle of rats fed a high-fat diet, Diabetes 31:232 (1982).

9. P.J.Randle, E.A.Newsholme and P.B.Garland, Regulation of glucose uptake by muscle. 8. Effects of fatty acids, ketone bodies and pyruvate, and alloxan-diabetes and starvation on the uptake and metabolic fate of glucose in rat heart and diaphragm muscles, Biochem. J. 93:652 (1964).

10. S.Ebner, A.F.Burnol, P.Ferré, M.A. De Saintaurin and J.Girard, Effects of insulin and norepinephrine on glucose transport and metabolism in rat brown adipocytes. Potentiation by insulin of norepinephrine-induced glucose oxidation, Eur.J.Biochem. 170:469 (1987).

MATURATION OF THE SYMPATHETIC NERVOUS SYSTEM:
ROLE IN NEONATAL PHYSIOLOGICAL ADAPTATIONS AND
IN CELLULAR DEVELOPMENT OF PERIPHERAL TISSUES

Theodore A. Slotkin, Elizabeth M. Kudlacz,
Qang-Chang Hou and Frederic J. Seidler

Department of Pharmacology
Duke University Medical Center
Durham, North Carolina 27710, U.S.A.

INTRODUCTION

It obvious that the development of the nervous system is marked by
the progression from immaturity of function to the acquisition of
neurotransmission and integrated control of synaptic activity. However,
there is an increasing realization that neural function in the fetus and
neonate also serve specialized needs which are particular to development,
and that the onset of maturity equally represents the loss of these unique
patterns of neural activity. Nowhere is this more evident than in the
sympathetic nervous system and its endocrine counterpart, the adrenal
medulla. This review will detail recent work which demonstrates how
catecholamines released first by the adrenal, and later by neurons, mediate
the transition from fetal to neonatal physiological function as well as the
subsequent programming of postsynaptic reactivity and cellular differentiation in target tissues. We have chosen the rat for study as this
species is altricial, and therefore develops neural function relatively
late. Thus, in the rat many fetal neural characteristics persist into the
postnatal period.

CATECHOLAMINES AND THEIR ROLE AT BIRTH

Hypoxia-induced Catecholamine Release From the Adrenal Medulla

Although release of adrenal catecholamines in the mature organism is
usually associated only with situations of extreme stress, studies from a
number of laboratories indicate that the adrenal medulla of the fetal and
neonatal organism plays a more important role in adaptation to the external
environment than it does in the adult. In many species, including the rat,
and to a somewhat lesser extent, man, sympathetic innervation of autonomic
end-organs is absent or non-functional at birth, and in these cases the
neonate is predominantly dependent upon adrenomedullary catecholamines for
achieving adrenergic responses (1-3). Catecholamine release increases with
the approach of parturition, culminating in a profound surge of catecholamine release at delivery (1-3), evoked in part by hypoxia associated with
vaginal delivery (4-5).

It is thus of prime importance that a key characteristic of the

Endocrine and Biochemical Development of the Fetus and Neonate
Edited by J. M. Cuezva *et al.*
Plenum Press, New York, 1990

67

fetal/neonatal adrenal medulla is its unique ability to respond to hypoxia. In the mature organism, stressors such as hypoxia result in reflex activation of the splanchnic nerve, which then leads to discharge of adrenal catecholamines. However splanchnic control of adrenomedullary function is generally absent at birth, appearing only toward the end of the first postnatal week in the rat, and becoming fully mature by 10 days of age (5-8). Nevertheless, the chromaffin cells do contain all the machinery necessary to synthesize, store and release catecholamines upon appropriate stimulation (9-11); indeed, despite the absence of functional neural connections to the immature adrenal medulla, the tissue is not passive to stressful stimuli. Instead, catecholamine release proceeds by a non-neurogenic mechanism which is not present in mature animals. The differences between neurogenic and non-neurogenic release mechanisms can be illustrated by the acute effects of hypoxia (5,7). In mature animals, stress-induced catecholamine secretion is reflexly-mediated and can therefore be blocked by chlorisondamine, a nicotinic antagonist which prevents splanchnic nerve signals from reaching the chromaffin cells of the adrenal medulla. In neonatal rats, hypoxia produces a catecholamine secretory response, but chlorisondamine cannot block the effect (Figure 1). Additionally, the proportion of norepinephrine and epinephrine released by non-neurogenic stimulation exactly duplicates their ratio in the immature adrenal; in contrast, mature, neurogenic secretion is characterized by selective release of either amine, attendant upon excitation of specific populations of descending spinal neurons which provide afferent input to the splanchnic nerve.

The unique mechanism for catecholamine release from immature adrenal disappears concurrently with the onset of neurogenic control (5-10). This is not simply a coincidence: the onset of nerve stimuli is directly responsible for terminating the phase in which non-neurogenic responses predominate. Thus, acceleration of the onset of splanchnic nerve function caused either by repeated maternal stress during late gestation, or by neonatal hyperthyroidism, is invariably accompanied by the loss of non-neurogenic catecholamine secretory responses (5). A definitive proof that it is the development of neural competence which initiates the switchover of secretory mechanism has been provided by recent studies in which neonatal rats were subjected to surgical denervation of the adrenals at 3 days of age, prior to the onset of splanchnic function (7). The denervated group maintained the ability to release catecholamines non-neurogenically in response to hypoxia, whereas the response was entirely neurogenic in the innervated cohort. Finally, rats denervated in adulthood also showed the reappearance of non-neurogenic response capabilities after 3 weeks without neural input to the adrenal. Thus, the development of neural stimulation itself appears to be responsible for the ontogenetic loss of the unique response pattern in the fetus and the neonate.

Physiological Role of Adrenal Catecholamine Release

The non-neurogenic release of catecholamines in response to hypoxia is of critical significance to the organism. Indeed, if animals are adrenalectomized, even with corticosteroid replacement they fail to survive hypoxia (5). In contrast, prevention of catecholamine release from sympathetic nerves (bretylium administration) has no effect, a finding in concert with the relative immaturity of sympathetic neural function. Similarly, chlorisondamine, which blocks neuronal input to the adrenal, does not influence hypoxic survival time until the development of neural competence after the first postnatal week.

We have recently explored the physiological role of catecholamines in enabling the newborn to survive during hypoxic stress (5,12,13). Blockade

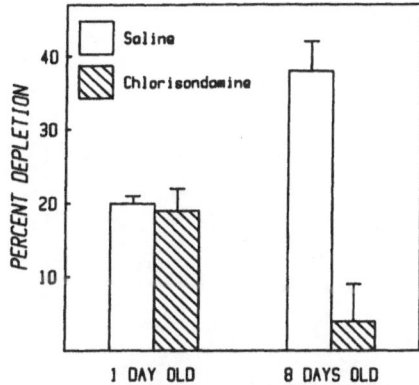

Fig. 1. Hypoxia-induced release of catecholamines.
Non-neurogenic release of adrenal catecholamines by hypoxia in neonatal rats. At one
day of age, hypoxia causes release which
cannot be blocked by chlorisondamine, a
nicotinic antagonist. By 8 days of age the
response is purely neurogenic and is thus
blocked by chlorisondamine (5).

of the effects of circulating catecholamines with either α or β2-selective
adrenergic antagonists, leads to death under low oxygen conditions.
Somewhat surprisingly, maintenance of cardiac function during hypoxia is
dependent upon catecholamine actions at α-, not β-receptors. The fetal and
neonatal myocardium is particularly enriched in α-receptors which are
postsynaptic and which disapear with the onset of innervation (14). So long
as catecholamine release and α-receptor function remain intact, hypoxia
does not cause any significant alteration of heart rate, and cardiac
conduction characteristics are well-maintained (Figure 2) (12). However,
administration of phenoxybenzamine, an α-receptor blocking agent, results
in loss of hypoxia tolerance, marked by a progressive decline in sinus
rhythm and appearance of marked atrioventricular conduction defects,
culminating in frank cardiac failure. Again these responses are not present
by 8 days of age, when non-neurogenic catecholamine secretory mechanisms
have disappeared and cardiac receptor populations lose their immature
characteristics. Thus, in addition to a specialized mechanism for
catecholamine secretion in response to hypoxia, the fetus and neonate
possess a unique adrenergic receptor population linked to a specialized
set of physiological responses which maintain cardiac conduction under
conditions in which it would otherwise fail.

In addition to cardiac components in the role of catecholamines in
neonatal survival during hypoxia, unusual respiratory adjustments are
critical and are mediated through β2-receptors. Normally, as birth
approaches, lung liquid needs to be reabsorbed and surfactant synthesized
and released to prepare the respiratory system for the demands of air-
breathing. Although the fetal and neonatal lung has a relatively low
concentration of β-receptors (15), the sites that are present become
coupled efficiently to lung liquid reabsorption (Figure 3) and to
surfactant synthesis on the day before birth (13). Immediately afterwards,
despite the fact that the number of receptor sites increases, the coupling
to liquid reabsorption disappears and β-receptors are then most highly
linked to enzymes involved in alveolar development. This illustrates that,
in addition to situations where unique receptor populations are present to

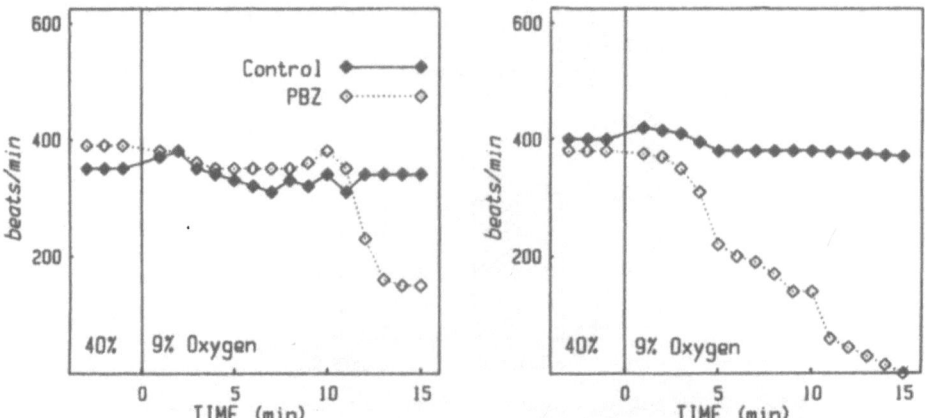

Fig. 2. Role of α-adrenergic stimulation in preserving
        cardiac function (heart rate) during hypoxia in 1
        day-old rats. Although phenoxybenzamine (PBZ) alone
        does not affect heart rate, function is severely
        impaired during hypoxia in drug-treated rats, leading
        to cardiac arrest (12).

receive the catecholamine signal at birth (cardiac α-receptors), existing
receptor populations may be linked transiently to selective processes
which are needed during the transition from intrauterine to extrauterine
life (lung β-receptors).

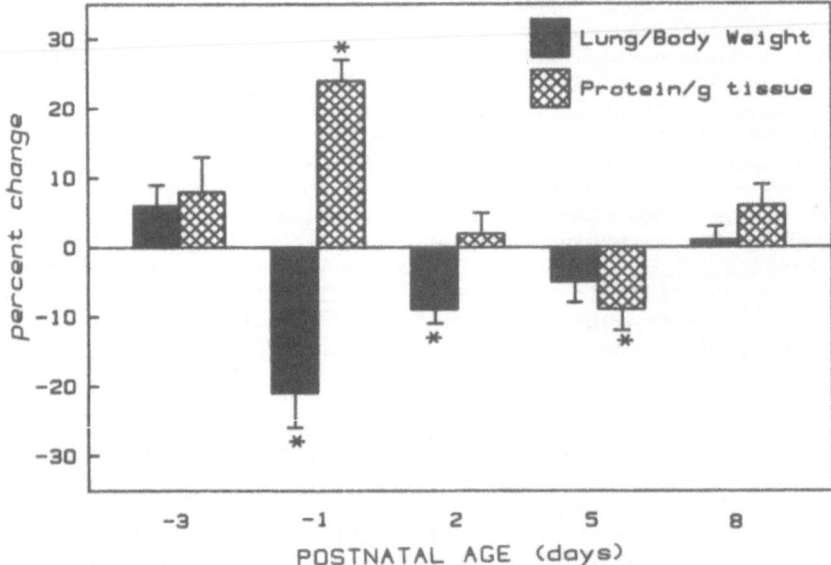

Fig. 3. Effects of β-adrenergic agonists on lung weigth and
        protein. Ability of β-receptor stimulation to evoke
        lung liquid reabsorption, as assessed by the
        consequent reduction in lung/body weight ratio or by
        the increase in lung protein concentration. Effects
        were assessed 4 hr after challenge with terbutaline
        (before birth) or isoproterenol (postnatally) (13).

These findings raise the issue of potential teratologic alteration of development and disappearance of specialized fetal/neonatal response mechanisms. Because of the crucial role played by adrenal catecholamines in neonatal survival, perinatal factors which affect development of sympathetic nerves, ontogeny of central regulation of sympatho-adrenal outflow, or ontogeny of receptor subtypes and their link to specific cellular events, all can have adverse impact on survival. The separation of neonatal adrenergic function into non-neurogenic and neurogenic phases produces a corresponding difference in vulnerability. In general, situations which accelerate the onset of nerve competence lead to the premature loss of non-neurogenic response mechanisms and their protective effect toward neonatal hypoxia (5-8,16,17). These include neonatal hyperthyroidism, exposure to heavy metals, maternal drug abuse and maternal stress. Consequently, acceleration of neural development is neither beneficial, nor innocuous because the premature onset of innervation leads to loss of non-neurogenic secretory mechanisms, unique receptor sub-populations and their selective linkage to fetal/neonatal physiological requirements. These animals are thus more susceptible to hypoxia-induced death. In practical terms, these neural effects may contribute to the increased perinatal mortality and Sudden Infant Death Syndrome which have been noted in the offspring of opiate addicts.

SYMPATHETIC NEURONAL CONTROL OF SYNAPTIC SENSITIVITY AND CELL MATURATION

The development of sympathetic neuronal function occupies two distinct phases during the postnatal period (20). Initially, while terminals are forming in their target tissues, neural activity and transmitter turnover are measurable, but quite low. After the establishment of fully-functional physiological connections of neurons to their targets (second to fourth postnatal week in the rat), there is a period of pronounced hyperactivity, in which impulse frequency and turnover may reach twice the adult value (19,20). Recent work has suggested that the surge of neuronal activity plays a crucial role in programming of cellular responsiveness and in controlling differentiation of sympathetic target tissues.

In the mature organism, neuronal hyperactivity leads to desensitization of receptor sites and thus one would ordinarily expect loss of receptors during the developmental surge of sympathetic activity. However, receptor sites actually increase and postsynaptic sensitivity rises during this phase (21,22), indicating that the adult-type transsynaptic relationship is not present at this time. Indeed, we have found that β-receptors in the neonatal heart do not desensitize with repeated agonist challenge (21). Recent work from our laboratory has shown conclusively that, in fact, a positive, rather than negative, trophic relationship exist between the development of neuronal input and the establishment of postsynaptic receptor sensitivity (23,24). Early denervation with 6-hydroxydopamine does not compromise the development of receptor binding sites themselves, nor does it lead to up-regulation of receptor numbers as it would in the adult. However, in these animals the linkage of the receptors to specific cellular events does not develop properly. In the heart, for example, the ontogenetic rise in β-adenergic sensitivity of heart rate control which ordinarily occurs during the surge in sympathetic tone, lags behind by several weeks if the tissue is denervated at birth (Figure 4,top). More drastic functional impairment is evident at the cellular level: the denervated hearts never develop a fully-efficient link between β-receptor activation and ornithine decarboxylase activity. As stimulation of this enzyme is a prerequisite for compensatory tissue growth, animals whose hearts were denervated at birth are permanently incapable of evoking reactive cardiac hypertrophy in response to increased cardiac work (Figure 4,bottom). The surge in

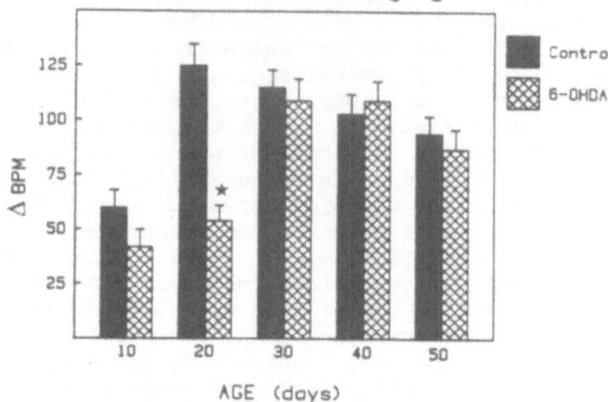

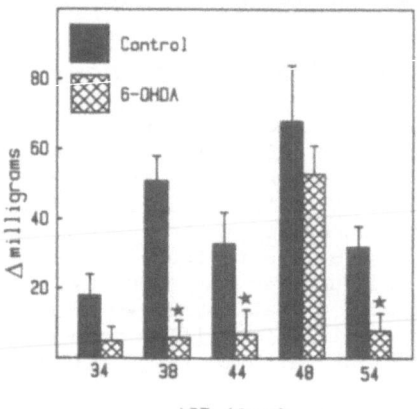

Fig. 4. Effects of neonatal denervation with 6-hydroxy-
        dopamine on the development of the heart rate
        response to isoproterenol (top) and on the
        ability of repeated challenge with isoproterenol
        to evoke cardiac hypertrophy, as assessed by
        tissue protein levels (bottom). Hypertrophy was
        evoked by daily administration of isoproterenol
        begun on postnatal days 30, 40 or 50, followed
        by sacrifice 4 or 8 days after initiating
        treatment (24).

sympathetic neuronal activity thus serves to "program" the future
relationship of β-receptors to cellular functions.

    Sympathetic neurons also appear to coordinate cell replication and
differentiation in a manner linked to the surge of activity. During
sympathetic hyperactivity, there is a switchover of postsynaptic cells
from the replication phase to the differentiation phase (15), thus
enhancing the developmental decline in DNA synthesis. These events are
coupled to the maturation of sympathetic neural activity and are mediated
through β-receptors; the specific linkage of the receptors to DNA

synthesis disappears immediately after the period of sympathetic hyper-activity. Consequently, premature development of sympathetic synapses and/or elevation of sympathetic tone also have adverse effects on cellular development. Indeed, administration of isoproterenol early in development has been shown to duplicate the shut-down of cell replication which ordinarily attends sympathetic hyperactivity (15). Even more dramatic is the demonstration that abnormalities of kidney growth caused by a teratogen (methylmercury) can be prevented by neonatal sympathectomy (16), confirming that premature arrival of neuronal signals and elevated sympathetic activity caused by this heavy metal compound are responsible for the effects on end-organ development. Thus, alterations in ontogeny of sympathetic neuronal function can have drastic actions on cellular development and responsiveness of peripheral tissues.

CONCLUSIONS

The neonatal nervous system possesses unique patterns of activation and signal transduction which provide for the needs of the immature organism. Release of catecholamines from the adrenal medulla in response to the hypoxia associated with birth, proceeds by specialized, non-neurogenic mechanisms and provides a critical signal necessary to survive the transition to extrauterine life. In a corresponding fashion, adrenergic target tissues possess unusual receptor sub-populations and transiently-expressed linkages of receptors to response patterns which are required only during this period. The catecholamine release mechanism and end-organ responses lose their fetal/neonatal characteristics and acquire the adult pattern as a result of the onset of mature patterns of sympathetic neuronal function, and thus both interference with, or promotion of, sympathetic activity have a net deleterious effect on adaptability of the neonate. After neuronal competence is established, sympathetic projections are transiently hyperactive, a condition necessary to the programming of postsynaptic responsiveness and in establishing the timing of cell replication and differentiation. Accordingly, factors which interfere with neuronal development can evoke permanent changes in physiological response patterns and in structure of innervated tissues.

ACKNOWLEDGMENTS

Supported by USPHS NS-06233, HD-09713 and EPA CR-813769.

REFERENCES

1.   H. Lagercrantz and P. Bistoletti, Catecholamine release in the newborn infant at birth, Pediat.Res. 11:889 (1973).
2.   H. Lagercrantz and T. A. Slotkin, The "stress" of being born, Sci. Am. 254 (4):100 (1986).
3.   T. A. Slotkin and F. J. Seidler, Adrenomedullary catecholamine release in the fetus and newborn: secretory mechanisms and their role in stress and survival, J. Devl. Physiol. 10:1 (1988).
4.   P. Bistoletti, L. Nylund, H. Lagercrantz, P. Hjemdahl and H. H. Strom, Fetal scalp catecholamines during labor, Am. J. Obset. Gynecol. 147:785 (1983).
5.   F. J. Seidler and T. A. Slotkin, Adrenomedullary function in the neonatal rat: Responses to acute hypoxia, J. Physiol. 358:1 (1985).
6.   T. A. Slotkin, P. G. Smith, C. Lau and D. L. Bareis, Functional aspects of development of catecholamine biosynthesis and release in the sympathetic nervous system, in: "Biogenic Amines in Development", H. Parvez and S. Parvez, eds., Elsevier, Amsterdam (1980).

7. F.J.Seidler and T.A.Slotkin, Ontogeny of adrenomedullary reponses to hypoxia and hypoglycemia: Role of splanchnic innervation, Brain Res.Bull. 16:11 (1986).

8. T.A.Slotkin, Development of the sympatho-adrenal axis, in: "Developmental Neurobiology of the Autonomic Nervous System", P.M.Gootman, ed., Humana Press, Clifton, N.J.(1986).

9. T.A.Slotkin, Maturation of adrenal medulla.I. Uptake and storage of amines in isolated storage vesicles of the rat, Biochem.Pharmacol. 22:2023 (1983).

10. T.A.Slotkin, Maturation of adrenal medulla. II. Content and properties of catecholamine storage vesicles of the rat, Biochem.Pharmacol. 22:2033 (1983).

11. R.N.Rosenthal and T.A.Slotkin, Development of nicotinic responses in the rat adrenal medulla and long-term effecs of neonatal nicotine administration, Brit.J.Pharmacol. 60:59 (1977).

12. F.J.Seidler, K.K.Brown, P.G.Smith and T.A.Slotkin, Toxic effects of hypoxia on neonatal cardiac function in the rat: α-Adrenergic mechanisms, Toxicol.Lett. 37:79 (1987).

13. E.M.Kudlacz, H.A.Navarro, J.P.Eylers, S.S.Dobbins, S.E.Lappi and T.A.Slotkin, Selective linkage of β-receptors to functional responses in developing rat lung and liver: Phosphatidic acid phosphatase, ornithine decarboxylase and lung liquid reabsorption, J.Devl.Physiol. (in press).

14. T.A.Slotkin, R.J.Kavlock, T.Cowdery, L.Orband, M.Bartolome, W.L.Whitmore and J.Bartolome, Effects of neonatal methylmercury exposure on adrenergic receptor binding sites in peripheral tissues of the developing rat, Toxicology 41:95 (1986).

15. T.A.Slotkin, W.L.Whitmore, L.Orband-Miller, K.L.Queen and K.Haim, Beta adrenergic control of macromolecule synthesis in neonatal rat heart, kidney and lung: Relationship to sympathetic neuronal development, J.Pharmacol.Exp.Ther. 243:101 (1987).

16. T.A.Slotkin and J.Bartolome, Biochemical mechanisms of developmental neurotoxicity of methylmercury, Neurotoxicology 8:65 (1987).

17. T.A.Slotkin, Perinatal exposure to methadone: How do early biochemical alterations cause neurofunctional disturbances?, Prog.Brain.Res. 73:265 (1988).

18. T.A.Slotkin, Endocrine control of synaptic development in the sympathetic nervous system: The cardiac-sympathetic axis, in: "Developmental Neurobiology of the Autonomic Nervous System", P.M.Gootman, ed., Humana Press, Clifton, N.J. (1986).

19. P.G.Smith, T.A.Slotkin and E.Mills, Development of sympathetic ganglionic transmission in the neonatal rat: Pre- and postganglionic nerve response to asphyxia and 2-deoxyglucose, Neuroscience 7:501 (1982).

20. F.J.Seidler and T.A.Slotkin, Development of central control of norepinephrine turnover and release in the rat heart: Responses to tyramine, 2-deoxyglucose and hydralazine, Neuroscience 6:2081 (1981)

21. C.Lau, S.P.Burke and T.A.Slotkin, Maturation of sympathetic neurotransmission in the rat heart.IX. Development of transsynaptic regulation of cardiac adrenergic sensitivity, J.Pharmacol.Exp.Ther. 223:675 (1982).

22. F.J.Seidler and T.A.Slotkin, Presynaptic and postsynaptic contributions to ontogeny of sympathetic control of heart rate in the preweanling rat, Brit.J.Pharmacol. 65:531 (1979).

23. Q.C.Hou, F.E.Baker, F.J.Seidler, M.Bartolome, J.Bartolome and T.A.Slotkin, Role of sympathetic neurons in development of β-adrenergic control of ornithine decarboxylase activity in peripheral tisues: Effects of neonatal 6-hydroxydopamine treatment, J.Devl.Physiol. (in press).

24. Q.C.Hou, F.J.Seidler and T.A.Slotkin, Development of the linkage of β-adrenergic receptors to cardiac hypertrophy and heart rate control: Neonatal sympathectomy with 6-hydroxydopamine, J.Devl.Physiol. (in press).

EFFECT OF THYROXINE AND CORTISOL ON

BRAIN CATECHOLAMINES IN NEONATAL RATS

Luis Goya, Cecilia Aláez, Francisco Rivero,
Maria J.Obregón* and Ana M. Pascual-Leone

Instituto de Bioquímica, Facultad de Farmacia
C. Mixto C.S.I.C.- Universidad Complutense
28040 Madrid, Spain

*Instituto de Investigaciones Biomédicas, C.S.I.C.
Facultad de Medicina, Universidad Autónoma
28029 Madrid, Spain

INTRODUCTION

There is strong evidence for an important role for both thyroxine and corticosterone as coordinators of maturational events. The role of thyroxine in the postnatal maturation of mammalian brain is well known (1). The influence of thyroid status or glucocorticoids on the maturation of the Autonomic Nervous System, particularly the Catecholamine System, has also been studied (2), and both thyroxine and glucocorticoids apparently accelerate the development of sympathetic neuronal interconnections in the short term. However, central biogenic amines are also regulated by steroids and thyroid hormones (3). Because adrenergic activity in the periphery is governed trans-synaptically by suprasegmental mechanisms in the central nervous system, it is important to determine the exact role of these monoamines in adrenomedullary maturation and the possible influence of hormonal status on these central mechanisms. The rat is the species of choice for these studies because innervation of the adrenal medulla develops after birth. Thus, the sympatho-adrenomedullary reflex evoked by insulin-induced hypoglycemia is absent in the neonate and must await the development of splanchnic innervation toward the end of first postnatal week (4). In young rats adrenal CA release has thus been used as a marker for the development of functional innervation.

Rats treated at birth with high doses of thyroxine (neo-T4 syndrome) or cortisol exhibit reduced brain and body weights and permanent derange- ment in behavior and cerebral cortical morphology. A decrease in pituitary TSH and in pituitary acidophilic cells, and consequently in pituitary GH are found in these animals, along with profound alterations in carbohydrate metabolism (5). The adaptation to fasting of neo-T4 rats suggested a possible perturbance in adrenomedullary secretion. Nevertheless neonatal treatement with thyroxine or cortisol on the maturation of adrenomedullary secretion and on the specific role of biogenic amines in the brain have not been completely elucidated. Indeed, retardation of development of the adrenomedullary sympathetic reflex has been identified in these populations. Accordingly, the current study examines the cerebral content

Endocrine and Biochemical Development of the Fetus and Neonate
Edited by J. M. Cuezva *et al.*
Plenum Press, New York, 1990

of dopamine (DA), norepinephrine (E) and total catecholamines (CA) in the neo-T4 and cortisol-treated rats as compared with hypothyroid and normal rats, in an attempt to correlate derangement of brain development with retardation of sympatho-adrenal development.

We have assessed the percentage of DA and NE vs. total CA at 6, 8 and 23 days of age in control, neo-T4, cortisol-treated and PTU (propylthio-uracil) treated rats. Finally we have contrasted these effects with those in the hypothalamus in order to evaluate potential involvement of biogenic amines in selective brain regions.

RESULTS

Four groups of neonatal rats were studied: a) T4 (20ug per day given in the first 5 days (neo-T4 syndrome), b) cortisol (0.5 mg on the day after birth), c) PTU by gastric intubation (50 mg/day) begin on the 18th day of gestation (hypothyroid rats) and d) control. All these animals were killed by decapitation at 6, 8 and 23 days postnatally and brain and hypothalamus were removed rapidly, frozen, weight, homogenized and assayed for catechol-amines by the COMT-radioenzymatic method (6). This method has a sensitivity in the femtomole range.

In normal development, there was a gradual increase in the percentage of DA in the total CA and a decrease in the percentage of NE (Figure 1).

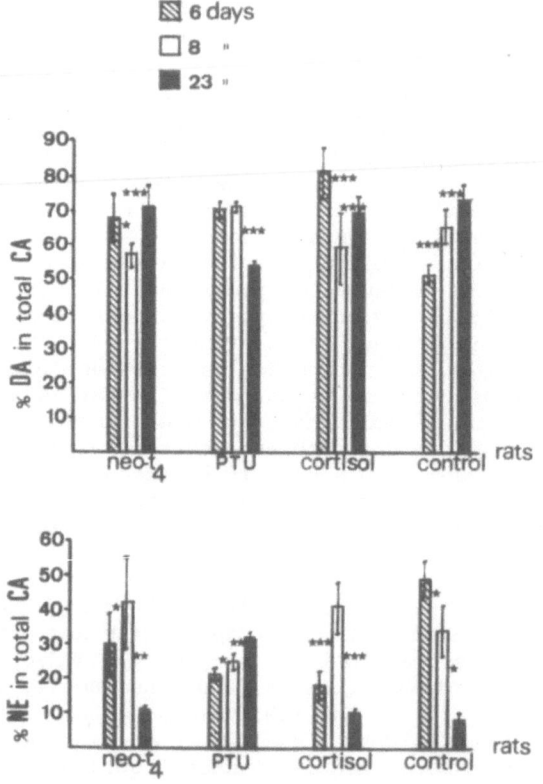

Fig. 1. Brain catecholamines. % DA and % NE in total CA in neo-T4, cortisol treated, PTU treated mother from gestation (hypothyroid rats) and control rats at 6,8 and 23 days of life. Number of rats: 8-10.

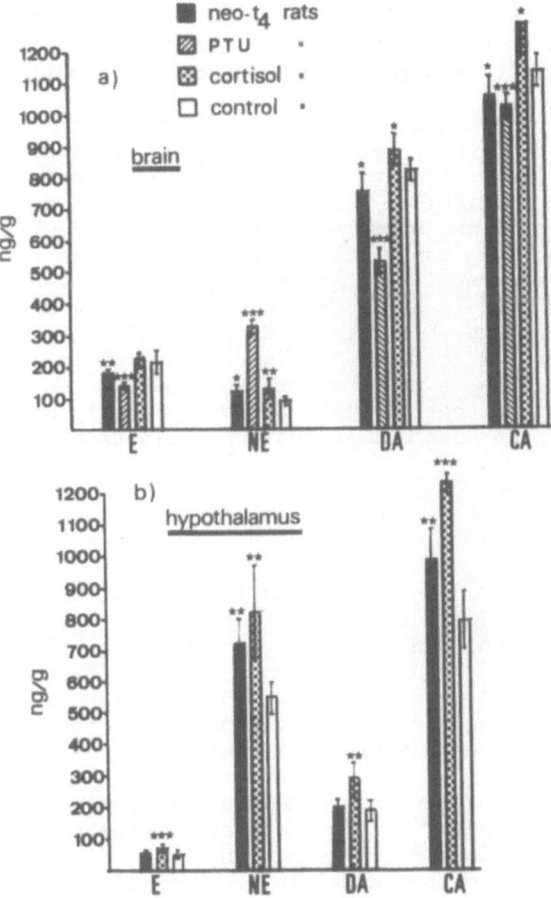

Fig. 2. a) E, NE, DA and total CA in the brain (ng/g tissue)
neo-T4, cortisol-treated, hypothyroid and control rats
at 23 days. Number of animals: 8-10. b) E, NE, DA and
total CA hypothalamus content in neo-T4, cortisol-
treated and control rats. Number of animals: 8.

Treatment with T4 or cortisol led to a decrease in percentage of DA between
6 and 8 days and an increse between 8 and 23 days. Thus, with respect to
control rats, the percentage of DA was higher in neo-T4 and cortisol-
treated rats at 6 days ($p<0.001$) smaller at 8 days ($p<0.05$) and at 23 days
is equal to that of controls. In contrast in hypothyroid animals the
percentage of DA remained the same between 6 and 8 days, and then decreased
thereafter. Thus, in this group the percentage was above control values at
6 and 8 days ($p<0.001$ and $p<0.05$ respectively) but lower at 23 days
($p<0.001$).

Accompanying these developmental changes in DA were reciprocal changes
in NE. The percentage of NE increased between 6 and 8 days in neo-T4 rats,
cortisol-treated and hypothyroid rats. Between 8 and 23 days the % NE
increased in PTU treated rats but decreased in neo-T4 and cortisol-treated
rats.

But the % NE in total CA in neo-T4 and cortisol-treated rats is higher
than in control rats at 8 and 23 days of life ($p<0.05$ for neo-T4 rats and

p<0.01 for cortisol-treated rats) (Figure 1). The % NE is smaller in hypothyroid animals than in controls at 8 and 6 days of life (p<0.05 and p<0.001). At 23 days it is higher than controls (p<0.001).

Figure 2 shows the brain and hypothalamic content E, NE, DA and total CA per g tissue at 23 days of age in the four populations. The finding can be summarized as follows: in part a) NE is higher than controls in animals treated with T4 and cortisol or PTU at 23 days old. E and DA are smaller than controls in neo-T4 and hypothyroid rats. But in cortisol-treated animals DA is higher than controls, and E is equal. The total CA in neo-T4 and hypothyroid rats is smaller than controls. It is higher than controls in cortisol-treated animals at 23 days old. Figure 2b) shows the hypo-thalamus content of E, NE, D and total CA per g of tissue for neo-T4 and cortisol-treated rats. NE content is higher than controls in the animals treated by hormones. The DA and E are higher than control rats in the hypothalamus of animals treated with cortisol. Finally at 23 days of life total CA is higher than controls in the hypothalamus of animals treated by T4 and cortisol.

DISCUSSION

In the fetal period, norepinephrine is the most abundant catechol-amine in the adrenal medulla and its replacement by epinephrine, and the development of secretion, can be used as a marker for the functional maturation of this tissue. In earlier data obtained by our experiment we had found an increase of E in plasma and adrenal gland on 8 days postpartum and it was therefore suggested that day 8 could be important in central nervous system regulation of adrenal maturation.

The central catecholamine neuron systems (dopamine, norepinephrine and epinephrine) showed markedly different developmental patterns. Dopamine neuron systems are the most numerous and in the first days of life, there is a faster development of brain DA. In the current study, we found that the gradual increase in brain DA typical of normal rats was perturbed by T4 or cortisol treatements and affected differently by hypothyroidism.

Because the perinatal period presents highly synchronized sequence of metabolic and endocrine events, it is very important to maintain the normal developmental sequence. Our findings suggest that thyroid status controls development of DA, NE, and E differentially so that either hyper or hypo-thyroidism discoordinates CNS maturation. Because there are some similarities in the eventual effects of the two treatment, these results are in accord with the fact that the neo-T4 animals have been called "pseudohypothyroid" animals. The increase of NE found in the brain of both neo-T4 and hypothyroid animals (albeit with different time courses) agrees with the relationship between thyroxine and the development of locus coeruleus neurons, which receive noradrenergic projections. Although all these changes may be indicative of overall endocrine derangements, more experiments are needed to correlate the levels of catecholamines with potential endocrine imbalances.

In cortisol-treated rats, the sequence of alterations was also initially similar to that the neo-T4 group. However, at 23 days, cortisol-treated rats showed an increase in the NE and DA content in brain and hypothalamus in agreement with effects of hydrocortisone (7). Similarities between neo-T4 and cortisol were also seen in the hypothalamus, where NE levels and total CA was found. Again, in this region our results show that the chronological sequence of the development of the 3 transmitters seems to be altered by cortisol or hypothyroidism. The later increase in NE versus control was also found at 23 days.

Our finding indicate that the entire pattern of catecholaminergic development in the CNS are carefully coordinated by thyroid and steroid status. It is thus interesting to consider these conclusions together with our previous work demostrating that the neo-T4 and cortisol-treated rats exhibit a retardation in the ontogeny of sympatho-adrenal reflexes. This retardation has been also in the hypothyroid rats (8). As a working hypothesis, we therefore postulate that these hormones influence sympatho-adrenal function through their actions on CNS catecholamines. In keeping with this idea, an accelerated ontogeny of the reflex response to insulin-induced hypoglycemia has been reported in neonatal rats treated with 8-hydroxidopamine (8-OHDA), which destroys catecholaminergic neurons (9). The destruction of supraspinal catecholaminergic neurons thus appears to facilitate the maturation of reflex competence, suggesting the imputs of central catecholamines exert an overall inhibitory influence of CNS catecholamine pathway on the maturation of the adrenal medulla and on the onset of secretory function, our results correlate well with these conclusions. In the three populations studied (neo-T4, cortisol-treated and hypothyroid animals) we found an increase in brain NE content through time, most evident at 23 days. In these three populations a retardation of sympathetic reflex development has indeed been found. Thus, although other authors have suggested a net excitatory function for CNS catecholaminergic neurons (10), our results are more consonant with an inhibitory one. In either case, the importance of CNS catecholamines as mediators of the developmental consequences of altered endocrine status should be seriously considered.

ACKNOWLEDGMENTS

Supported by D.G.I.C.Y.T. grant PM88-0021, Spain.

REFERENCES

1. J.H.Dussault and J.Ruel, Thyroid and brain development, Ann.Rev. Physiol. 49:321 (1987).
2. T.A.Slotkin, Thyroid hormone regulations of sympathetic nervous system development, Trends Autonomic Pharmac. 3:147 (1985).
3. A.O.Davies and L.J.Lefkowitz, Regulation of adrenergic receptors by steroid hormones, Ann.Rev.Physiol. 46:119 (1985).
4. F.J.Seidler and T.A.Slotkin, Adrenomedullary function in neonatal rat, response to acute hypoxia, J.Physiol. 358:1 (1985).
5. F.Escrivá and A.M.Pascual-Leone, Decrease of blood glucose, liver glycogen and insulin and changes in glucose tolerance in suckling rats treated neonatally with high L-thyroxine doses, Endocrinology 108:336 (1981).
6. M.Da Prada and G.Zürcher, Simultaneous radioenzymatic determination of plasma and tissues adrenaline, noradrenaline and dopamine within the fentomole range, Life Sciences 19:1161 (1976).
7. R.Ulrich, A.Yuwiler and E.Geller, Effects of hydrocortisone on biogenic amine levels in the hypothalamus, Neuroendocrinology 19: 259 (1975).
8. D.Gripois, M.Valens and A.Diarra, Adrenal-medullary response to insulin-induced hypoglycemia in the young rat. Influence of thyroid hormones, J.Autonomic Nervous System 15:165 (1975).
9. C.Lau, L.L.Ross, W.L.Whitmore and T.A.Slotkin, Regulation of adrenal chromaffin cell development by the central monoaminergic system: Differential control of norepinephrine and epinephrine levels and secretory responses, Neurosciences 22:1067 (1987).
10. J.P.Gagner, S.Gauthier and F.C.Sourkes, Participation of spinal monoaminergic and cholinergic systems in the regulation of adrenal thyrosine hydroxylase, Neuropharmacology 22:45 (1983).

G-PROTEINS IN NEONATAL LIVER: ONTOGENY AND RELATIONSHIP

TO ACTIVATION OF ADENYLATE CYCLASE

Ifeanyi J.Arinze and Yumiko Kawai

Department of Biochemistry
Meharry Medical College
Nashville, TN 37208, U.S.A.

INTRODUCTION

In mammals a variety of metabolic processes are initiated after birth to enable the newborn to adjust to living in an extra-uterine environment. Such metabolic processes include, but are not limited to, gluconeogenesis, glycogenolysis, fatty acid oxidation, ketogenesis, ketone body utilization, and myelination. Other than myelination, the majority of the metabolic processes in this list occur in the liver, albeit with different time frames. Thus, the liver is a key organ for metabolic homeostasis in mammalian newborns. This organ accumulates glycogen as well as tri-glycerides prior to birth (1-5) such that at birth the concentration of glycogen in various species is generally twofold higher than the level of glycogen found in liver of the adult of the same species (5,6). At least two decades of research dating back to the work of Shelley (5) have delineated enzymatic and metabolite profiles which engender initial synthesis of liver glycogen prior to birth and its breakdown postnatally. This information is available in numerous reviews (7-12).

Because hormonal regulation is an important feature of virtually all aspects of metabolism it is not surprising that hormonal influences on the initial activation of metabolic processes in the newborn have also been studied. The changing ratio of insulin:glucagon (13,14) and the postnatal increase in catecholamine levels in plasma (15) favor the production of hepatic cAMP during the immediate postnatal period. If one considers the activation of glycogen breakdown in neonatal liver as a model system, one observes that the influence of this change in hormonal milieu is particularly dramatic. According to Patel et al. (12), "the initiation of hepatic glycogenolysis in the newborn rat follows the changes in cAMP concentration". In newborn guinea pigs the onset of glycogenolysis is delayed until 3-4 hours after birth and the liver is less responsive to

---

Abbreviations: G protein, guanine nucleotide regulatory protein; Gs, the stimulatory G protein; Gi, the inhibitory G protein; Go, the other G protein; Gs-α and Gi-α , α-subunit of Gs and Gi; cyc$^-$, Gs-deficient variant of the S49 lymphoma cell; TED, 20 mM Tris-HCl, 1 mM EDTA, 1 mM dithiothreitol; SDS-PAGE, sodium dodecyl sulfate-polyacrylamide gel electrophoresis; Gpp(NH)p, guanyl-5'-yl imidodiphosphate; GTP-γ-S, guanosine 5'-(3-thiotriphosphate); Bmax, maximum binding capacity.

Endocrine and Biochemical Development of the Fetus and Neonate
Edited by J. M. Cuezva *et al.*
Plenum Press, New York, 1990

glucagon or catecholamine, as glycogenolytic stimuli, than in the adult
(16). No delay is exhibited when cAMP is applied as a direct stimulus
(16), suggesting developmentally associated regulatory events at the level
of the plasma membrane.

The mammalian adenylate cyclase system which is responsible for the
production of cAMP minimally consists of a receptor, the adenylate cyclase
catalytic unit and G-protein(s) which mediate activation or inhibition of
the catalytic unit (17). In this chapter we will present data which are
consistent with our view that responsiveness of the hepatocyte to hormonal
activation of the adenylate cyclase system in the neonate is predicated on
developmental changes in the G-protein component. For this purpose we chose
to study the beta-adrenergic receptor-adenylate cyclase system because this
receptor system has been a prototype for understanding mechanisms of a
variety of hormone-induced transmembrane signalling events.

EXPERIMENTAL PROCEDURES

Animals. New Zealand white male (1.2-2.0 kg, body weight) and pregnant
(24-28 days of gestation) rabbits were purchased from Myrtle's rabbitry,
Thompson Station, TN, and fed ad libitum. The animals had free access to
water at all times. Term fetuses were allowed to be delivered naturally
and were kept in a nested box with dam.

Cholate extraction and reconstitution assays. Liver plasma membranes
were extracted with 1% sodium cholate/TED, pH 8.0, at a protein
concentration of 10 mg/ml. The mixture was stirred on ice for 1 h and
centrifuged at 60,000 x g for 90 min. The supernatant was collected and
stored at $-80^{\circ}C$ until used. Immediately before use the extracts were
incubated at $30^{\circ}C$ for 10 min to inactivate endogenous adenylate cyclase
activity. Cholate extracts thus obtained (5 μg protein) were reconstituted
with cyc⁻ membranes (40 μg protein) as described by Sternweis et al. (18).
NaF-stimulated adenylate cyclase activity was then assayed. Under the
reconstitution conditions the enzyme activity was proportional to the
amounts of extracts added.

Purification of G proteins. G proteins and their subunits were
purified from bovine brain by the procedure of Sternweis and Robishaw (19)
except that octyl-Sepharose (Pharmacia) was used instead of hepthylamine-
Sepharose. Partially purified G proteins from the octyl-Sepharose column
were then applied to a Mono Q HR5/5 column which had been equilibrated
with 0.5 % Lubrol PX/TED pH 7.6, and the column was eluted with a linear
gradient of NaCl using a Pharmacia FPLC instrument. Major protein peaks
were pooled and each peak was re-chromatographed on the Mono Q column under
the same conditions. The purified proteins, Gi, Go, and subunits thereof
were identified by GTP-γ-^{35}S binding activity (19), pertussis toxin-
dependent ADP-ribosylation and SDS-PAGE.

Preparation of antibodies. New Zealand white rabbits were injected
intradermally with purified Gi, Go, or βγ-subunit in complete Freund's
adjuvant (2 ml total, 100 μg protein per animal), followed by two
subsequent injections with 50 μg protein in incomplete Freund's adjuvant
at 2-week intervals. Animals were bled weekly and the antisera collected
were subjected to heat inactivation.

Immunoblot analysis. Liver plasma membranes (10-20 μg of protein) or
purified G proteins (20-50 ng) were subjected to SDS-PAGE in 10% acrylamide
gel and electrophoretically transferred to nitrocellulose using a Bio-Rad
Mini Trans-Blot apparatus. The nitrocellulose was then incubated with
blocking solution containing 3% gelatin followed by diluted rabbit antisera

and immunostained using alkaline phosphatase-conjugated goat anti-rabbit IgG as a second antibody. Immunostained bands were scanned with an LKB laser densitometer and the amounts of G proteins or their subunits in membranes were quantified using purified proteins as standards.

Other experimental methods. Methods for preparation of partially purified liver plasma membranes (20), quantitation of β-adrenergic receptor through binding assays with [^{3}H] dihydroalprenolol (20,21), assays for adenylate cyclase (21) and for bacterial toxin-dependent ADP-ribosylation of membrane proteins (22) have been previously described.

RESULTS

Ontogeny of Cholera Toxin and Pertussis Toxin Substrates in Rabbit Liver Membranes

In a previous publication (21) we showed that the coupling efficiency in the β-adrenergic receptor-adenylate cyclase system in rabbit liver is altered during neonatal development such that an initial apparently coupled system becomes uncoupled within 5-6 hours after birth and recoupled after 6 hours postpartum. The data suggested that changes in the amount/properties of Gs and/or Gi may be responsible, in part, for the alterations observed. Gs and Gi are substrates for ADP-ribosylation by the bacterial toxins cholera toxin and pertussis toxin, respectively (17). Cholera toxin enhances adenylate cyclase activity by inhibiting GTPase activity associated with Gs. This activating effect of the toxin is concomitant with ADP-ribosylation of Gs peptides. Table 1 shows that pretreatment of neonatal and adult liver membreanes with cholera toxin resulted in differential levels of enhancement of adenylate cyclase activity. Whereas the difference in activity in the presence of the toxin was 13-16 pmol/min/mg in membranes from newborn (0-6 h) animals, it was about 67 pmol/min/mg in membranes from the adult. Intermediate values were seen with membranes from 2-week-old animals, suggesting a development-dependent change in Gs.

Table 1. Development-dependent effect of cholera toxin on GTP-stimulated adenylate cyclase activity

Source of membranes	Cholera toxin		Difference
	−	+	
	pmol cAMP/min/mg protein		
Newborn			
0 h	9.4±1.2	25.6±0.0	16.2±1.1
6 h	7.8±1.4	20.8±4.3	13.0±2.9
Neonate			
2 week	20.6±0.8	52.2±9.5	31.6±8.7
Adult	25.2±1.7	92.0±4.6	66.8±3.0

Membranes (200 µg) were incubated for 30 min at 30°C with 1 mM NAD in the absence or presence of pre-activated cholera toxin (20 µg/ml). Enzyme activity in the washed membranes (<30 µg) was measured in the presence of 100 µM GTP. Results are means ± S.E. of two experiments. Data are taken from Ref. 21 with permission from the publisher.

In order to assess the relative amounts of Gs in membranes from neonatal and adult animals, we decided to label membranes with [^{32}P]NAD in the presence of cholera toxin. Figure 1 shows that there was very little ^{32}P incorporation into cholera toxin-dependent bands in membranes isolated from term animals. Thereafter the two major cholera toxin-dependent ADP-ribosylated proteins corresponding to 45 and 52 kDa-peptides became gradually evident, reaching maximal levels in membranes from 4- to 6-week-old animals. In contrast, the pertussis toxin substrate (a doublet of 41/42 kDa-peptide) was clearly evident even in membranes from term animals and in all age groups studied. Densitometric analysis of the data showed that membranes isolated from term animals contained only 8% of the amount of Gs (45 + 52 kDa peptides) and about 50% of that of Gi detected in membranes from the adult (Table 2). The pattern for Gs did not change for up to 24 h after birth at which time Gi amounts already equaled adult values. When membranes from newborn animals were mixed with membranes from the adult the extent of labeling of either Gs or Gi in the mixed samples was additive, suggesting the absence of inhibitor(s) or no influence from potential endogenous activator(s) for the ADP-ribosylation reaction in either membranes. Increased amounts of both cholera and pertussis toxins or higher concentrations of NADP$^+$ did not change the ADP-ribosylation patterns from those shown in Table 2.

Table 2 shows that alterations in the amount of Gi-α during neonatal development were not as marked as for Gs-α. For example, there were not significant changes in Gi-α after one day of life. A caveat, however, is that the availability of β- and γ-subunits during the period studied can influence the results obtained since the presence of $\beta\gamma$ is required to ADP-ribosylate the α-subunit of Gi (23). Thus the amount of α-subunit of Gi could increase markedly, but if $\beta\gamma$ did not keep pace, the incorporation of [^{32}P]ADP-ribose into Gi-α would be constant. To test this possibility, membranes from newborn and adult animals were incubated with purified $\beta\gamma$-subunits and then ADP-ribosylated with pertussis toxin. The results did not show any significant differences in the developmental pattern of ^{32}P-labeling of Gi-α (data not shown), suggesting that the amount of $\beta\gamma$ in the membranes was probably not limiting for the ADP-ribosylation of Gi.

Changes in Gs Activity during Neonatal Development

Because cholera toxin-catalyzed ADP-ribosylation of Gs-α in some

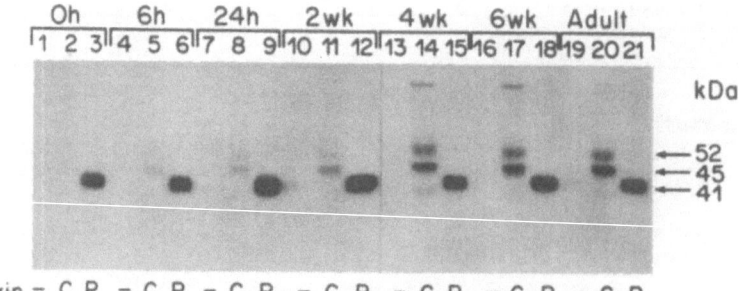

Fig. 1. Development-dependent [^{32}P]ADP-ribosylation of cholera toxin and pertussis toxin substrates in rabbit liver membranes. Membranes (100 μg protein) from various age groups of animals were ADP-ribosylated in the absence (-) or presence of cholera toxin (C) or pertussis toxin (P). [^{32}P]ADP-ribosylated proteins were separated by SDS-PAGE and detected by autoradiography. A representative autoradiogram is shown.

Table 2. Densitometric analysis of cholera toxin- and
pertussis toxin-labeled bands in liver
membranes from various age groups of rabbits

| Source of membranes | Gs | | Gi |
	45 kDa	52 kDa	41/42 KDa
	pmol ADP-ribose/mg protein		
Newborn			
0 h	0.13±0.02*	0.09±0.03*	1.51±0.41*
6 h	0.21±0.03*	0.15±0.03*	1.75±0.39*
24 h	0.22±0.04*	0.17±0.03*	3.09±0.65
Neonate			
2 week	0.50±0.08*	0.24±0.05*	3.88±0.46
4 week	1.06±0.29*	0.61±0.17	2.97±0.22
6 week	1.48±0.28	0.77±0.15	3.37±0.31
Adult	1.98±0.37	0.90±0.10	3.17±0.34

Membranes from various age groups of animals were
ADP-ribosylated as described in the legends to
Figure 1. Autoradiograms were scanned and analyzed
using an LKB laser densitometer. Results are means
± S.E. for six different membrane preparations.
* Significantly different from adult values (p<0.05,
Mann-Whitney test).

systems appears to require endogenous soluble (24) and membrane-bound
(25,26) protein factors the use of the ADP-ribosylation reaction to
estimate the amounts of Gs-α in membranes may be problematical. It is
generally accepted that Gs activity determined by reconstitution assay
using cyc⁻ membranes reflects directly the amount of Gs present in the
test systems (27,28). Therefore in order to obviate the potential effects
of soluble or membrane-bound factors on the extent of ADP-ribosylation,
we also assayed Gs by using a recombination system deficient in these
proteins.

Table 3 shows that the Gs activity in cholate extracts of membranes
from term animals, as measured by reconstitution with cyc⁻ membranes, is
about 40% of the adult value; this low activity was unchanged in the first
6 h after birth, and increased gradually with age of the animals, reaching
maximal levels in extracts of membranes from 6-week-old animals. This
developmental pattern is similar to that of the cholera toxin-dependent
[32]P-labeling of Gs in membranes (Table 2), indicating that the cholera
toxin-catalyzed ADP-ribosylation of membranes was a good index of the
relative abundance of Gs in this system. However, the relative increase
in the Gs activity during development is less than that in the cholera
toxin-mediated labeling of Gs. This may reflect, in part, the difficulties
associated with quantitation of Gs activity with this reconstitution
technique, as reported elsewhere (18,28).

Use of Antibodies to Quantify G-protein Subunits

In order to provide additional quantitative information on the G-
protein subunits, we have developed several polyclonal antibodies against
subunits of Gi and Go, purified from bovine brain. Figures 2 and 3 are
inmunoblot analyses of the specificity of some of these antibodies and the
quantitation of G protein subunits in liver plasma membranes. So far we
have not generated antibodies against the α-subunit of Gs. However, using
our available antibodies it is clear that the developmental pattern for

Table 3. Ontogeny of Gs activity in rabbit liver membranes

Source of membranes	Adenylate cyclase activity
	pmol cAMP/min
Newborn	
0 h	0.86±0.08
6 h	0.94±0.05
12 h	1.18±0.08
24 h	1.24±0.00
Neonate	
2 week	1.05±0.10
4 week	1.57±0.32
6 week	2.05±0.32
Adult	2.19±0.34

Cholate extracts (5 μg protein) of membranes from
various age groups of animals were reconstituted with
cyc⁻ memmbranes (40 μg protein) at 30°C for 10 min. The
concentration of cholate in the reconstitution phase
was 0.1%. Adenylate cyclase activity in the mixture
was then determined in the presence of 10 mM NaF.
Results are means ± S.E. for two different membrane
preparations. Cyclase activity in the absence of
cholate extracts was 0.03±0.00 pmol/min.

Gi-α determined by immunoblot procedures (Figure 4) parallels that
determined by ADP-ribosylation assay. The level of the β-subunit remains
at 28-30 pmol/mg membrane at 0-6 hours after birth, increasing gradually
to 45-50 pmol/mg at 2 weeks before subsequently doubling to adult levels

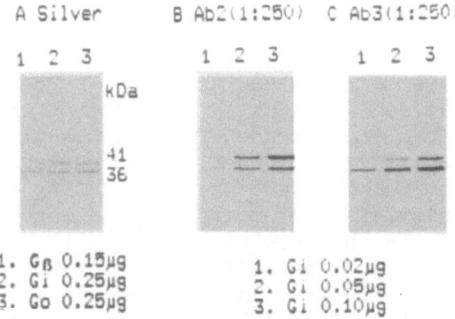

1. Gβ 0.15μg
2. Gi 0.25μg
3. Go 0.25μg

1. Gi 0.02μg
2. Gi 0.05μg
3. Gi 0.10μg

Fig. 2. Specificity of antisera to G protein subunits.
Purified G proteins were separated by SDS-PAGE
and either stained with silver (A) or transferred to nitrocellulose (B and C). The trans-
blots were incubated with the indicated
antisera and immunostained with alkaline
phosphatase-conjugated goat anti-rabbit IgG.
Rabbit antisera Ab2 and Ab3 were raised against
Gi and Go, respectively. G proteins used as
antigens (A) were purified from bovine brain
using FPLC as a final step.

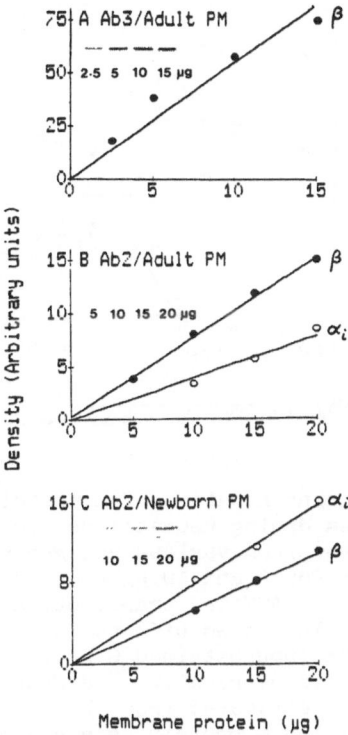

Fig. 3. Activity of antisera as a function of increasing
membrane protein concentration. Rabbit liver
plasma membranes (5-20 μg protein) from adults (A
and B) and 0 h-old newborn (C) were separated by
SDS-PAGE and immunoblotted as described. The
immunostained G protein bands (inset) were
quantified by densitometry. Antisera used were
same as those in Figure 2.

(Figure 4). To our knowledge this is the first quantitation of these
subunits in neonatal liver of any species.

Density of β-adrenergic Receptor and Coupling Efficiency between R-G and G-C Components

In a previous publication (21) we showed that maximally measurable
β-adrenergic receptors are already present in liver plasma membranes of
newborn rabbits at term. The Bmax for [3H]dihydroalprenolol binding in
membranes is about 200-300 fmol/mg protein during neonatal development. In
binding experiments using antagonist ligands, it is generally accepted that
the guanine nucleotide-dependent shift in agonist competition curves to
lower affinities (29) is indicative of receptor coupling to the guanine
nucleotide regulatory protein, i.e. Gs. Comparison of isoproterenol
competition of [3H]dihydroalprenolol binding in the presence and absence
of Gpp(NH)p showed that the degree of shift was age-dependent (21). We
were interested in providing a physiological correlation of this shift
in terms of activatability of the cyclase by a number of known activators
of the enzyme, especially guanine nucleotides.

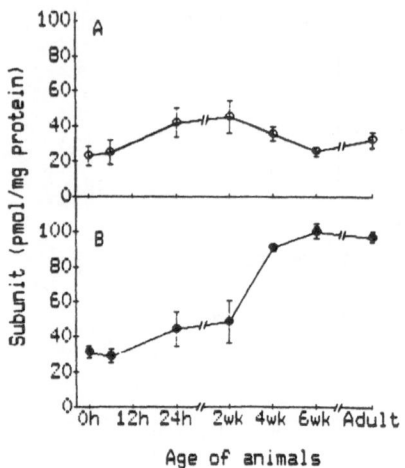

Fig. 4. Estimation of concentration of G protein subunits in
liver membranes during neonatal development. Liver
plasma membranes from various age groups of rabbits
(20 μg protein for A and 10 μg protein for B) were
separated by SDS-PAGE and transblotted as described.
The blots were incubated with antiserum Ab2 (A) or
Ab3 (B) and the immunostained G protein bands were
quantified densitometrically. The amounts of αi- and
β-subunits were estimated from the density of bands,
using the purified G protein as a standard.

Figure 5 shows that the ability of Gpp(NH)p, GTP, GTP plus
l-isoproterenol, or NaF to activate the cyclase was also age-dependent.
In all cases the stimulation of the enzyme was biphasic, i.e. a decrease
followed by an increase, the least activation occurring at 4-6 h postpartum. Mixing of membranes from different neonatal (< 6 h) and adult age
groups resulted in simple additivity of activity of the enzyme (data not

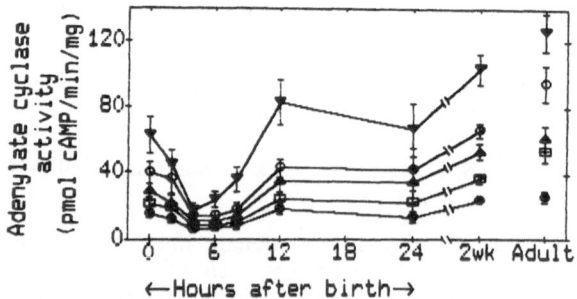

Fig. 5. Profile of hepatic adenylate cyclase activation
during development. Liver plasma membranes (< 30
μg) were incubated with 100 μM Gpp(NH)p (o), 100
μM GTP (□), 100 μM GTP plus 10 μM isoproterenol
(▲), or 10 mM NaF (▼) and with no additions for
basal cyclase activity (●) for 5 min at 30°C.
[^{32}P]cAMP formed was measured according to Salomon
et al.(30). Points plotted are means ± S.E. for
three to six experiments. Data are taken from Ref.
21 with permission from the publisher.

shown), thus negating the possibility that unknown inhibitors/activators, if present, might influence the results obtained. The temporal change in the ability of Gpp(NH)p to modulate adenylate cyclase was highly correlated with the shift in agonist (l-isoproterenol) competition curves. The former parameter is an index of the interaction between Gs and the catalytic unit of the cyclase whereas the latter is an index of interaction between the receptor and Gs since it can be concluded from studies in a variety of systems that guanine nucleotide-dependent stimulation of adenylate cyclase can be used as an index of coupling between Gs and the catalytic unit of the cyclase (31-33). Unlike the results obtained with effectors which act through Gs (Figure 5) there was no overall increase in the Mn^{2+}-dependent activity from newborn to adult, suggesting that the intrinsic activity of the catalytic component is, at least, not increased during the developmental period studied. These results are interpreted to suggest that there is decreased efficiency in the coupling of the β-adrenergic receptor to hepatic adenylate cyclase in early neonatal life.

DISCUSSION

The experimental approach used in this study was to quantitate at the protein level the concentrations of the subunits of heterotrimeric G-proteins, specifically Gs and Gi, in rabbit liver plasma membranes during neonatal development. The results demonstrate changing levels of the β-subunit as well as Gs-α and Gi-α. Because the γ-subunit was not quantified in our studies we cannot make conclusions regarding this subunit. Nevertheless, results of parallel studies on the pattern of activation of the catalytic unit of the adenylate cyclase as well as an associated receptor, i.e. the β-adrenergic receptor, suggest a pattern of uncoupling early after birth followed by re-coupling (Figure 5). The molecular basis for this phenomenon is as yet unclear.

Gs and Gi are known to share identical β- and γ-subunits (17,34,35). Since it is believed that dissociation of α-subunits of Gs or Gi complex results in stimulation or inhibition of adenylate cyclase (17), it is intriguing to speculate that the formation of Gs complexes in neonatal membranes, in situ, can be greatly influenced by competition for β- and/or γ-subunit present. In fact using data from our immunoblotting and ADP-ribosylation analyses one can calculate $\beta\gamma/\alpha$-s ratios of 78 and 34 for neonatal (4-6 hr postpartum) and adult liver membranes, respectively, suggesting an excess of $\beta\gamma$ over α-subunit or an abundance of free $\beta\gamma$. It should be noted that at 4-6 hrs postpartum the system is least coupled with respect to G-C communication (Figure 5) and little Gs activity is detectable at 0-6 hrs postpartum, as judged by reconstitution assay with cyc$^-$ membranes (Table 3). Based on immunoblotting data only, one can calculate $\beta\gamma/\alpha$-i ratios of only 1.1 and 3.0 for neonatal (4-6 hours postpartum) and adult liver membranes, respectively. Quantitation of Gs-α using the ADP-ribosylation reaction has been recently argued to give lower values than can be detected by immunoblotting methods (36). Since the concentration of $\beta\gamma$ can be expected to influence the coupling phenomenon, conclusions as to how far in excess the $\beta\gamma$ subunits are over the α-subunit probably need to be based on the use of one method, e.g. immunoblotting, to quantify both components.

Another possibility which is presently being explored in our laboratory is that the uncoupling-recoupling phenomenon may be predicated on the phosphorylation state of one or more G-protein subunits or the β-receptor component. This possibility is akin to the idea that phosphorylation events underlie the mechanism for both homologous and heterologous desensitization (37). Also of interest is whether such changes in coupling characteristics also occur in liver membranes of species other than the rabbit.

The results from the reconstitution study with cyc⁻ membranes (Table 3) corroborate the suggestion that activation of adenylate cyclase by appropriate hormones during neonatal development may be hampered by lack of competent Gs-components. The analysis at the protein level described in this paper is consistent with the finding, using molecular cloning methods, that the expression of Gs and Gi is programmed by different genetic regulatory processes (38). By using molecular probes Luetje et al. (39) have recently shown differences in the expression of G protein subunits in rat atria and ventricles during cardiac development as well as regulation of the levels of the subunits. Their analyses also show an excess or the presence of free β-subunits in chick heart during development (39). Our studies provide a groundwork for further analysis of the genetic expression of G protein subunits during development in mammalian liver.

SUMMARY

The coupling efficiency in the β-adrenergic receptor-adenylate cyclase system in rabbit liver is altered during neonatal development such that an initial apparently coupled system becomes uncoupled within 5-6 hours after birth and recoupled after 6 hours postpartum. Changes in the amount/ properties of Gs and/or Gi have been suggested to be responsible, in part, for these alterations. Since Gs and Gi are substrates for ADP-ribosylation by the bacterial toxins, cholera toxin and pertussis toxin, respectively, the ontogeny of trimeric G proteins and their subunits in rabbit liver during neonatal development was studied, using bacterial toxin-catalyzed ADP-ribosylation of membrane proteins to quantitate Gs-α and Gi-α, reconstitution of Gs activity with cyc⁻ membranes, and inmunoblotting analysis to quantitate Gi-α and β-subunits. Under optimal conditions of ADP-ribosylation little cholera toxin substrate, Gs-α, was detected in membranes from liver of neonatal animals up to 2 weeks of age. Thereafter ribosylatable Gs-α peptides (45 and 52 kDa) were increasingly evident, reaching maximal levels in membranes from animals aged 4-6 weeks. Gs activity in cholate extracts of membranes, measured by reconstitution assay, exhibited similar development pattern as that of cholera toxin-mediated labeling. In contrast, the pertussis toxin substrate, Gi, was clearly evident even in membranes from term animals and in all age groups studied. Its developmental pattern paralleled that of Gi-α determined by immunoblotting analysis. The concentration of the β-subunit of the G-proteins also showed an age-dependent increase. These results are part of continuing studies that suggest that the activation of adenylate cyclase in neonatal liver may be predicated on the amount/properties of G proteins and their subunits.

ACKNOWLEDGMENTS

This work was supported by grant No. HD 08792 from the National Institutes of Health, and by grant #1-598 from the March of Dimes-Birth Defects Foundation. We thank Gale Beamer for typing the manuscript and Dr. Craig C. Malbon, Department of Pharmacological Sciences, State University of New York at Stony Brook, for providing S49 cyc⁻ membranes.

REFERENCES

1.   M.J.R.Dawkins, Glycogen synthesis and breakdown in fetal and newborn rat liver, Ann.N.Y.Acad.Sci. 111:203 (1963).
2.   R.Kornfeld and D.H.Brown, The activity of some enzymes of glycogen metabolism in fetal and neonatal guinea pig liver, J.Biol.Chem. 238:1604 (1963).

3. F.J.Ballard and I.T.Oliver, Glycogen metabolism in embryonic chick and neonatal rat liver, Biochim.Biophys.Acta, 71:578 (1963).

4. P.Devos and H.-G.Hers, Glycogen metabolism in the liver of foetal rat, Biochem.J. 140:331 (1974).

5. H.J.Shelley, Glycogen reserves and their changes at bith and in anoxia, Br.Med.Bull. 17:137 (1961).

6. H.J.Shelley, Carbohydrate metabolism in the foetus and the newly born, Pro.Nutr.Soc. 28:42 (1969).

7. M.J.R.Dawkins, Biochemical aspects of developing function in newborn mammalian liver, Br.Med.Bull. 22:27 (1966).

8. G.S.Dawes and H.J.Shelley, Physiological aspects of carbohydrate metabolism in the foetus and newborn, in: "Carbohydrate Metabolism and Its Disorders," F.Dikens, P.J.Randle and W.J.Whelan, eds., Acad. Press, N.Y. (1968).

9. D.G Walker, Developmental aspects of carbohydrate metabolism, in: "Carbohydrate Metabolism and Its Disorders," F.Dikens, P.J.Randle and W.J.Whelan, eds., Acad. Press, N.Y. (1968).

10. R.W.Hanson, L.Reshef and F.J.Ballard, Hormonal regulation of hepatic P-enolpyruvate carboxykinase(GTP) during development, Fed.Proc. 34:166 (1975).

11. M.W.Haymond, A.S.Pagliara and D.M.Bier, Endocrine and metabolic aspects of fuel homeostasis in the fetus and neonate, in: "Endocrinology," L.J.DeGroot, ed., W.B.Saunders Company 3:2215 (1989).

12. M.S.Patel, P.Van Lelyveld and R.W.Hanson, The development of the pathways of glucose homeostasis in the newborn, in: "Biochemical Development of the Fetus and Neonate," C.T.Jones, ed., Elsevier Biochemical Press (1982).

13. J.R.Girard, G.S.Cuendet, E.B.Marliss, A.Kervran, M.Rieutort and R. Assan, Fuels, hormones, and liver metabolism at term and during the early postnatal period in the rat, J.Clin.Invest. 52:3190 (1973).

14. P.N.DiMarco, A.V.Ghisalberti, C.E.Martin and I.T.Oliver, Perinatal changes in liver corticosterone, serum insulin and plasma glucagon and corticosterone in the rat, Eur.J.Biochem. 87:243 (1978).

15. J.M.Cuezva and M.S.Patel, Plasma catecholamine concentrations in the newborn rat during the first six postnatal hours, Biochem.Soc.Trans. 10:521 (1982).

16. Y.Kawai and I.J.Arinze, Activation of glycogenolysis in neonatal liver, J.Biol.Chem. 256:853, ibid 12612 (1981).

17. A.G.Gilman, G proteins: Transducers of receptor-generated signals, Ann.Rev.Biochem. 56:615 (1987).

18. P.C.Sternweis, J.K.Northup, M.D.Smigel and A.G.Gilman, The regulatory component of adenylate cyclase: Purification and properties, J.Biol. Chem. 256:11517 (1981).

19. P.C.Sternweis and J.D.Robishaw, Isolation of two proteins with high affinity for guanine nucleotides from membranes of bovine brain, J.Biol.Chem. 259:13806 (1984).

20. Y.Kawai and I.J.Arinze, β-Adrenergic receptors in rabbit liver plasma membranes: Predominance of β_2-receptors and mediation of adrenergic regulation of hepatic glycogenolysis, J.Biol.Chem. 258:4364 (1983).

21. Y.Kawai, S.M.Graham, C.Whitsel and I.J.Arinze, Hepatic adenylate cyclase: Development-dependent coupling to the β-adrenergic receptor in the neonate, J.Biol.Chem. 260:10826 (1985).

22. Y.Kawai, C.Whitsel and I.J.Arinze, $NADP^+$ enhances cholera and pertussis toxin-catalyzed ADP-ribosylation of membrane proteins, J.Cyclic Nucleotide and Prot.Phos.Res. 11:265 (1986).

23. S.-C.Tsai, R.Adamik, Y.Kanaho, E.L.Hewlett and J.Moss, Effects of guanyl nucleotides and rhodopsin on ADP-ribosylation of the inhibitory GTP-binding component of adenylate cyclase by pertussis toxin, J.Biol.Chem. 259:15320 (1984).

24. K. Enomoto and D. M. Gill, Cholera toxin activation of adenylate cyclase: Roles of nucleoside triphosphate and a macromolecular factor in the ADP ribosylation of the GTP-dependent regulatory component, J. Biol. Chem. 255:1252 (1980).

25. R. A. Kahn and A. G. Gilman, Purification of a protein cofactor required for ADP-ribosylation of the stimulatory regulatory component of adenylate cyclase by cholera toxin, J. Biol. Chem. 259:6228 (1984).

26. R. A. Kahn, C. Goddard and M. Newkirk, Chemical and immunological characterization of the 21-kDa ADP-ribosylation factor of adenylate cyclase, J. Biol. Chem. 263:8282 (1988).

27. P. C. Sternweis and A. G. Gilman, Reconstitution of catecholamine-sensitive adenylate cyclase: Reconstitution of the uncoupled variant of the S49 lymphoma cell, J. Biol. Chem. 254:3333 (1979).

28. J. Codina, W. Rosenthal, J. D. Hildebrandt, L. Birnbaumer and R. D. Sekura, Purification of Ns and Ni, the coupling proteins of hormone-sensitive adenylyl cyclase without intervention of activating regulatory ligands, Methods Enzymol. 109:446 (1985).

29. R. J. Lefkowitz, J. M. Stadel and M. G. Caron, Adenylate cyclase-coupled beta-adrenergic receptors: structure and mechanisms of activation and desensitization, Annu. Rev. Biochem. 52:159 (1983).

30. Y. Salomon, C. Londos and M. Rodbell, A highly sensitive adenylate cyclase assay, Anal. Biochem. 58:541 (1974).

31. R. S. Kent, A. De Lean and R. J. Lefkowitz, A quantitative analysis of beta-adrenergic receptor interactions: Resolution of high and low affinity states of the receptor by computer modeling of ligand binding data, Mol. Pharmacol. 17:14 (1980).

32. A. De Lean, J. M. Stadel and R. J. Lefkowitz, A ternary complex model explains the agonist-specific binding properties of the adenylate cyclase-coupleed β-adrenergic receptor, J. Biol. Chem. 255:7108 (1980).

33. L. E. Limbird, Activation and attenuation of adenylate cyclase: The role of GTP-binding proteins as macromolecular messengers in receptor-cyclase coupling, Biochem. J. 195:1 (1981).

34. J. Codina, D. Stengel, S. L. C. Woo and L. Birnbaumer, β-Subunits of the human liver Gs/Gi signal-transducing proteins and those of bovine retinal rod cell transducin are identical, FEBS Lett. 207:187 (1986).

35. H. K. W. Fong, J. B. Hurley, R. S. Hopkins, R. Miake-Lye, M. S. Jonhson, R. F. Doolittle and M. I. Simon, Repetitive segmental structure of the transducin β subunit: Homology with the CDC4 gene and identification of related mRNAs, Proc. Natl. Acad. Sci. USA. 83:2162 (1986).

36. L. A. Ransnäs and P. A Insel, Quantitation of the guanine nucleotide binding regulatory protein Gs in S49 cell membranes using antipeptide antibodies to αs, J. Biol. Chem. 263:9482 (1988).

37. D. R. Sibley, J. L. Benovic, M. G. Caron and R. J. Lefkowitz, Phosphorylation of cell surface receptors: A mechanism for regulating signal transduction pathways, EndocrineRev, 9:38 (1988).

38. P. L. Ashley, J. Ellison, K. A. Sullivan, H. R. Bourne and D. R. Cox, Chromosomal assignment of the murine Giα and Gsα genes: Implications for the obese mouse, J. Biol. Chem. 262:15299 (1987).

39. C. W. Luetje, K. M. Tietje, J. L. Christian and N. M. Nathanson, Differential tissue expression and developmental regulation of guanine nucleotide binding regulatory proteins and their messenger RNAs in rat heart, J. Biol. Chem. 263:13357 (1988).

40. C. W. Luetje, P. Gierschik, G. Milligan, C. Unson, A. Spiegel and N. M. Nathanson, Tissue-specific regulation of GTP-binding protein and muscarinic acetycoline receptor levels during cardiac development, Biochemistry, 26:4876 (1987).

POSTNATAL DEVELOPMENT OF β-ADRENERGIC RECEPTOR

KINASE ACTIVITY IN DIFFERENT TISSUES OF THE RAT

Irene García-Higuera and Federico Mayor, Jr.

Departamento de Biología Molecular
Centro de Biología Molecular
Universidad Autonóma de Madrid
28049 Madrid, Spain

INTRODUCTION

Recent studies have suggested a key role for the phosphorylation of plasma membrane receptors in the modulation of the sensitivity of cells to extracellular signals. Receptor phosphorylation has been proposed to regulate either receptor function or intracellular distribution, and has been shown to be implicated in the molecular mechanisms of desensitization (1,2). Desensitization is a widespread phenomenon characterized by the fact that the intensity of a response wanes over time despite the presence of a stimulus of constant intensity. Several recent investigations, utilizing the β-adrenergic receptor (βAR) system as a model, have demonstrated that desensitization of adenylate cyclase (AC) coupled receptors is associated with receptor phosphorylation by different protein kinases. Whereas the phosphorylation of βAR by cAMP-dependent protein kinase and protein kinase C contribute to different forms of "heterologous" regulation (1), Lefkowitz and col. have recently identified a novel protein kinase, termed β-adrenergic receptor kinase (βARK) that has been implicated in the molecular mechanisms of "homologous" or agonist-specific desensitization. βARK is a ubiquitously distributed cytosolic enzyme that phosphorylates only the agonist-occupied form of the βAR, up to 9 mol of phosphate per mol of receptor (3). βARK may also be capable of agonist-dependent phosphorylation of a variety of receptors coupled to either stimulation or inhibition of AC, thus emerging as an enzyme of general regulatory significance (4,5). However, very little is known about the physiological role of βARK.

In this regard, we have initiated a study on the development of βARK activity during the first hours after birth. This period of time is characterized by important alterations of messenger levels, and particularly by an increase in plasma catecholamine concentrations (6).Catecholamines control a wide variety of critical metabolic processes at this stage. Our results show a marked increase in βARK activity during the first 2 hours of extrauterine life. This increase is transient in all tissues examined, except for the brain. Interestingly, kinase levels in the lung, liver and heart of rat neonates are elevated 3 to 5-fold with respect to adult values. These results suggest an active process of receptor regulation in the early postnatal period of the rat.

Endocrine and Biochemical Development of the Fetus and Neonate
Edited by J. M. Cuezva *et al.*
Plenum Press, New York, 1990

RESULTS AND DISCUSSION

βARK is a very specific kinase that only phosphorylates the agonist-occupied form of the receptor (3). In order to determine βARK activity in crude tissue preparations, we have taken advantage of the ability of βARK to phosphorylate rhodopsin, the retinal receptor, in a light-dependent fashion (7). Thus, cytoplasmic fractions prepared from brain, lung, liver or heart homogenates were incubated with purified rhodopsin under phosphorylating conditions in the presence of light. The phosphorylated rhodopsin was resolved and quantitated by SDS-PAGE and autoradiography (4). Figure 1 shows a representative autoradiogram of such experiments. Figure 1 also illustrates the light dependence of the reaction, thus indicating that agonist-dependent phosphorylation by βARK or a closely related enzyme is being observed. Similar results were obtained with all tissues tested (data not shown).

Figure 2 summarizes the main findings of this study. First, βARK activity increases markedly in brain, liver, heart and lung between 0 and 2 hours of extrauterine life. A rapid decline in βARK levels is subsequently detected in liver and heart, whereas in the lung (where βARK activity is very high at birth) a slower decrease is observed. No significant changes are apparent in brain. Secondly, it is worth noting that, except for the brain, βARK activity in the first hours after birth is 2.5 to 7-fold higher than that detected in the corresponding adult tissues. At 2 h. of life, kinase levels attain values 7, 6.2 and 5.2 fold higher than adult activity in heart, liver and lung, respectively. On the contrary, adult brain displays a high βARK activity. The results are not due to the presence of an inhibitor of the assay in adult tissues (data not shown) and is also

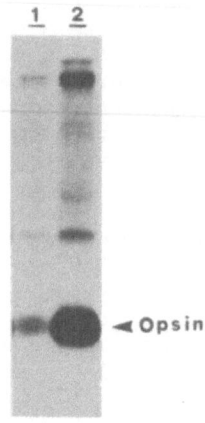

Fig. 1. Light-dependent phosphorylation of rhodopsin as an assay for βARK activity in different tissues of the newborn rat. The autoradiogram shows opsin phosphorylation in the absence (lane 1) or presence (lane 2) of light when a supernatant isolated form the brain of adult rats was incubated with a rhodopsin preparation. Similar light-dependence was obtained with other tissues at all ages examined. Crude supernatants of different rat tissues and rhodopsin kinase-free purified rod outer segments (approx. 95% rhodopsin) were obtained as described (4,7). Phosphorylation assays, SDS-PAGE and autoradiography were performed as described elsewhere (3,4).

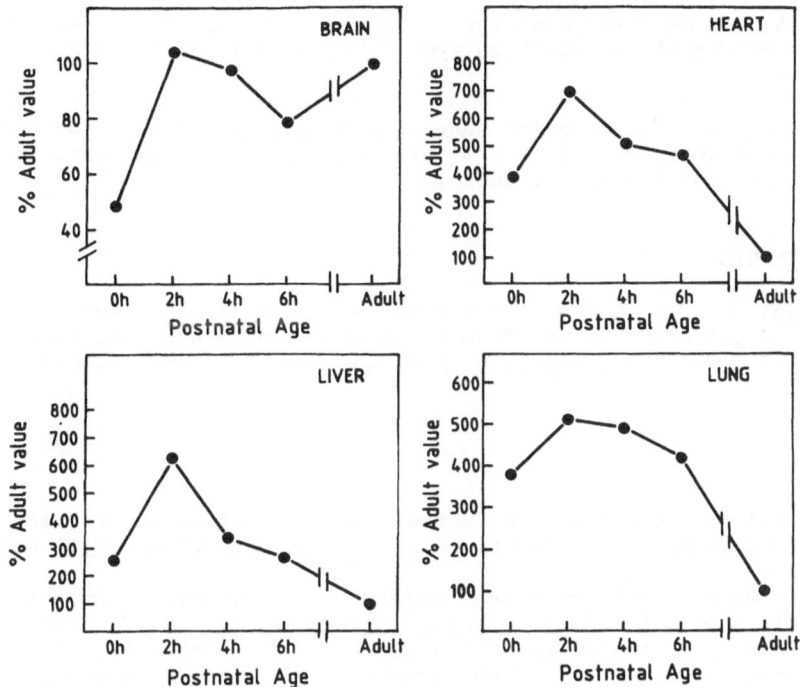

Fig. 2. Postnatal development of βARK activity in different
tissues of the rat. Kinase activity was determined
as described in Figure 1 and the text. βARK levels
were expressed in pmol of phosphate incorporated in
rhodopsin per mg of supernatant protein and referred
to adult values, taken as 100% for each tissue.
Adult activities were 47.4±6, 4.6±0.1, 2.2±0.4 and
3.6±0.4 pmol phosphate/mg of protein for brain,
lung, heart and liver, respectively (average ± SEM
of 5-6 determinations). Tissues from 8 newborn rats
were pooled for each determination.

apparent when values are expressed as pmol of phosphate incorporated in
rhodopsin per gram of wet tissue instead of per mg of protein (data not
shown). Our adult βARK values are comparatively similar to those previously
reported in rat or bovine tissue (3,8).

The results presented herein would indicate the potential for receptor
regulation in the first hours of the life of the rat. Since βARK has been
shown to be an agonist-dependent kinase and given the high levels of
catecholamines at birth (6), it is tempting to suggest that adrenergic
receptors would be a major target of phosphorylation and regulation by this
kinase. In fact, β- and α2-adrenergic receptors have been shown to be phosphorylated and desensitized by βARK *in vitro* (1,5). In this regard, it has
been described that a refractoriness to AC stimulation by β-adrenergic
agonists and other AC activators occurs in liver of the rabbit newborn
early after birth (9). Moreover, βAR levels in rat livers membranes
diminish during the first postnatal day (10). Thus, it is suggested that
the dramatic surge of catecholamines that takes place after birth could
promote a β-adrenergic receptor regulation process that would be reflected
by an increase in βARK activity. However, since βARK has been reported to
be able to modulate other non-adrenergic, AC-coupled receptors, changes in
the levels of other messengers could also trigger the increase in kinase

activity. A more detailed description of the postnatal development of the components of signal transduction systems is needed to shed light on this issue.

In conclusion, our results suggest a physiological role for βARK in an active process of receptor regulation that would be important in the adaptation of neonates to extrauterine life.

ACKNOWLEDGMENTS

We thank Dr. J.M. Cuezva for helpful discussions. This work was partially supported by grants from Fundación Ramón Areces and Boehringer Ingelheim. I.G.H. is a recipent of a Basque Community predoctoral fellowship.

REFERENCES

1. D.R.Sibley, J.L.Benovic, M.G.Caron, and R.J.Lefkowitz, Regulation of transmembrane signaling by receptor phosphorylation, Cell 48:913 (1987).
2. R.L.Huganir, and P.Greengard, Regulation of receptor function by protein phosphorylation, Trends Pharmacol.Sci. 8:472 (1987).
3. J.L.Benovic, F.Mayor jr., C.Staniszewski, R.J.Lefkowitz, and M.G. Caron, Purification and characterization of the β-adrenergic receptor kinase, J.Biol.Chem. 262:9026 (1987).
4. F.Mayor jr., J.L.Benovic, M.G.Caron, and R.J.Lefkowitz, Somatotatin induces translocation of the β-adrenergic receptor kinase and desensitizes somatostatin receptors in S49 lymphoma cells, J.Biol. Chem. 262:6468 (1987).
5. J.L.Benovic, J.W.Regan, H.Matsui, F.Mayor jr., S.Cotecchia, L.M.F. Leeb-Lundberg, M.G.Caron, and R.J.Lefkowitz, Agonist-dependent phosphorylation of the α2-adrenergic receptor by the β-adrenergic receptor kinase, J.Biol.Chem. 262:17251 (1987).
6. J.M.Cuezva, E.S.Burkett, D.S.Kerr, H.M.Rodman, and M.S.Patel, The newborn of diabetic rat. I. Hormonal and metabolic changes in the postnatal period, Pediatr.Res. 16:632 (1982).
7. J.L.Benovic, F.Mayor jr., R.L.Somers, M.G.Caron, and R.J.Lefkowitz, Light dependent phosphorylation of rhodopsin by the β-adrenergic receptor kinase, Nature 322:869 (1986).
8. R.L.Somers, and D.C.Klein, Rhodopsin kinase activity in the mammalian pineal gland and other tissues, Science 226:182 (1984).
9. Y.Kawai, S.M.Graham, C.Whistel, and I.J.Arinze, Hepatic adenylate cyclase. Development-dependent coupling to the β-adrenergic receptor in the neonate, J.Biol.Chem. 260:10862 (1985).
10. K.Snell, and C.A.Evans, Characterization of rat liver β-adrenorecep-tors during perinatal development as determined by ^{125}I-iodopindolol radioligand binding assays, Br.J.Pharmacol. 93:817 (1988).

MOLECULAR AND FUNCTIONAL ASPECTS OF MITOCHONDRIAL DEVELOPMENT

PERINATAL DEVELOPMENT OF LIVER

MITOCHONDRIAL FUNCTION

June R. Aprille

Mitochondrial Physiology Unit
Department of Biology
Tufts University
Medford, MA 02155, U.S.A.

INTRODUCTION

Just before birth, liver mitochondria are relatively undeveloped with
respect to a capacity for aerobic ATP synthesis. They also lack the ability
to perform certain reactions compartmentalized in the matrix that are an
essential part of metabolic pathways such as gluconeogenesis, ureagenesis,
and fatty acid oxidation. The postnatal development of these mitochondrial
activities is a prerequisite for independent metabolism in the neonate.
Morphologically, compared to postnatal mitochondria, prenatal mitochondria
are: larger; more variable in shape; have fewer cristae; are fewer in
number per cell; and have a higher equilibrium density. Shortly after birth
the mitochondria begin to proliferate. Details of all these enzymatic,
functional and morphological aspects of mitochondrial development in the
perinatal period were reviewed recently (1). This paper will focus on some
interesting regulatory events that occur immediately after birth.

At term the adenine nucleotide content (ATP+ADP+AMP) of rat liver
mitochondria is much lower than in adult mitochondria. In response to
physiological signals elicited by parturition, mitochondria accumulate
adenine nucleotides from the cytoplasm so that the matrix ATP+ADP+AMP
content increases to adult levels within a few hours (2-5). During the
same time interval the rates of oxidative phosphorylation (2,4-6), pyruvate
carboxylation (7-10), citrulline synthesis (11), and intramitochondrial
protein synthesis (5,12,13) increase as well (reviewed in reference 1). The
purpose of this discussion is to consider the hypothesis that the rapid
postnatal development of metabolic functions in mitochondria is mediated
in part by the increased mitochondrial adenine nucleotide content. Particular attention will be given to oxidative phosphorylation and to pyruvate
carboxylation because the development of these activities is necessary
for the immediate onset of postnatal gluconeogenesis (9,10).

POSTNATAL ACCUMULATION OF ADENINE NUCLEOTIDES IN MITOCHONDRIA

A number of investigators (2-5) have noticed that the adenine
nucleotide content of mitochondria isolated from term or just-born rat
liver is very low (3 nmol/mg mitochondrial protein), and that it increases
to near adult levels (12 nmol/mg mitochondrial protein) soon after birth

Endocrine and Biochemical Development of the Fetus and Neonate
Edited by J. M. Cuezva *et al.*
Plenum Press, New York, 1990

(Figure 1, left). Similar findings have been noted in rabbit (9,14), except that the adenine nucleotide content at term is a little higher than in rat, and in the rabbit there is a large overshoot in matrix adenine nucleotide content that peaks at 6 hours postnatal (14) (Figure 1, right).

There is no change in the adenine nucleotide content of the whole tissue (3,10,15-18) while the mitochondrial adenine nucleotide content is increasing, thus it is clear that the increase occurs as net uptake from the cytoplasm into the matrix compartment. There is no increase in matrix volume (under phosphorylating conditions) (5), and so the increase in mitochondrial adenine nucleotide content results in a real increase in the absolute concentration of ATP+ADP (also AMP, but to lesser extent) in the matrix compartment. The mitochondria occupy only about 10-13% of hepatocyte volume (1,19); simple algebra shows that the large postnatal increase (3-4 fold) in mitochondrial adenine nucleotide content decreases cytoplasmic adenine nucleotide concentration by only 20-25%. As we shall see, the significance of this result is that the higher concentration of ATP and ADP in the matrix may stimulate the activity of enzymes in that compartment that involve one of the adenine nucleotides as substrate, whereas the activity of adenine nucleotide-requiring enzymes in the cytoplasm may remain relatively less affected (reviewed in reference 20).

The transport mechanism by which liver mitochondria are able to take up adenine nucleotides from the cytoplasm against a concentration gradient has been reviewed in detail elsewhere (20). Net uptake does *not* occur via the ADP/ATP translocase, which can transport adenine nucleotides only in a one-for-one exchange. The new carrier that accounts for net uptake is called the ATP-Mg/Pi carrier, because the substrate for transport is ATP-Mg and net uptake occurs when cytoplasmic ATP-Mg moves over the carrier in exchange for matrix Pi (Figure 2). Matrix Pi is not depleted because the normal Pi gradient re-equilibrates continuously over the much faster Pi/OH

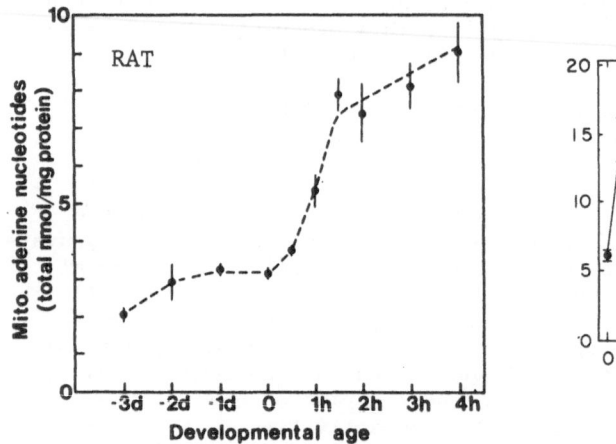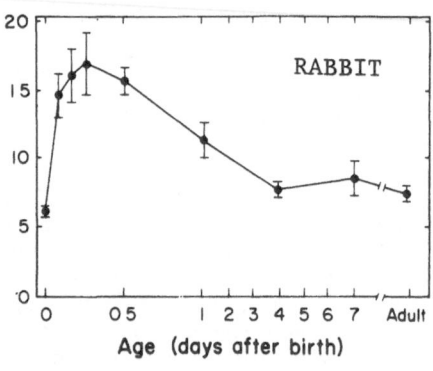

Fig. 1. Total adenine nucleotides (ATP+ADP+AMP) in rat (left panel) or rabbit (right panel) liver mitochondria isolated at various developmental ages. For rat, x-axis indicates days (d) before full term at 22 days gestation or hours (h) after birth. For rabbit, x-axis indicate days after birth. Adenine nucleotides were determined enzymatically in neutralized perchloric acid extracts of mitochondria. Data are from references 30 and 14 respectively, where other details may be found.

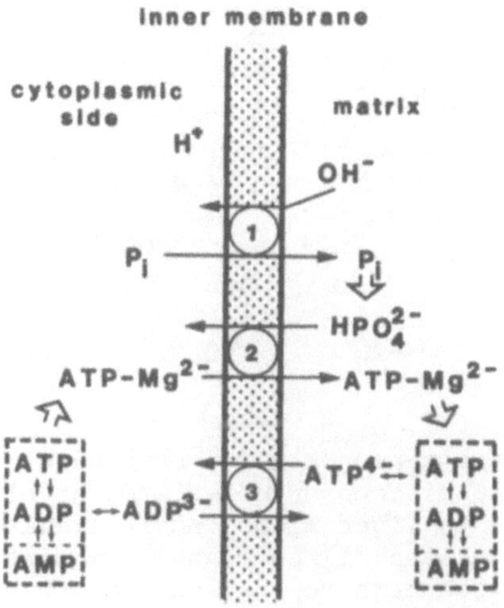

Fig.2. Model for regulation of the adenine nucleotide pool
size by the ATP-Mg/Pi carrier in liver mitochondria.
ATP-Mg exchanged for Pi over this carrier (2)
equilibrates rapidly into the ATP+ADP+AMP pools
(boxes) in the matrix and cytoplasm. The normal Pi
concentration gradient is not perturbed by ATP-Mg/Pi
exchange, because Pi reequilibrates over the much
faster Pi/OH carrier (1). The ATP-Mg/Pi carrier is
distinct from ATP-ADP translocase (3). Diagram from
reference 20.

carrier. In general, the ATP-Mg/Pi carrier is *very* slow compared to other
metabolic carriers and activities in mitochondria, so that ATP-Mg brought
into the matrix cannot contribute directly to flux though matrix reactions.
Instead the accumulated ATP-Mg rapidly equilibrates into the matrix ATP+ADP
pool (and eventually also to AMP) to cause proportional increases in the
absolute concentration of all the nucleotides. The steady state ATP/ADP
ratio is determined independently by the much faster reactions of the
ADP/ATP translocase and ATP synthetase. The ATP-Mg/Pi carrier can affect
the ATP/ADP ratio indirectly, but only to the extent that variations in
the matrix adenine nucleotide content affect the kinetics of oxidative
phosphorylation as discussed below and in references 1 and 20.

DEVELOPMENT OF OXIDATIVE PHOSPHORYLATION RATES

For many years it was supposed that isolated fetal liver mitochondria
were "leaky", or even uncoupled (1). This is clearly not the case. Electron
transport is in fact well-coupled to the phosphorylation of ADP; that is,
there is a very low rate of substrate oxidation in the absence of ADP or
after added ADP has been phosphorylated. Reasons for earlier confusion on
this point have been reviewed (1); investigators seem to have been misled
by the observed low respiratory control ratio (RCR=state 3/state 4) which
is commonly used as a relative index of coupling. However in this case, the
low RCR is due to a low state 3 rate, not a high state 4 rate. Maximally
uncoupled rates equivalent to those in adult mitochondria can be elicited
with chemical uncouplers, showing that electron transport is not rate-
limiting (4) (Figure 3). Thus the fundamental finding with respect to
oxidative phosphorylation in fetal or just-born liver mitochondria is not
uncoupling, but simply a very *slow* coupled phosphorylation rate. There

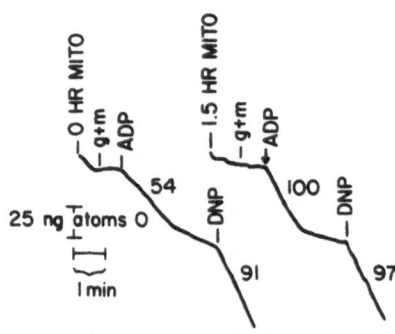

Fig. 3. Polarographic measurement of respiratory function
in rat liver mitochondria isolated at birth (left)
as compared to liver mitochondria isolated after
1.5 h of postnatal life (right). Sequential
additions were made to the assay chamber at the
times indicated as follows: mito, mitochondria; g+m,
glutamate+malate; ADP (150 nmol); DNP, 2,4-dinitro-
phenol (f.c. 40 μM). Numbers indicate rates
calculated as ng atoms O/min/mg protein for the
adjacent region of the recording. Data are from
reference 4 where other details may be found.

are conflicting reports (e.g. reference 4 vs. 5) as to whether there is
developmental improvement in ADP/O ratios, perhaps because it is somewhat
difficult to measure ADP/O when state 3 is as low as it is in newborns.
Recent work showing a normal protonmotive force in newborn mitochondria
(5) further supports the idea that coupling is intact and that state 3
respiration is simply slow.

State 3 respiration develops to nearly adult rates in rat liver
mitochondria as the mitochondrial adenine nucleotide content increases from

Table 1. Respiratory function in mitochondria
isolated from rat liver at birth compared
to those isolated at 60 min postnatal

Postnatal age (min)	0	60
RCR (State3/State4)	3.6±0.5(6)	6.0 ±0.7*(3)
State 3 respiration (ng atoms O/min/mg)	68.1±8.4(9)	108.9±9.4*(3)
State 4 respiration (ng atoms O/min/mg)	19.0±3.0(6)	19.6±3.5(3)
ADP/O	3.0±0.2(6)	2.8±0.1(3)

Respiratory rates were measured polarographically as
described under Figure 3. ADP/O ratios were calculated
by measuring total oxygen consumed in State 3 after
the addition of 120 nmol ADP to the 1 ml assay. Each
value shown is the mean±SE for the number of
experiments indicated in parentheses. Data are from
reference 4. *Difference from the values at 0 min to
at least p < 0.005.

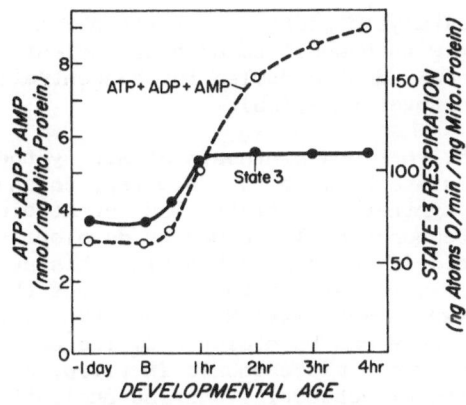

Fig. 4. Adenine nucleotide content and state 3
respiration in rat liver mitochondria
during postnatal development. This
figure is reproduced from reference 1.
Methods are under Figures 2 and 3.

about 3 nmol/mg mitochondrial protein at birth to 5-6 nmol/mg mitochondrial
protein during the first hour of postnatal life (2,4-6) (Table 1). The
adenine nucleotide content increases further during the next hour or two,
but state 3 respiration remains constant (Figure 4). This suggests that
only a small initial increase in adenine nucleotides is sufficient for
maximum development of oxidative phosphorylation rates. Two lines of
evidence support this idea. First, we have shown that if isolated newborn
rat liver mitochondria are allowed to accumulate modest amounts of adenine
nucleotide from the medium, state 3 "matures" just as it does *in vivo* (21).
Second, if the adenine nucleotide content of isolated adult rat liver
mitochondria is depleted to values found in the newborn, state 3 rates are
also decreased to newborn values, but an effect on state 3 is not seen
until the adenine nucleotide content is lowered to a threshold of 5-7
nmol/mg mitochondrial protein (1,22) (Figure 5). It is interesting that
rabbit liver mitochondria, which at birth already have an adenine
nucleotide content of 5-6 nmol/mg mitochondrial protein, also have
maximally developed rates of state 3 respiration (14).

The mechanism by which matrix adenine nucleotide concentration affects
state 3 is not known for sure. The activity of ADP/ATP translocase varies
as a linear function of the matrix adenine nucleotide pool size, and
probably is not rate-limiting for state 3 *in vitro* at least when adenine
nucleotides are greater than 6-7 nmol per mg mitochondrial protein
(4,14,22,23) (Figure 5). In contrast, the activity of the ATP synthetase
complex is affected by variations in the matrix adenine nucleotide content
in a pattern that mirrors the effect on state 3 respiration, showing
saturation when adenine nucleotide content exceeds 6-7 nmol/mg mitochondrial protein (Figure 5). Because translocase activity is an intermediate step in the ATP synthetase reaction in intact mitochondria,
we cannot tell from these results whether synthetase or translocase
limits state 3 when matrix adenine nucleotides fall below 6-7 nmol/mg
mitochondrial protein. More definitive flux control studies of oxidative
phosphorylation in newborn liver mitochondria support the idea that the
rate-limiting step is ATP synthetase (24). We have suggested that the
activity of the ATP synthetase might be stimulated postnatally by a simple
mass action effect of the increase in matrix adenine nucleotides, in other
words, due to a higher absolute concentration of substrate (ATP+ADP) for

the enzyme. Also the catalytic mechanism of the ATP synthetase exhibits positive cooperativity with respect to adenine nucleotide substrates; modest increases in local adenine nucleotide concentrations can have dramatic effects on turnover rate (25).

One study suggests that new synthesis of ATP synthetase contributes to the development of state 3 rates (5). However, in another report (26), it was found that ATP synthetase activity was low at birth but only when measured in intact mitochondria. There was no developmental increase in enzyme activity measured in lysed mitochondria. This suggests that it is the availability of ATP substrate in the matrix, not the amount of enzyme, that limits newborn activity. Moreover, the fact that the development of state 3 rate can be mimicked by increasing the adenine nucleotide content of newborn mitochondria *in vitro*, and that it can be reversed by lowering the mitochondrial adenine nucleotide content (see above), argues strongly for a direct role of the adenine nucleotide pool as the overall initiating event for development of state 3. Subsequent increases in the amount of ATP synthetase enzyme probably contribute to further development of oxidative phosphorylation rates. In fact, intra-mitochondrial protein synthesis and ATP synthetase complex assembly could even be dependent on the postnatal increase in matrix adenine nucleotides, since these processes are energy-requiring and therefore might benefit from a higher matrix ATP concentration (1).

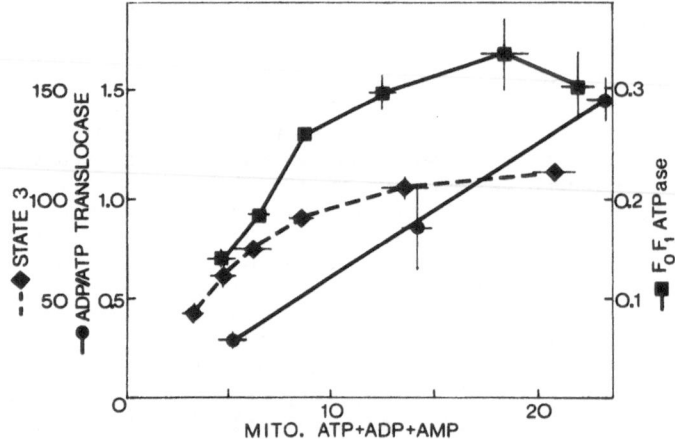

Fig. 5. State 3 respiration, ADP/ATP translocase, and ATP synthetase activities in relation to the matrix adenine nucleotide content in adult rat liver mitochondria. State 3 (ng atoms 0/min/mg mitochondrial protein) was measured as described under Figure 3. ADP/ATP translocase (nmol 0/min/mg mitochondrial protein) was assayed as forward exchange of ^{14}C-ADP (22). ATP synthetase was assayed as uncoupler-stimulated, oligomycin-sensitive ATP hydrolysis (µmoles/hour/mg mitochondrial protein). The matrix adenine nucleotide content (nmol/mg mitochondrial protein) was varied by incubating under conditions that promoted adenine nucleotide uptake or depletion (32). The figure is from reference 20. Bars are S.E.

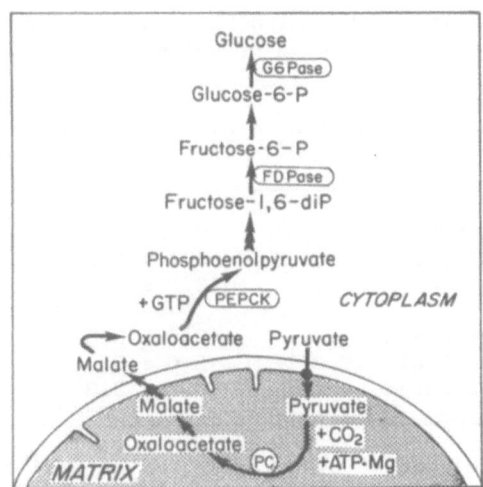

Fig. 6. In the gluconeogenic pathway, pyruvate carboxylase
(PC) is compartmented within the mitochondria and is
subject to regulation by the ATP concentration in
the matrix. This regulation has the potential to
control gluconeogenic rates. Other gluconeogenic
enzymes (glucose-6-phosphatase, (G6Pase); fructose-
1,6-biphosphatase, (FDPase); and phosphoenolpyruvate
carboxykinase, (PEPCK)) are located in the
cytoplasm, although in certain species some PEPCK is
found in mitochondria. The figure is from reference 1.

DEVELOPMENT OF PYRUVATE CARBOXYLATION RATES

Pyruvate carboxylase is located exclusively in the matrix and it is
a key enzyme in the gluconeogenic pathway (Figure 6). Many years ago Snell
(7) reported that pyruvate carboxylase activity in intact mitochondria
increases several-fold soon after birth, but this result was seen only if
activity was measured in intact mitochondria. There was no developmental
change if enzyme activity was assayed in lysed mitochondria. The enzyme
requires ATP, and Snell's results prompted us to think that developmental
increases in pyruvate carboxylation rates in intact mitochondria might be
a direct result of the postnatal increase in the matrix adenine nucleotide
content. This idea was investigated in the newborn rat an rabbit (8-10).
Crossover studies showed that pyruvate carboxylase was rate-limiting for
gluconeogenesis for the first two hours after birth (10). Pyruvate
carboxylation rates did increase in concert with the increased adenine
nucleotide content (Figure 7). Furthermore, the developmental increase in
pyruvate carboxylation rates could be mimicked by increasing the matrix
adenine nucleotide content of newborn rat or rabbit liver mitochondria *in
vitro*, and it could be reversed by lowering the adenine nucleotide content
of older liver mitochondria to newborn values (8,9) (Table 2). Any
physiological signal (various hormones, pO2, see below) which stimulated or
inhibited the uptake of adenine nucleotides into mitochondria *in vivo* also
stimulated or inhibited the rate of pyruvate carboxylation in like fashion
(10).

All of these results suggest that the matrix adenine nucleotide
concentration has a direct effect on the activity of pyruvate carboxylase,
probably by mass action of increased availability of ATP as a substrate,
but perhaps also by allosteric effects of ATP and ADP. The advantage to the
neonate is a mechanism that promotes the rapid onset of gluconeogenesis to

Table 2. Alteration of the mitochondrial adenine nucleotide
pool size: effect on pyruvate carboxylation

Postnatal Age	Adenine Nucleotide Manipulation	Mitochondrial ATP+ADP+AMP	Pyruvate carboxylation
Zero hr	untreated	6.4 ± 0.4	42.8 ± 4.9
	control	5.4 ± 0.6	37.5 ± 2.0
	loaded	12.3 ± 1.4	79.9 ± 9.0
1 hr	untreated	8.9 ± 0.3	88.7 ± 9.0
	control	8.6 ± 0.9	79.6 ± 5.3
	depleted	5.3 ± 0.3	43.5 ± 9.1

Mitochondria isolated at birth (zero hr) or at 1 hr of
postnatal life were incubated to either load or deplete the
mitochondrial adenine nucleotide pool. Controls were treated
similarly except that conditions were chosen so as to
maintain a constant matrix adenine nucleotide pool size.
Mitochondria were then pelleted, washed, and resuspended for
adenine nucleotide and pyruvate carboxylation determinations
(units: nmol/mg mitochondrial protein and nmol HCO_3^-/min/mg
mitochondrial protein, respectively). Data are from reference 9, where other experimental details may be found. Each
value is the mean ± S.E. for three separate experiments.

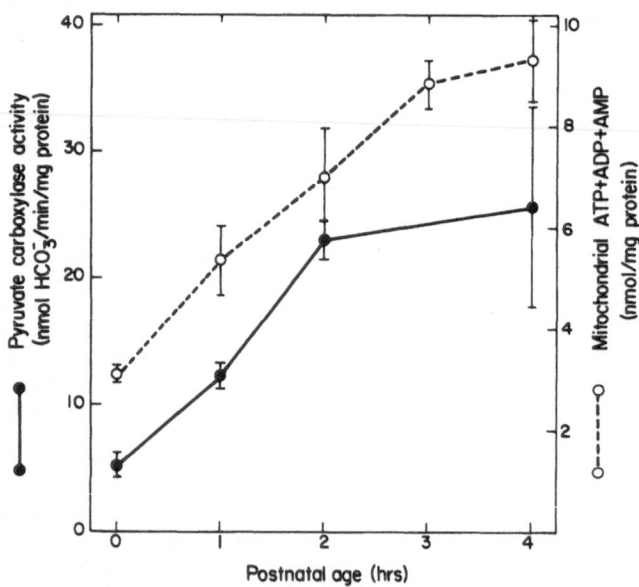

Fig. 7. Pyruvate carboxylation in intact rat liver mitochondria
in relation to the mitochondrial adenine nucleotide
content during early postnatal life. Pyruvate
carboxylation was measured as $H^{14}CO_3^-$ incorporation into
pyruvate under phosphorylating conditions with no added
ATP or ADP. Adenine nucleotides were determined as under
Figure 1. The data are from reference 8 where further
details may be found. Bars are S.E.

correct the postnatal hypoglycemia that ensues once the umbilicus is severed. This initial 2-3 fold increase in gluconeogenesis occurs within 2 hours and is independent of new enzyme synthesis (10); later synthesis of gluconeogenic enzymes (phosphoenolpyruvate carboxykinase and pyruvate carboxylase especially) over a period of days enables much larger increases in gluconeogenic rates (reviewed in reference 1).

DEVELOPMENT OF OTHER METABOLIC FUNCTIONS AND MITOCHONDRIAL BIOGENESIS

In theory, any matrix reaction that requires adenine nucleotides may be susceptible to regulation by the postnatal increase in the matrix adenine nucleotide concentration. Citrulline synthesis, an essential reaction in ureagenesis, is a candidate for such regulation. Indeed, citrulline synthesis is low in newborn mitochondria and does increase in concert with the postnatal increase in matrix adenine nucleotide content (11). The rate of citrulline synthesis in adult liver mitochondria is increased or decreased when the matrix adenine nucleotide content is increased or decreased by in vitro manipulation (27).

It is intriguing to consider that the rapid onset of mitochondrial biogenesis (DNA replication, RNA and protein synthesis) might depend on the postnatal accumulation of adenine nucleotides into the matrix (1,20). These reactions are all ATP dependent, and thus might very well be stimulated by the postnatal increase in the matrix adenine nucleotide concentration. Intramitochondrial synthesis of DNA and RNA does increase as proliferation ensues postnatally (28).

PHYSIOLOGICAL REGULATION OF ADENINE NUCLEOTIDE ACCUMULATION INTO NEWBORN LIVER MITOCHONDRIA

The net uptake of adenine nucleotides from the cytoplasm into the mitochondria is exquisitely controlled by pO_2 and by hormones (10,20,29, 30). The design of fetal circulation ensures the preferential delivery of oxygenated blood from the placenta to the ascending aorta and cephalic branches. Blood delivered to the caudal half of the fetus contains relatively less oxygen, about 30% saturation. At birth, as umbilical flow ceases and the lungs inflate, changes in circulation occur that result in sudden perfusion of the liver with well-oxygenated blood. This is an important permissive condition for the uptake of adenine nucleotides into mitochondria, presumably because with oxygen present, the cytoplasmic ATP/ADP ratio increases which provides the substrate (ATP-Mg) for adenine nucleotide transport into the matrix. If animals are subjected to prolonged postnatal hypoxia, the normal increase in mitochondrial adenine nucleotide does not occur, and the related development of metabolic functions is suppressed (10,20,30).

At birth, catecholamines and glucagon increase while insulin decreases (18,31). Calcium and cAMP are second messengers that mediate the effect of these hormones on adenine nucleotide uptake by mitochondria (20,29). The ATP-Mg/Pi carrier has an absolute requirement for micromolar calcium, and cAMP serves to stimulate glycolysis which supports an increase in the cytoplasmic ATP/ADP ratio (20). In pups from diabetic mothers, the postnatal insulin/glucagon ratio is abnormally high and the uptake of mitochondrial adenine nucleotides is supressed (18,30). This condition impairs the activation of pyruvate carboxylation and delays the onset of gluconeogenesis, which probably contributes to prolonged hypoglycemia in this situation (10,18,20,30).

109

SUMMARY

    A subcellular shift of adenine nucleotides from cytoplasm to the
mitochondrial compartment does appear to be an important factor in the
rapid postnatal development of independent metabolic function (Figure 8)
The uptake of adenine nucleotides into mitochondria is signaled by
oxygenation and hormones via an increased cellular ATP/ADP ratio, and via
direct stimulation of the ATP-Mg/Pi carrier by calcium (20). The effect of
increased mitochondrial adenine nucleotide concentrations is stimulation
of matrix enzyme activities that require adenine nucleotides as a
substrate. Of matrix reactions affected by adenine nucleotides, oxidative
phosphorylation is initially the most important because this leads to
positive feedback for further adenine nucleotide uptake (secondary to
increased cellular ATP/ADP ratios), and to a faster rate of ATP supply
for metabolic activities (Figure 8).

    Oxygenation is permissive for the hormone effects, probably because
electron transport is essential to establish ATP/ADP ratios that favor net
uptake (20). In hypoxic newborns, cytoplasmic ATP/ADP ratios remain low,
so no net uptake can occur even though the carrier might be stimulated by
hormones. Net uptake of adenine nucleotides is completely reversible. If
newborns are first normoxic, but then later suffer an hypoxic event, the
lower cytoplasmic ATP/ADP ratio will favor adenine nucleotide efflux from
mitochondria via the ATP-Mg/Pi carrier. This could selectively shut down
certain metabolic pathways (such as gluconeogenesis and ureagenesis) by
decreasing the activities of their matrix reactions, thus preserving any
available ATP for more essential cellular functions. This reversible
regulation may provide protection of hepatocyte integrity during hypoxic
episodes in the perinatal period, in that systemic demands for glucose can
be ignored at times when supplying ATP to support gluconeogenesis would
jeopardize life-support activities within the liver cell.

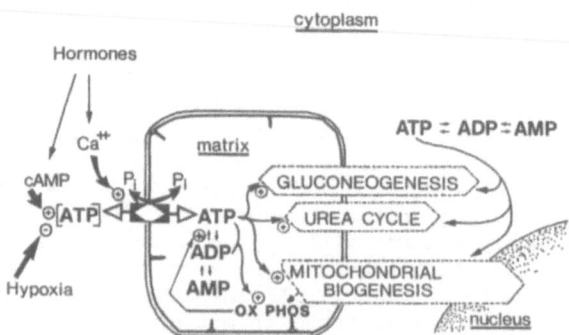

Fig. 8. The postnatal increase in mitochondrial adenine
         nucleotide content is regulated by hormones and
         pO2 via second messenger effects on ATP-Mg/Pi
         transport activity. An increased matrix adenine
         nucleotide content stimulates (probably by mass
         action) rates of matrix reactions that involve
         adenine nucleotide substrates. Stimulated
         reactions include oxidative phosphorylation;
         intramitochondrial steps in the gluconeogenic
         pathway and urea cycle; and probably also the
         intramitochondrial steps of mitochondrial
         biogenesis. Figure adapted from reference 20.

110

ACKNOWLEDGMENTS

This work was supported by N.I.H. grant number HD16936 from the U.S. Public Health Service. I thank Valerie Ricciardone and Carol Valente for preparation of the manuscript, I am grateful to the Fundación Ramón Areces, to the Perinatal Biochemical Group of the Spanish Biochemical Society, and to Drs. J.M. Cuezva and A.M. Pascual-Leone for the opportunity to participate in this symposium.

Figures and Tables are reprinted with publishers permission as follows: Table 2 is from ref. 9, Pergamon Press; Figure 3 and Table 1 are from ref. 4, Academic Press; Figure 1B is from ref. 14 and Figure 7 is from ref. 8, Elsevier Publ.; Figures 4 and 6 are from ref. 1, Van Nostrand Reinhold; Figure 1A is from ref. 30, Pediatric Research; Figures 2, 5 and 8 are from ref. 20, FASEB.

REFERENCES

1. J.R.Aprille, Perinatal development of mitochondria in rat liver, in: "Mitochondrial Physiology and Pathology", G.Fiskum, ed., Van Nostrand Reinhold, New York, pp. 66 (1986).
2. T.Nakazawa, K.Asami and O.Yukawa, Appearance of energy conservation system in rat liver mitochondria during development. The role of adenine nucleotide translocation, J.Biochem. 73:397 (1973).
3. R.Sutton and J.K.Pollak, The increasing adenine nucleotide concentration and the maturation of rat liver mitochondria during the neonatal development, Differentiation 12:15 (1978).
4. J.R.Aprille and G.K.Asimakis, Postnatal development of rat liver mitochondria: State 3 respiration, adenine nucleotide translocase activity, and the net accumulation of adenine nucleotides, Archiv.Biochem.Biophys. 201:564 (1980).
5. C.Valcarce, R.M.Navarrete, P.Encabo, E.Loeches, J.Satrústegui and J.M.Cuezva, Postnatal development of rat liver mitochondrial functions: The roles of protein synthesis and of adenine nucleotides, J.Biol.Chem. 263:7767 (1988).
6. J.K.Pollak, The maturation of the inner membrane of foetal rat liver mitochondria. An example of a positive feedback mechanism, Biochem.J. 150:477 (1974).
7. K.Snell, Pyruvate carboxylation in liver mitochondria of the developing neonatal rat, Internat.J.Biochem. 5:463 (1974).
8. J.R.Aprille, P.Yaswen, and J.Rulfs, Acute postnatal regulation of pyruvate carboxylase activity by compartmentation of mitochondrial adenine nucleotides, Biochim.Biophys.Acta 675:143 (1981).
9. W.A.Brennan, Jr. and J.R.Aprille, Pyruvate carboxylase and the acute regulation of hepatic gluconeogenesis in the newborn rabbit, Comp.Biochem.Physiol. 77B:35 (1984).
10. W.A.Brennan, Jr. and J.R.Aprille, Regulation of hepatic gluconeo-genesis in newborn rabbit: Controlling factors in presuckling period, Am.J.Physiol. 249:E498 (1985).
11. R.T.Kelley and J.R.Aprille, Postnatal accumulation of adenine nucleotides in rabbit liver mitochondria and the development of citrulline synthesis, Pediatr.Res. 18:140A (1984).
12. J.K.Pollak, Mitochondrial heterogeneity in foetal and suckling rat liver: Differential leucine incorporation into the proteins of two mitochondrial populations, Biochem.Biophys.Res.Commun. 69:823 (1976)
13. M.Hallman, Effect of intraperitoneal chloramphenicol on some mitochondrial enzymes in neonatal rats, Biochem.Pharmacol. 20:1797 (1971).

14. J. Rulfs and J. R. Aprille, Adenine nucleotide pool size, adenine nucleotide translocase activity and respiratory activity in newborn rabbit liver mitochondria, Biochim. Biophys. Acta 681:300 (1982).

15. P. H. Van Lelyveld and F. A. Hommes, Adenine nucleotides in foetal rat liver cells. Compartmentation and variation with age, Biochem. J. 174:527 (1978).

16. J. M. Cuezva and J. M. Medina, Adenine nucleotide concentrations in liver of fetal rats. Neonatal changes in the premature newborn, Rev. Esp. Fisiol. 38:161 (1982).

17. F. J. Ballard, Adenine nucleotides and the adenylate kinase equilibrium in livers of foetal and newborn rats, Biochem. J. 117:231 (1970).

18. J. M. Cuezva C. I. Chitra, and M. S. Patel, The newborn of diabetic rat. II. Impaired gluconeogenesis in the postnatal period, Pediatr. Res 16:638 (1982).

19. J. Vassy, M. Kraemer, M. T. Chalumeau, and J. Foucrier, Development of the fetal rat liver: ultrastructural and stereological study of hepatocytes, Cell Differentiation 24:9 (1988).

20. J. R. Aprille, Regulation of the mitochondrial adenine nucleotide pool size in liver: mechanism and metabolic role, FASEB. J. 2:2547 (1988).

21. J. R. Aprille, Net uptake of adenine nucleotides by newborn rat liver mitochondria, Archiv. Biochem. Biophys. 207:157 (1981).

22. G. K. Asimakis and J. R. Aprille, In vitro alteration of the size of the liver mitochondrial adenine nucleotide pool: Correlation with respiratory functions, Archiv. Biochem. Biophys. 203:307 (1980).

23. R. Grunwald and J. J. Lemasters, Control of oxidative phosphorylation by ADP-ATP translocation: modulation by matrix adenine nucleotide, Fed. Proc. 43:1877 (1984).

24. L. Baggeto, D. C. Gautheron and C. Godinot, Effects of ATP on various steps controlling the rate of oxidative phosphorylation in newborn rat liver mitochondria, Archiv. Biochem. Biophys 203:670 (1984).

25. H. S. Penefsky, Molecular mechanism of action of mitochondrial ATPase in: "Achievements and Perspectives of Mitochondrial Research, Vol I: Bioenergetics", E. Quagliariello, E. C. Slater, F. Palmieri, C. Saccone, and A. M. Kroon, eds., Elsevier, Amsterdam (1985). pp. 35-41.

26. M. Hallman, Changes in mitochondrial respiratory chain proteins during perinatal development. Evidence of the importance of environmental oxygen tension, Biochim. Biophys. Acta 253:360 (1971).

27. F. G. Goldstein and J. R. Aprille, Citrulline synthesis: regulation by alterations in the total mitochondrial adenine nucleotide content, Arch. Biochem. Biophys. 213:7 (1982).

28. P. Cantatore, P. L. Polosa, F. Fracasso, Z. Flagella and M. N. Gadaleta, Quantitation of mitochondrial RNA species during rat liver development: the concentration of cytochome oxidase subunit I (CoI) mRNA increases at birth, Cell Differentiation 19:125 (1986).

29. P. C. Tullson and J. R. Aprille, Regulation of mitochondrial adenine nucleotide content in newborn rabbit liver, Am. J. Physiol. 253:E530 (1987).

30. J. R. Aprille and M. T. Nosek, Neonatal hypoxia or maternal diabetes delays postnatal development of liver mitochondria, Pediatr. Res. 21:266 (1987).

31 J. M. Cuezva, E. S. Burkett, D. S. Kerr, H. M. Rodman, and M. S. Patel, The newborn of diabetic rat. I. Hormonal and metabolic changes in the postnatal period, Pediatr. Res. 16:632 (1982).

32. J. Austin and J. R. Aprille, Carboxyatractyloside-insensitive influx and efflux of adenine nucleotides in rat liver mitochondria, J. Biol. Chem. 259:154 (1984).

POSTNATAL MITOCHONDRIAL DIFFERENTIATION

IN THE NEWBORN RAT

José M. Cuezva, Carmen Valcarce, Ana M. Luis, José M. Izquierdo, Agustín Alconada and Margarita Chamorro

Departamento de Biología Molecular
Centro de Biología Molecular (U.A.M.-C.S.I.C.)
Universidad Autónoma de Madrid
28049 Madrid, Spain

INTRODUCTION

During mammalian development an important aspect of cell metabolism is that related with the pathway of energy provision, because, in one way or another, the rest of metabolic pathways and cellular functions depend on an efficient supply of energy. It is in the mitochondria where energy is generated by the oxidation of cellular substrates into its useful form of ATP. Cells devoid of, or which contain a poorly developed mitochondria, rely on the less efficient anaerobic glycolysis for harnessing their ATP needs.

The late fetal and early neonatal stages of mammalian development are characterized by sudden changes in the pathways and substrates relevant for energy provision. During late fetal life, the fetal liver meets its energy demands by anaerobic glycolysis (1-4), because its respiratory capacity is low (2,4,5), due to the low number of mitochondria per cell (6-9) and the low activity of these mitochondria (10-15). During this stage of development, the main substrate oxidized by the fetal liver is the glucose made available by the transplacental transfer from maternal circulation . The non-restrained transfer of glucose from the maternal to the fetal compartment supports the high glycolytic activity of the fetal rat tissues and introduces, mainly towards the end of gestation, an active lactate cycle between the fetal producing tissues and the maternal consuming ones (16), a Cori cycle that is mainly supported by the increased gluconeogenic capacity of the maternal liver (16-18). In addition, the fetal mitochondria lack the enzymatic machinery that allows fatty acid oxidation during this stage of development (19).

At birth, the supply of nutrients is abruptly interrupted and, at the same time, a number of energy-demanding physiological processes are initiated or increased. These are the well known cases of thermoregulation, respiration and muscular activity, and metabolic pathways such as gluconeogenesis, ureogenesis and protein synthesis. Also, an abrupt change in the main mitochondrial oxidative substrates takes place, that is, the switch from glucose to fatty acid oxidation for the fetal and neonatal periods, respectively (20,21). These circumstances trigger the development

Endocrine and Biochemical Development of the Fetus and Neonate
Edited by J. M. Cuezva *et al.*
Plenum Press, New York, 1990

113

of an efficient energy generating system, i.e., mitochondrial differentiation or mitochondrial maturation (9-15,22).

An excellent review of the different enzymatic and functional aspects of mitochondrial development has been recently published (9). In this paper, we will focus our attention on a very short time period of mitochondrial development, that is, the first hours of neonatal life. This time-period is unique because the process of mitochondrial differentiation occurs in just a matter of minutes (9,10,14,15,22). This process only happens physiologically once in the life span of a mammalian organism, and endows specific tissues in which it occurs with the functional machinery to harness their energy needs. Further, and because of clear clinical implications of postnatal mitochondrial differentiation for neonatal adaptation to extrauterine life, we will briefly summarize those pathophysiological situations which we consider at present to be high risk in terms of postnatal impairments in mitochondrial differentiation.

PROLIFERATION AND DIFFERENTIATION OF RAT LIVER MITOCHONDRIA

The fetal liver contains a much lower number of mitochondria per liver cell than the adult liver (6-9). Figure 1 illustrates the differences between these two stages of liver development in the rat. There are two main aspects that should be pointed out: (i) the low number of mitochondria per hepatocyte in term fetal liver when compared to adults and (ii) the different sizes of the organelle in both tissues, being much bigger in fetal than in adult liver, most likely as a result of the physiological swelling experienced by mitochondria under a hypoxic condition (23). These

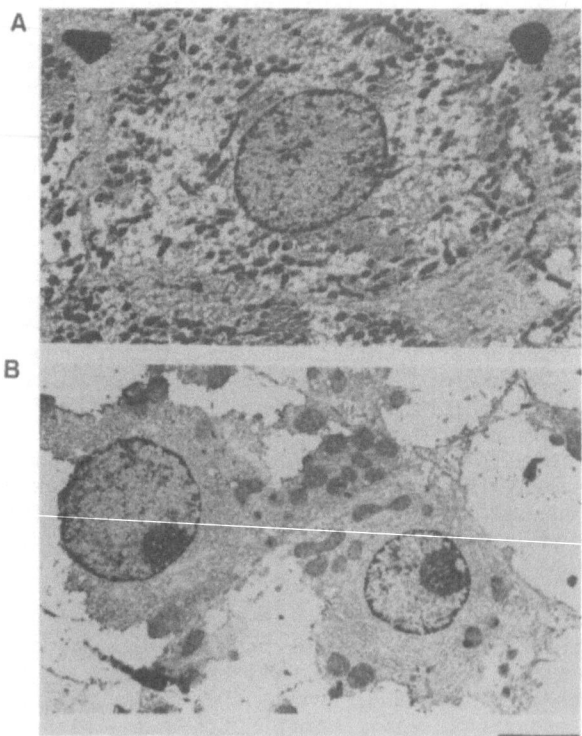

Fig. 1. Electron microscopy of thin sections from adult (A) and term fetal (B) livers. Bar, 5 μm.

findings imply that during rat liver development at least two different changes should be accomplished. *First*, the proliferation of the organelle in order to achieve the number of mitochondria per hepatocyte found in the adults and *second*, the acquisition of the ultrastructural, molecular and functional characteristics that define mitochondrial function. As reviewed recently (9), proliferation of mitochondria is a continous process that occurs during the entire developmental period. This results in an increase of the number of mitochondria per liver cell, in principle identical to the parent population, as a result of a faster rate of division for the organelle than for the liver cell. In contrast, the acquisition of the ultrastructural, molecular and functional features of adult liver mitochondria, that is, mitochondrial differentiation, is a very rapid process that is accomplished during the first hours of extrauterine life (9, 10, 14, 15, 22). Mitochondrial differentiation applies for those pre-existing mitochondria and does not necessarily have to be reflected by an increase in the number of mitochondria per cell, but rather by an increase in any of the parameters tested when they are expressed per mitochondrial unit (9, 15). However, the most feasible alternative occuring during this stage of development is that mitochondrial proliferation is concurrent with mitochondrial differentiation. Thus, we find a rapid increase in both the number of mitochondria per cell and in the functional feature of the organelle.

At present time, we should envision differentiation of fetal $\longrightarrow$ neonatal rat liver mitochondria as being brought about by the synergistic action of two main processes that occur soon after birth in the neonatal liver (15): (i) those related with the synthesis of proteins for the mitochondria, and (ii) those dependent on the postnatal enrichment in mitochondrial adenine nucleotides.

THE ROLE OF PROTEIN SYNTHESIS IN MITOCHONDRIAL DIFFERENTIATION

Development of Rates of Respiration

Previous studies (13-15) of the respiratory activity of mitochondria isolated from fetal rat liver have clearly established that their respiratory rates were lower than those observed in adults. Figure 2 represents the developmental changes in the rates of respiration of isolated mitochondria from fetal, neonatal and adult rat liver. The most remarkable findings are that both State 3 and FCCP-uncoupled rates of respiration show a two-fold increase in just 1 h after birth. Also, State 4 respiration rates show a two-fold increase during the same time period. The postnatal surge in State 3 rates of respiration has been suggested to be solely mediated by the increase in intramitochondrial adenine nucleotides experienced by the neonatal mitochondria during the first hours following birth (14, 22, 24). However, several lines of evidence show that the main postnatal increase in State 3 rates of respiration, as well as in many other parameters of mitochondrial function (15), are dependent on the coordinated induction of the rates of cytosolic and mitochondrial protein synthesis (15). When term newborns were administered cycloheximide at birth to inhibit cytosolic protein synthesis, the surge in both State 3 and FCCP-uncoupled rates of respiration by their liver mitochondria isolated at 2 h postnatal were almost blunted, and only a 30% increase for both parameters was noticed to be resistant to the antibiotic treatment (Figure 2). However, cycloheximide treatment to the neonates did not prevent: the postnatal enrichment in intramitochondrial adenine nucleotides (Table 1), the postnatal increase in State 4 respiratory rates (Figure 2), or the dramatic reduction of mitochondrial matrix volume that occurs soon after birth (Table 1). These findings will be considered below under "the role of adenine nucleotides in mitochondrial differentiation" section.

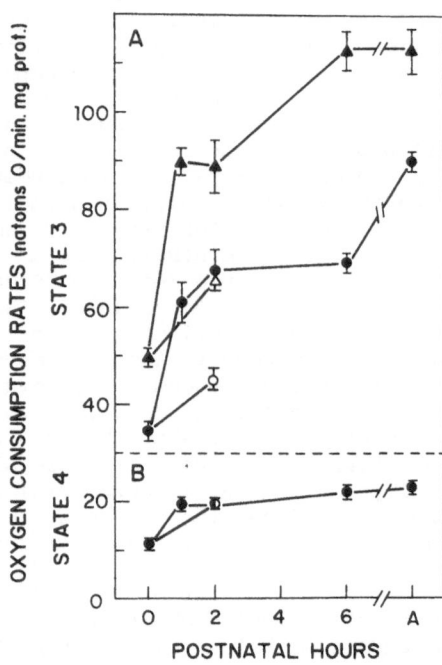

Fig. 2. Developmental changes in oxygen consumption
rates by rat liver mitochondria. Respiratory
rates were measured polarographically using 8
mM sodium succinate as respiratory substrate.
(A). State 3 (●) and FCCP-uncoupled (▲) rates
of respiration were determined in the presence of 0.1 mM NaADP and 1.2 μM FCCP,
respectively. (B). State 4 (●) rates of
respiration. The open circles and triangle in
both A and B, are the same parameters when
assayed in isolated mitochondria from 2 h-old
neonates, that were administered 10 mg/Kg
body weight of cycloheximide at the time of
delivery, to inhibit cytosolic protein
synthesis. Results of this figure are from
reference (15).

Development of Oxidative Phosphorylation

        Fetal mitochondria are remarkably less efficient in the generation
of ATP, i.e., in oxidative phosphorylation, than adult or neonatal
mitochondria (Figure 3). It has been argued that the conflicting reports in
the literature (12, 15 versus 9, and the preceding chapter) as to whether
there is a postnatal increase in the ADP/O ratio (Figure 3), could be due
to the difficulty of estimating this parameter for fetal mitochondria.
However, the mitochondrial transmembrane potential ($\Delta\psi$), the main driving
force in oxidative phosphorylation, is similar under State 3 and 4
conditions for the fetal mitochondria (15), adult levels for this parameter
develop during the first postnatal hour (15); and the addition of ADP to
adult or neonatal mitochondria promotes a reduction of $\Delta\psi$ up to a value
similar to that found under State 4 in the fetal mitochondria (15), clearly
indicating the lower phosphorylating capability of the fetal mitochondria.
As shown for State 3 and FCCP-uncoupled rates of respiration (Figure 2),
administration of cycloheximide to the newborns prevented the postnatal

Table 1. Effect of cytosolic and mitochondrial protein synthesis inhibitors on intramitochondrial ATP and ADP concentrations and in matrix volume

Postnatal h + treatment	ATP(mM)	ADP(mM)	Matrix volume (μl/mg)
0+none	0.6±0.1	1.4±0.3	1.00±0.17
2+NaCl	2.4±0.2**	6.0±1.2*	0.43±0.05**
2+Cycloh.	1.8±0.1	6.3±0.8	0.49±0.05
2+Strept.	2.2±0.4	6.7±0.8	0.42±0.05

Cytosolic or mitochondrial protein synthesis was inhibited in newborn rats at the time of delivery by the intraperitoneal administration of 10 or 100 mg/Kg body weight of cycloheximide (cycloh.) or streptomycin (strepto.), respectively. Mitochondria were isolated at 0 and 2 h after the treatment and the ATP and ADP concentrations measured. The mitochondrial matrix space was determined under State 4 conditions using Na-succinate (8 mM) as respiratory substrate (15). Values are means ±S.E. *, $p<0.0125$ and **, $p<0.0005$ compared to values from 0 h versus 2 h NaCl treated.

increase in the ADP/O ratio (Figure 3). A similar finding has been observed when mitochondrial protein synthesis was inhibited at birth by administration of streptomycin (15), suggesting the cooperation of both cytosolic and mitochondrial protein synthesis in order to achieve functional energy conserving mitochondria in the neonatal period.

Development of Inner Mitochondrial Membrane Complexes

The developmental increase in FCCP-uncoupled and State 3 rates of respiration, in addition to the increase in $\Delta\psi$ (15) and in the ADP/O ratio, suggests a rapid postnatal increase in the enzymatic capacity of the

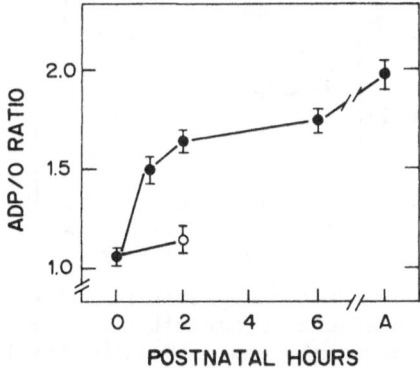

Fig. 3. Developmental changes in the ADP/O ratio in rat liver mitochondria. ADP/O ratios were calculated by measuring total oxygen consumed in State 3 after the addition of 100 nmol ADP. Results of this figure are from reference (15). For symbols, see legend to Figure 2.

mitochondrial respiratory chain and oxidative phosphorylation. In fact, determination of the specific activity of complexes I, II and IV of the respiratory chain and of the F_1-ATPase complex in isolated mitochondria during rat liver development confirmed the previous hypothesis (Figure 4). It should be noted that most of the activities assayed increase almost two-fold in their specific activity during the first postnatal hours, indicating that the fetal mitochondria have an incomplete set of respiratory chain and ATP synthetase proteins. Changes in enzyme activities were not observed in previous reports (13,14) possibly because they were measured in liver homogenates (14), where the bulk of total liver protein could mask any postnatal surge of mitochondrial enzyme activities. Moreover, a two-fold increase in cytochrome oxidase activity was, in fact, found (13), but was not taken into consideration; thus, leading to the

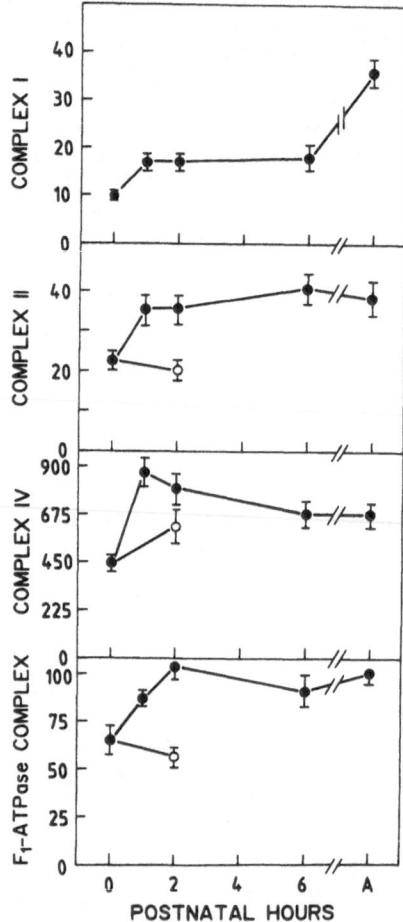

Fig. 4. Developmental changes in rat liver mitochondrial enzyme activities of the respiratory chain and oxidative phosphorylation. The enzyme activities were expressed as mU/mg of mitochondrial protein. The results are means ± S.E. for six to eight experiments. Results of this figure are taken in part from reference (15). For symbols, see legend to Figure 2.

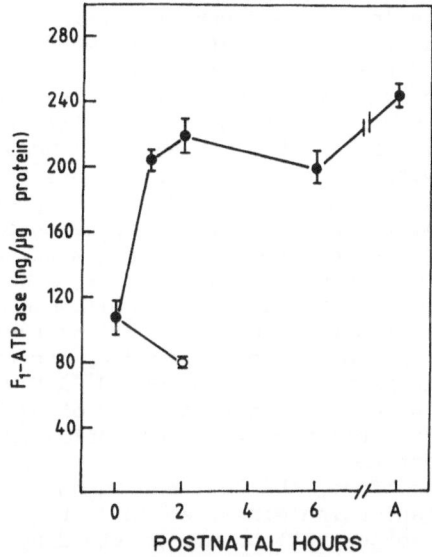

Fig. 5. Developmental changes in the amount of
F_1-ATPase complex in rat liver mitochondria. The amount of F_1-ATPase
proteins were determined by *elisa* in
sonicated mitochondria with a rabbit
antiserum raised against pure rat
liver F_1-ATPase (15). The antibody
recognized the major α and β subunits
of the complex. The results of this
figure are taken from reference (15).
For symbols, see legend to Figure 2.

erroneous suggestion that the postnatal enrichment in adenine nucleotides
(14,15,24) was the sole trigger factor promoting postnatal mitochondrial
differentiation (9,14,22).

The postnatal increase in mitochondrial respiratory and oxidative
phosphorylation enzymes (Figure 4) is due to a rapid and preferential
increase in the rates of protein synthesis for mitochondrial proteins,
occuring in the neonatal liver and maintained for at least 2 h following
birth (15). In fact, quantitation of the amount of mitochondrial F_1-ATPase
proteins by immunological methods during liver development (Figure 5) with
an antibody raised against purified F_1-ATPase from inner mitochondrial
membrane vesicles prepared from adult rat liver (15), showed that the
amount of these proteins increase two-fold during the first postnatal hour
(Figure 5). Further, administration of cytosolic protein synthesis
inhibitors to the neonates at birth (15) showed, in isolated mitochondria
from the 2 h-old treated neonates, that the increase in activity of both
complex II of the respiratory chain and the F_1-ATPase complex (Figure 4) as
well as the amount of F_1-ATPase protein (Figure 5), were arrested (15). In
contrast, inhibition of mitochondrial protein synthesis by streptomycin
administration to the newborns at birth did not prevent the postnatal
increase in the specific activities of Complex II and F_1-ATPase or in the
amount of F_1-ATPase proteins (15), in agreement with their known synthesis
on cytosolic ribosomes (25,26).

It has been proposed (14,27) that adenine nucleotide translocase
activity controls respiration during postnatal development. However, it is

unclear whether the translocase performs this role since, during postnatal development, there is a strict correlation between ADP/O ratios and the mitochondrial content of F_1-ATPase (15). This is in agreement with studies showing that the H^+-ATP synthetase, and not the translocase, limits respiration in the newborn (28). The only situation in which an increase in F_1-ATPase did not correlate with the measured ADP/O ratios was when neonates were treated with streptomycin (15). Since this treatment is unable to block translocase expression because the protein is coded by the nuclear genome (29), the absence of efficient coupling between respiration and phosphorylation should be attributed to the lack of complete H^+-ATP synthetase proteins (15). Several subunits of the F_1-ATPase, those that constitute the inner mitochondrial membrane H^+--channel of the H^+-ATP synthetase, are encoded for by mitochondrial DNA (30). Thus, the low ADP/O ratios obtained by streptomycin treatment (15) are most likely the result of incomplete H^+-ATP synthetase and not the translocase. In addition, during the first hour of extrauterine life, newborn rat liver mitochondria undergo a net influx of adenine nucleotides (14,15,24) by a transport mechanism independent of the adenine nucleotide translocase (31,32). Therefore, ADP delivery through translocase would not be expected to limit respiration since, at this stage of development, intramitochondrial ADP concentrations are provided by another transport mechanism (31,32).

A Translational Mechanism for Postnatal Mitochondrial Differentiation?

Regulation of the amount of a particular protein in any given experimental system could involve a combination of different mechanisms at both pre- and post-transcriptional and/or translational levels. Thus, we have recently explored (33) the possibility of a rapid postnatal increase in liver mRNA levels for the β-subunit of the F_1-ATPase complex that could result in the doubling of the amount of β-F_1-ATPase protein per mitochondrial unit in just 1 h postnatal. However, the absence of changes for liver β-F_1-ATPase mRNA levels during postnatal development (data not shown) suggests that mitochondrial differentiation is regulated at a translational level, most likely by the existence of a limitation in the translational machinery of the fetal liver (33).

THE ROLE OF ADENINE NUCLEOTIDES IN MITOCHONDRIAL DIFFERENTIATION

Ultrastructural Morphology of Isolated Mitochondria during Rat Liver Development

Since the pioneering work of Hackenbrock (34,35), on "ultrastructural basis for metabolically linked mechanical activity in mitochondria" and further similar observations by others (36,38), it is well known that adenine nucleotides, mainly ADP, promote ultrastructural changes in isolated mitochondria which result in a reduction of their matrix volume. Term fetal and 2 h-old neonatal isolated rat liver mitochondria (Figure 6, A and B) are larger than adult mitochondria (Figure 6C), in agreement with previous reports (39) and with the ultrastructural observation of the organelle in liver thin sections (Figure 1). We have recently showed (15) that three different mitochondrial morphologies could be observed during rat liver development. These mitochondrial morphologies have been defined as follows: orthodox (predominant morphology in 6A), condensed (predominant morphology in 6C), and highly condensed (predominant morphology in 6E). As previously mentioned, addition of ADP to mitochondrial preparations promoted significant ultrastructural changes at all time points of development studied, shifting the predominant conformation at that time to the highly condensed conformation (Figure 6, A, B and C versus D, E, F). A quantitative estimation of the percent contribution of each mitochondrial

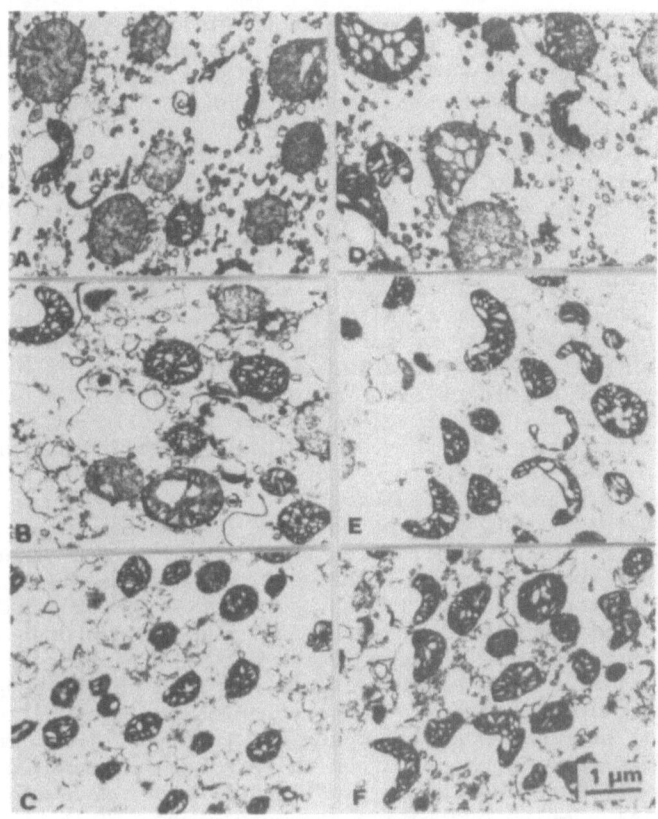

Fig. 6. Effect of ADP on the morphology of mitochondria
        isolated from livers of fetal, 2 h-old neonatal,
        and adult rats. Liver mitochondria from term fetus
        (A,D), 2 h-old neonatal (B,E), and adult (C,E) rats
        incubated under State 4 (A,B,C) and State 3 (D,E,F)
        conditions are shown at a magnification x 18,000.

morphology at the different developmental stages studied and the effect of
ADP is shown in Figure 7. In the absence of ADP, the orthodox morphology
was the most abundant "conformation" for fetal mitochondria (70%). In
contrast, adult and 2 h-old mitochondria showed the condensed (100%) and
highly condensed (70%) morphologies, respectively (Figures 6 and 7).

Developmental Changes in Mitochondrial Matrix Volume and Adenine Nucleotide Concentrations

The results obtained by electron microscopy studies are in agreement
with the values of the mitochondrial matrix space determined by biochemical
methods (15), which showed that isolated fetal mitochondria under State 4
conditions had a larger matrix space than in any of the other developmental
stages studied. Addition of ADP to mitochondrial preparations promoted a
shift in the mitochondrial morphologies present to a condensed or highly
condensed state (Figures 6 and 7). This was also manifested by the values
obtained from the measurements of the matrix volume under State 3
conditions (15). In this situation, no differences were observed in
mitochondrial volumes at any stage of development studied (15). The
existence of a close parallelism between the developmental changes in
intramitochondrial levels of adenine nucleotides and the matrix volume of

the organelle in rat liver was recently shown (15). There exist a significant linear correlation between both parameters up to a threshold value of 6-7 mM intramitochondrial ATP+ADP (Figure 8). A similar linear correlation could be observed when intramitochondrial ADP concentrations are considered against the mitochondrial volume (data not shown). Once the intramitochondrial ATP+ADP concentrations reach the threshold value, corresponding to the minimum volume of the organelle, no further increase in the nucleotide is able to promote a further contraction of the matrix space (Figure 8).

Regulation of mitochondrial matrix volume involves changes in the transport of K^+ through the inner mitochondrial membrane by two main mechanisms, an electrogenic K^+ uniporter, which mediates K^+ entry into the mitochondria, and a K^+/H^+ antiporter which pumps K^+ out of the mitochondria (40-44). Further, the existence of an anionic channel (23), with a putative function of restoring the matrix volume of the organelle after its pathological swelling has been suggested. The ultrastructural changes promoted both *in vivo* and *in vitro* by adenine nucleotides on mitochondrial morphology (Figures 6 and 7), resulting in a decrease in its matrix volume (15 and Figure 8), are thought to be mediated by the interaction of the nucleotides with the adenine nucleotide translocase of the inner membrane (36-38). This involves differential changes in the permeability of the membrane towards potassium ions. It has been suggested (45) that the interaction of adenine nucleotides with the translocase promotes a conformational change in the protein, which results in changes in the H^+ and K^+ permeabilities of the membrane. This has lead to the suggestion that the translocase could act as a "gated pore" or channel, specific for H^+ and K^+ ions. Thus, it seems reasonable to suggest that the higher matrix volume of fetal mitochondria (15) results from an unfavorable translocase conformation that allows the entrance of K^+ and water, because the low intramitochondrial adenine nucleotide concentrations (14,15,24) maintain

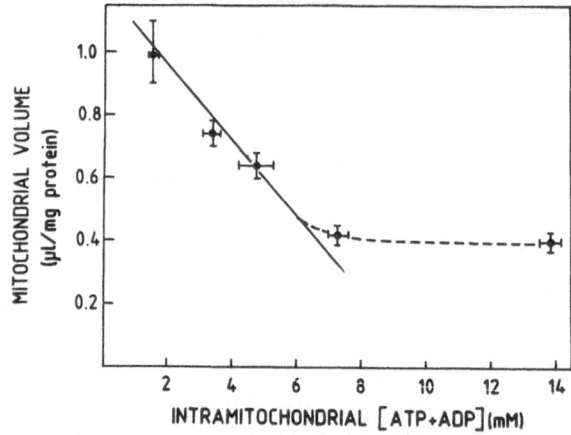

Fig. 7. Quantitative estimation of the different mitochondrial morphologies observed during rat liver development. Open, dotted and hatched bars represent the orthodox, condensed and highly condensed morphologies, respectively. The results were obtained by counting 12 different fields taken at random and at x 18,000 magnification. Figure 6 is one representative example. Results are means ± S.E.

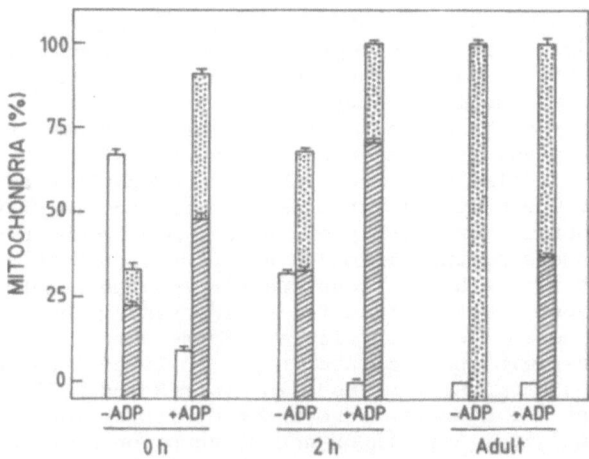

Fig. 8. Linear correlation between the intramitochondrial ATP+ADP concentrations and the mitochondrial matrix volume during postnatal development. The figure has been constructed with the
ATP+ADP concentrations for liver mitochondria
from 0-, 1-, 2-, and 6-h-old neonates and
adults and their corresponding State 4 values
of matrix volume. Data for this figure have
been taken from reference (15). The solid line
has the following linear parameters: slope =
0.098, Y intercept = 1.123, r = 0.988,
P<0.001. Dashed line shows the minimum matrix
volume of the organelle achieved during development in 2- and 6-h-old neonates, a time when
intramitochondrial nucleotide concentration
are even higher than in adult mitochondria
(15).

the "open" conformation of such a gated pore. During neonatal development,
the increase in adenine nucleotide concentrations (14,15,24) will promote
the conformational change in the translocase, closing the channel and thus
promoting the concurrent reduction in the mitochondrial matrix volume.

Effects on Respiratory Rates

The evidence that matrix volume regulation of the mammalian
mitochondria plays an important role in metabolic and bioenergetic
functions of the organelle has been suggested for a long time. A review on
this issue has recently appeared (46). Reduction of matrix volume has an
immediate positive effect on the intramitochondrial concentrations of the
metabolic intermediates and effectors of several key mitochondrial
reactions needed for postnatal development of metabolic pathways in the
neonatal liver (for reviews see 9 and the preceding chapter).

Whereas our results indicate that protein synthesis is required to
develop efficient oxidative phosphorylation, the postnatal accumulation of
adenine nucleotides (14,15,24) could play a complementary role on the rates
of respiration by triggering the reduction of the mitochondrial matrix
volume (Figures 6-8). Interestingly, protein synthesis inhibitors that were
unable to block the postnatal increase in intramitochondrial adenine
nucleotides (Table 1) were of course unable to prevent matrix volume
reduction (Table 1). In this situation, the increases in State 4 rates of
respiration were not affected (Figure 2); while a 30% increase in both

State 3 and FCCP-uncoupled rates of respiration (Figure 2) was still observed. Since administration of cytosolic protein synthesis inhibitors arrested the postnatal surge in inner membrane mitochondrial proteins (Figure 4), it is reasonable to suggest that the increases in the rates of respiration are due to the reduction of the matrix volume. A decrease of mitochondrial matrix volume might represent a reduction in the surface area of the inner mitochondrial membrane with increase packing of the different membrane proteins. The postnatal changes in membrane fluidity, estimated with the fluorescent apolar 1,6-diphenyl-1,3,5-hexatriene probe (15) indicated a continuous postnatal increase in the order parameter of inner mitochondrial membranes (15). A closer packing of the respiratory membrane proteins could favour electron transfer by increasing the number of effective collisions in the lipid bilayer of the respiratory chain proteins (47). This could explain the postnatal surge in State 4 rates of respiration and the marginal increases in State 3 and FCCP-uncoupled rates of respiration when the neonates were treated with cycloheximide at birth (15). These marginal increases in oxygen consumption rates are in the range of 7-17 natoms O/min/mg protein, and do not account for more than 10-15% of the maximum rates of respiration determined in the 6 h-old neonatal or adult mitochondria (15).

A SUMMARY OF THE MECHANISMS INVOLVED IN FETAL ⟶ NEONATAL MITOCHONDRIAL DIFFERENTIATION IN THE RAT LIVER

During the first hour of life, two main processes occur in the neonatal liver that trigger the sudden change in the metabolic pathways relevant for energy provision by promoting the acquisition of a functional energy conserving mitochondria (Figure 9). These two processes are (i) the preferential postnatal increase in the rates of protein synthesis for the neonatal mitochondria (15) and (ii) the postnatal increase in liver adenine nucleotides (48,49), which most likely promote subsequent enrichment in intramitochondrial nucleotides (14,15,24). These two processes, with a clear postnatal induction, acting in a synergestic way, promote differentiation of the fetal (non-energy conserving) to neonatal (energy conserving) mitochondria, by affecting different ultrastuctural, molecular and functional features of the pre-existing mitochondria.

The coordinated postnatal induction of cytosolic and mitochondrial protein synthesis promotes the increase in respiratory complexes of the inner mitochondrial membrane (Figure 9), allowing the development of adult respiratory rates and hence the membrane potential and proton motive force to be used in oxidative phosphorylation (15). The coordinated induction of protein synthesis also involves a doubling per mitochondrial unit of H^+-ATP synthetase complexes (Figure 9) which allows efficient utilization of the proton motive force generated by the respiratory chain. Further, we have recently shown using immunological techniques that protein synthesis is also involved in the rapid postnatal surge of the pyruvate dehydrogenase complex activity (50), by a coordinated increase in at least three of its main components, the E1α, E2 and E3 proteins. This latter result points out an event of utmost importance for mammalian adaptation to extrauterine life, that is, that the early postnatal synthesis of mitochondrial proteins involves those inner membrane proteins related to mitochondrial bioenergetic function, and matrix proteins directly involved in its metabolic functions.

The postnatal increase in intramitochondrial adenine nucleotides (14,15,24) promotes ultrastructural changes of the mitochondria (Figures 6 and 7) that result in a dramatic reduction of its matrix volume (Figure 8). This results in a sudden increase in the intramitochondrial concentrations of the different substrates and effectors of the metabolic and bioenergetic

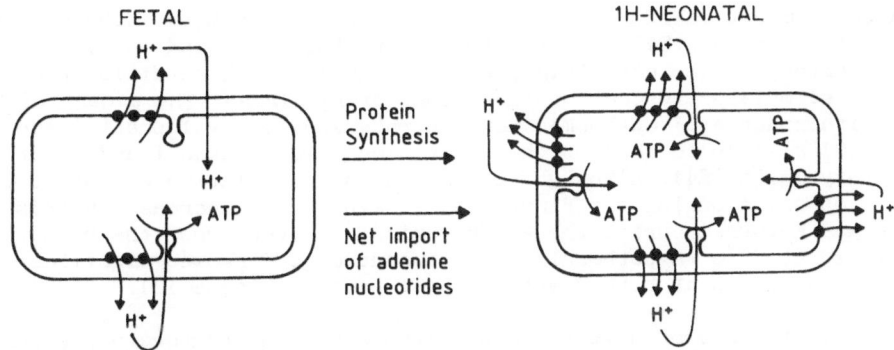

FETAL   1H-NEONATAL

Fig. 9. Mechanisms involved in the rapid postnatal differentiation
of rat liver mitochondria. Closed circles represent
respiratory complexes of the inner membrane, pumping H^+
out of the organelle as a result of electron transfer.
Reentry of H^+ through ATP synthetase complex couples
respiration to the synthesis of ATP. The reduction of the
matrix volume in 1 h-neonatal mitochondria has not been
represented to avoid clumping of the various elements of
the design. Reentry of H^+ not coupled to the synthesis of
ATP is shown in fetal mitochondria, due to the higher
passive proton permeability of this membrane.

pathways located within the organelle and in a closer packing of the
membrane proteins (15). These most likely contribute to an increase in
the metabolic flux of mitochondria and in marginal increases in their
rates of respiration. Adenine nucleotides also affect the passive proton
permeability of the inner membrane (Figure 9) (Valcarce and Cuezva,
manuscript in preparation), a property that defines the bioenergetic
quality of a membrane, contributing to the increased efficiency of
oxidative phosphorylation observed after birth.

POSTNATAL DIFFERENTIATION OF MITOCHONDRIA IN OTHER RAT TISSUES

Brown Adipose Tissue

The main physiological function of brown adipose tissue (BAT) in
newborns is the generation of heat in a process called "non-shivering
thermogenesis" (51,52). The specialized function of this tissue depends on
a mitochondrial inner membrane protein of 32 KDa called the "uncoupling
protein" (UCP) (51,53). The role of the 32 KDa protein in heat production
is to provide a short-circuit for the proton electrochemical gradient
generated by the respiratory chain (51). This unique protein (measured as
GDP-binding activity) of BAT mitochondria has been shown to increase during
development in several species, most likely by increasing the rate of
transcription of its gene (53). We have recently shown (54), using
immunological methods, the rapid postnatal increase of BAT mitochondrial
F_1-ATPase and UCP proteins, thus establishing the timing of induction of
the thermogenic capacity in BAT mitochondria and hence, its postnatal
differentiation.

Postnatal development of respiratory enzymes in isolated BAT
mitochondria of euthermic neonates takes place during the first 2-6
postnatal hours (54). However, when neonates were placed in a hypothermic
condition (22°C), the postnatal surge in both complexes I and IV was
accelerated (54). The amount of F_1-ATPase proteins in BAT mitochondria

increased by about 50% during the first 2 h of life in both euthermic and hypothermic neonates (54). At the sixth postnatal hour, and as shown for the respiratory enzyme activities, neonatal hypothermic condition promoted a further stimulatory effect on the amount of F_1-ATPase. The amount of UCP in BAT mitochondria of euthermic newborn rats show a developmental profile (Figure 10) similar to that of the activities of complexes I and IV of the respiratory chain (54). Although no changes in the relative amount of UCP could be observed during the first 2 h, a significant increase in this protein was measured during the 2-6 h period. However, when the neonates were placed at 22°C (hypothermia), the amount of UCP showed significant postnatal increases at both 2 and 6 h *post partum* (Figure 10).

The maintenance of newborn rats at two different temperatures, thermal neutrality (37°C) and nest temperature (22°C), a condition known to promote hypothermia and to enhance brown fat lipolysis in the immediate postnatal period of the newborn (55), have helped to define the time and conditions of natural onset of the thermogenic process and a putative mitochondrio-genic signal for BAT (54). The mitochondrial UCP/F_1-ATPase protein ratio (Figure 11), an index of BAT mitochondrial differentiation, showed that neonatal hypothermia significantly accelerated the process of mitochon-drial differentiation with a rapid postnatal onset after birth, while in BAT mitochondria of neonates held at thermoneutrality there was at least a 2-h delay in its onset (Figure 11). However, at the sixth postnatal hour this ratio was the same in both groups of neonates, although the amounts of UCP (Figure 10) and F_1-ATPase proteins present in BAT mitochondria of euthermic neonates were significantly lower (54). The finding that at the

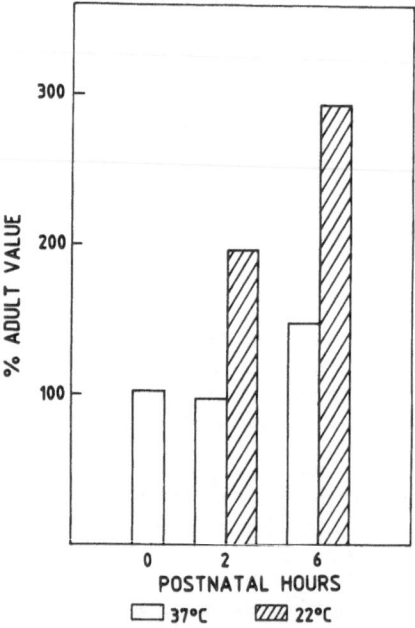

Fig. 10. Postnatal development of the uncoupling protein in isolated brown adipose tissue mitochondria of neonates held at thermo-neutrality (37°C) and under hypothermic conditions (22°C). The amount of protein was measured as indicated in reference (54), and expressed as percent of adult values.

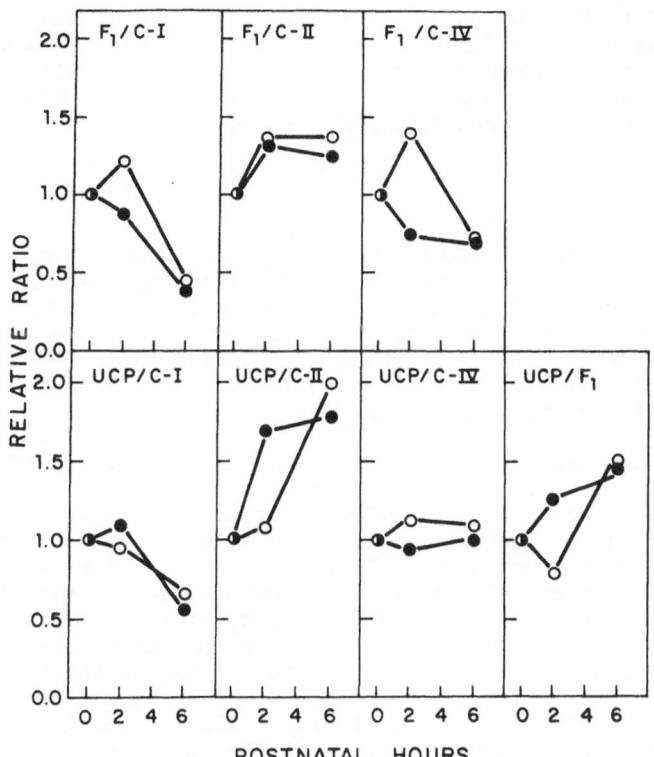

Fig. 11. Developmental changes in the relative ratios of
BAT inner mitochondrial membrane proteins from
euthermic (o) and hypothermic (●) neonates. The
ratios were obtained by using the protein amounts
and enzyme activities reported in reference (54).
F_1, F_1-ATPase complex; UCP, uncoupling protein;
C-I, C-II and C-IV are Complexes, I, II and IV of
the respiratory chain.

sixth postnatal hour the UCP/F_1-ATPase protein ratio was the same in both
groups of neonates also holds when the amount of any of the two proteins is
expressed relative to the activity of complexes I, II and IV of the respiratory chain (Figure 11). However, the developmental profiles for these
ratios (Figure 11) showed marked differences, ranging from preferential
increases, to no changes, to a decrease. These results (Figure 11), may
indicate the relative bioenergetic importance of Complex I of the respiratory chain as an electron acceptor for the oxidized substrates in the
early neonatal BAT. This finding is consistent with the high rates of
lactate utilization by this tissue (56).

As in newborn rat liver mitochondria (15), BAT mitochondria of both
euthermic and hypothermic neonates significantly increase their F_1-ATPase
content rapidly after birth (54), suggesting a common biological signal for
triggering of mitochondrial differentiation in both tissues. UCP synthesis,
however, is known to be under noradrenergic control, and the UCP gene is
acutely regulated at the level of transcription after activation of the
plasma membrane β-adrenoreceptors of the brown adipocyte (53,57,58). Thus,
the rapid postnatal increase in UCP protein found in BAT mitochondria of
neonates held at 22°C is most likely due to a rapid increase in the rates
of transcription of the UCP gene. Stimulation of sympathetic nervous system

activity of BAT is thought to be the major physiological regulator of the response of this tissue in animals exposed to a cold environment, suggesting that the concurrent postnatal increase in F_1-ATPase protein and in the activity of the respiratory complexes (54) observed in hypothermic neonates is a general mitochondriogenic response of BAT to the secreted noradrenaline.

Other Tissues

A major limitation in the study of mitochondrial differentiation at this stage of development is the low amount of tissue available to obtain a reliable mitochondrial preparation. Thus, we have studied the developmental changes in mitochondrial ATPase activity in rat homogenates of heart, lung and brain, to detect possible processes of mitochondrial proliferation and hence to initiate studies for possible identification of concurrent processes of mitochondrial differentiation.

Table 2 summarizes the developmental changes in mitochondrial ATPase activity in rat heart, lung and brain homogenates. In the three tissues studied, the specific activity of ATPase was lower in fetal and neonatal samples than in adult. Further, an increase in activity could be observed after the first or second postnatal hour in the three tissues. These results suggest that proliferation or differentiation of mitochondria may be occuring. At present, we have tested our hypothesis only in isolated brain mitochondria. Our results showed that respiratory complexes, I, II and IV and the F_1-ATPase complex did not change during the first 6 h following birth (Table 3). These findings most likely exclude differentiation of mitochondria in the rat brain, at least during this stage of neonatal development, but suggest that proliferation of the organelle is taking place.

NEONATES AT HIGH RISK FOR POSTNATAL IMPAIRMENTS IN MITOCHONDRIAL DIFFERENTIATION

The Premature Newborn

Premature birth is accompanied by a whole set of metabolic and func-tional impairments due to the abrupt interruption of gestation before a normal fetal stage of maturity has been attained. These alterations cover a broad spectrum of functional delays in the development of several organs, metabolic pathways and signal transducing elements, the degree of stress and rates of survival related to the length of the uncompleted period of gestation.

Table 2. Development of mitochondrial ATPase activity in rat heart, lung and brain homogenates

Postnatal hour	Heart	Lung	Brain
0	19	7	28
1	21	7	28
2	28	6	40
6	31	12	41
Adult	47	24	56

Mitochondrial ATPase activity is expressed as mU/mg of protein in the homogenate. The results are means of two experiments.

128

Table 3. Postnatal changes in rat brain mitochondrial
enzyme activities linked to the respiratory
chain and oxidative phosphorylation

Postnatal hour	Complex I	Complex II	Complex IV	F_1-ATPase complex
0	7±1	47±3	446±39	376±45
2	8±1	41±6	490±74	361±54
6	8±2	44±7	498±54	380±47
Adult	13±1	83±3	576±42	442±49

Brain mitochondria were isolated in a Ficoll gradient.
The enzyme activities are expressed as mU/mg of
mitochondrial protein. The results are means ± S.E.
for four to five experiments.

In the rat, premature delivery of the fetus only 24 h in advance of
natural delivery is accompanied by an extremely high rate of mortality
during the first postnatal hours (Figure 12). In these neonates, it is a
general characteristic to detect in their circulating blood higher lactate
and lower oxygen concentrations than in the full term neonates (59,60). A
linear correlation between plasma lactate and blood oxygen concentrations
has been observed (60), suggesting that lactate utilization depends on
blood oxygenation. Further, administration of oxygen to premature newborn
rats (61) accelerated the clearance of plasma lactate from circulation. The
metabolic fate of lactate during the first postnatal hours is its oxidation
to CO_2 (62), and it has been suggested to be an alternative substrate for
the neonatal tissues (63) before glucose and ketone bodies become available.

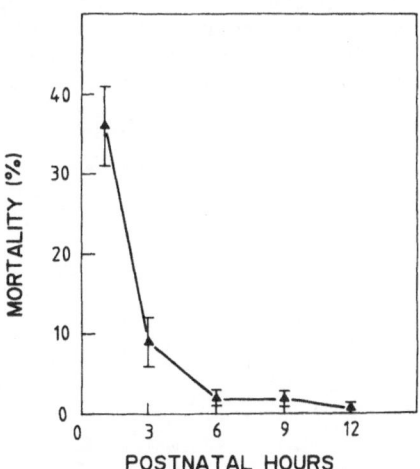

Fig. 12. Mortality rates in the premature newborn
rat during postnatal development. Preterm
neonates (4.3±0.1 g) were delivered by
rapid hysterectomy from cervically
dislocated pregnant rats on day 21 of
gestation and maintained without feeding
as described (59). The results shown are
means ± S.E. of 12 different litters.
Mortality rates in term newborns were
negligible during the same time period.

Although lung functional impairments, due to prematurity, could be an important physiological factor involved in the high mortality rates observed in the premature newborn rat (Figure 12), the fact that at the first postnatal hour both term and preterm neonates have similar blood oxygen concentrations (60) suggests that other alteration(s) are responsible for the high mortality rates observed up to the third postnatal hour in these neonates (Figure 12).

In the premature newborn rat, lactate oxidation is significantly reduced when compared to term neonates during the first two hours after birth (Figure 13A) (62). Further, lactate turnover rates (64) in premature newborn rats are 2-3 fold higher than in term neonates, even at 3 h after birth (Figure 13 B). These results point out that the premature newborn maintains the glycolytic pathway as a main route for harnessing ATP needs, and suggests that a delayed induction or impairment in mitochondrial differentiation could be taking place in some neonatal tissues. This could result in the lower postnatal rates of lactate oxidation and in the high lactate turnover rates (Figure 13). This hypothesis is currently being investigated in our laboratory.

The Hypothyroid Newborn

Thyroid hormones are known to regulate differentiation, development and growth of most mammalian tissues (65). The mechanism by which thyroid hormones exert their physiological role in different experimental systems is by promoting the accumulation of specific nucleus-directed mRNAs (66). It has been proposed that both T4 and T3 affect mitochondrial functions (67-69), number, morphology (70,71)and molecular structure (72,73) in a long term action of the hormones. Even an *in vitro* short-term action of T3 in several mitochondrial functions has been reported (69,74-76). In order to investigate the possible role of thyroid hormones in liver postnatal mitochondrial differentiation, we have studied the effect of fetal and neonatal hypothyroidism (77) on the developmental changes of the β subunit of mitochondrial F_1-ATPase complex, both at the protein and mRNA levels (33). Our results indicate that mitochondrial differentiation is impaired in these neonates, and that treatment of the hypothyroid neonates at birth

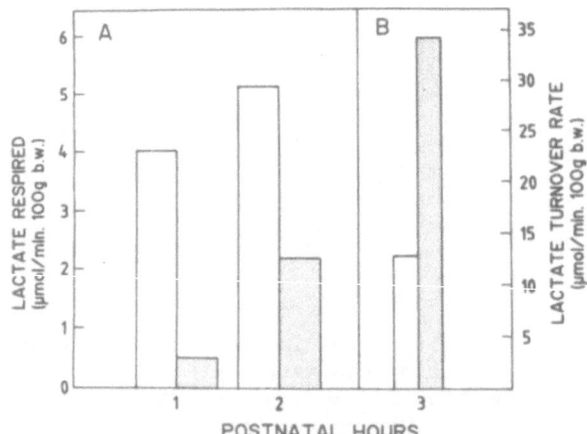

Fig. 13. Lactate oxidized and lactate turnover rates in the immediate postnatal period of term (open bars) and preterm (dotted bars) newborn rats. Data are from references (62) and (64), respectively.

with thyroid hormone restored the process of mitochondrial differentiation in their livers. Further, this study has allowed us to demonstrate that the β-subunit of the mitochondrial F_1-ATPase complex is accutely regulated by thyroid hormones at the level of transcription (33).

The Newborns of Diabetic Mothers

Newborns of insulin-treated diabetic rats had several clinical features including macrosomia, hyperinsulinemia, prolonged hypoglycemia and impaired glucagon secretory response, similar to those observed in the newborn infants of insulin-dependent diabetic women (78,79). Further, we have reported that the prolongued postnatal hypoglycemia experienced by these neonates results from alterations in the rates of gluconeogenesis (49) due to a delayed induction of hepatic cytosolic phosphoenolpyruvate carboxykinase activity (49) and alterations in the energy and the cytosolic redox state in liver of newborns of insulin-treated diabetic rats (49,80). The high ATP concentrations observed at birth in the liver of newborns of insulin-treated diabetic rats and during the first 3 postnatal hours, showed a significant decline after this time period (49), suggesting an alteration in hepatic mitochondrial metabolism. One might speculate that a yet unknown postnatal impairment in mitochondrial differentiation could be occuring in the liver of these neonates, resulting in the alterations of their liver energy state (49,80) and in the impaired postnatal development of normal rates of hepatic gluconeogenesis (49). A recent report has identified a delayed postnatal increase in intramitochondrial adenine nucleotides in newborns of severely diabetic rats (81).

ACKNOWLEDGEMENTS

The work performed in this laboratory and reported in this review was supported by grants PB85-0199 and PM88-0024 from the "Comisión Asesora de Investigación Científica y Técnica", and "Dirección General de Investigación Científica y Técnica", and by an institutional grant from "Fundación Ramón Areces". A.M.L., J.M.I. and A.A. are predoctoral fellows from Fondo de Investigaciones Sanitarias and Plan de Formación de Personal Investigador, MEC-CSIC. We would like to thank the contributions, made by our collaborators and colleagues over the years and reviewed in this article.

REFERENCES

1. H.B.Burch, D.H.Lowry, A.M.Kuhlman, J.Skerjance, E.J.Diamant, S.R. Lowry and P.Von Dippe, Changes in patterns of enzymes of carbohydrate metabolism in the developing rat liver, J.Biol.Chem. 238:2267 (1963).
2. F.A.Hommes, Energetic aspects of late fetal and neonatal metabolism, in: "Normal and Pathological Development of Energy Metabolism", F.A. Hommes and C.J.Van den Berg, eds. Academic Press, London, pp. 1 (1975).
3. F.A.Hommes, G.P.B.Kraan and R.Berger, The regulation of ATP synthesis in fetal rat liver, Enzyme 1:351 (1973).
4. E.Chico, J.S.Olavarria and I.Nuñez de Castros, Crabtree effect induced by fructose in hepatocytes isolated from developing rats, Enzyme 24: 209 (1979).
5. R.Berger and F.A.Hommes, Regulation of pyruvate oxidation in mitochondria isolated from fetal and adult rat liver, Biochim. Biophys.Acta. 314:1 (1973).
6. H.P.Rohr, A.Wirz, L.C.Henning, V.N.Riede and L.Bianchi, Morphometric analysis of the rat liver cell in the perinatal period, Lab.Invest. 24:128 (1971).

7. C.A.Lang and G.H.Herber, Quantitative comparison of the mitochondrial populations in the livers of newborn and weaning rats, Dev.Biol. 29: 176 (1972).

8. H.David, Quantitative and qualitative changes in the mitochondria in hepatocytes during postnatal development of male rats, Exp.Pathol. 17:359 (1979).

9. J.R.Aprille, Perinatal development of mitochondria in rat liver, in: "Mitochondrial Physiology and Pathology", G.Fiskum, ed., Van Nostrand Reinhold Co., New York, pp. 66 (1986).

10. J.K.Pollak and C.G.Duck-Chong, Changes in rat liver mitochondria and endoplasmic reticulum during development and differentiation, Enzyme 15:139 (1973).

11. M.Hallman, Changes in mitochondrial respiratory chain proteins during perinatal development. Evidence of the importance of environmental oxygen tension, Biochim.Biophys.Acta 253:360 (1971).

12. T.Nakazawa, K.Asami, H.Suzuki and O.Yukawa, Appearance of an energy conservation system in rat liver mitochondria during development, J.Biochem. 73:397 (1973).

13. J.K.Pollak, The maturation of the inner membrane of fetal rat liver mitochondria, Biochem.J. 150:477 (1975).

14. J.R.Aprille and G.K.Asimakis, Postnatal development of rat liver mitochondria: State 3 respiration, adenine nucleotide translocase activity, and the net accumulation of adenine nucleotides, Arch. Biochem.Biophys. 201:564 (1980).

15. C.Valcarce, R.M.Navarrete, P.Encabo, E.Loeches, J.Satrústegui and J.M.Cuezva, Postnatal development of rat liver mitochondrial functions: The roles of protein synthesis and of adenine nucleotides, J.Biol.Chem. 263:7767 (1988).

16. C.Valcarce, J.M.Cuezva and J.M.Medina, Increased gluconeogenesis in the rat at term gestation, Life Sciencies 37:553 (1985).

17. C.Valcarce, J.M.Cuezva and J.M.Medina, Phosphoenolpyruvate carboxykinase activity in the kidney of pregnant rats during late gestation, Biochem.Soc.Trans. 12:789 (1984).

18. J.M.Cuezva, C.Valcarce, M.Chamorro, A.Franco and F.Mayor, Alanine and lactate as gluconeogenic substrates during late gestation, FEBS Lett. 194:219 (1986).

19. P.C.Foster and E.Bailey, Changes in the activities of the enzymes of hepatic fatty acid oxidation during development of the rat, Biochem.J. 154:49 (1976).

20. J.Girard and P.Ferré, Metabolic and hormonal changes around birth, in: "The Biochemical Development of the Fetus and Neonate", C.T. Jones, ed., Elsevier Biomedical Press, Amsterdam, pp. 517 (1982).

21. F.Mayor and J.M.Cuezva, Hormonal and metabolic changes in the perinatal period, Biol.Neonate 48:185 (1985).

22. J.K.Pollak and R.Sutton. The differentiation of animal mitochondria during development, Trends Biochem.Sci. 5:23 (1980).

23. K.D.Garlid and A.D.Beavis, Evidence for the existence of an inner membrane anion channel in mitochondria, Biochim.Biophys.Acta 853: 187 (1986).

24. R.Sutton and J.K.Pollak, The increasing adenine nucleotide concentration of rat liver mitochondria during neonatal development, Differentiation 12:15 (1978).

25. R.Hay, P.Bohni and S.Gasser. How mitochondria import protein, Biochim. Biophys.Acta 779:65 (1984).

26. A.Tzagoloff and A.M.Myers, Genetics of mitochondrial biogenesis, Ann. Rev.Biochem. 55:249 (1986).

27. D.E.Hale and J.R.Williamson, Developmental changes in the adenine nucleotide translocase in the guinea pig, J.Biol.Chem. 259:8737 (1984).

28. L.Bagetto, D.C.Gautheron and C.Godinot, Effects of ATP on various steps controlling the rate of oxidative phosphorylation in newborn

rat liver mitocondria, <u>Arch.Biochem.Biophys</u>. 232:670 (1984).

29. N.Pfanner and W.Neupert, Transport of proteins into mitochondria: a potassium diffusion potential is able to drive the import of ADP/ATP carrier, <u>EMBO J</u>. 4:2819 (1985).

30. Y.Kagawa, Proton motive ATP synthesis, <u>in</u>: "Bionergetics", New Comprehensive Biochemistry, L.Ernster, ed., Elsevier/North-Holland Biomedical Press, Amsterdam, vol. 9, pp. 149 (1984).

31. J.K.Pollak and R.Sutton, The transport and accumulation of adenine nucleotides during mitochondrial biogenesis, <u>Biochem.J</u>. 192:75 (1980).

32. J.Austin and J.R.Aprille, Carboxyatractyloside-insensitive influx and efflux of adenine nucleotides in rat liver mitochondria, <u>J.Biol. Chem</u>. 259:154 (1984).

33. J.M.Izquierdo, A.M.Luis and J.M.Cuezva, Postnatal mitochondrial differentiation in rat liver. Regulation by thyroid hormones of the β-subunit of the mitochondrial F_1-ATPase complex, <u>J.Biol.Chem</u>. (in press) (1990).

34. C.Hackenbrock, Ultrastructural basis for metabolically linked mechanical activity in mitochondria. I. Reversible ultrastructural changes with change in metabolic steady state in isolated liver mitochondria, <u>J.Cell.Biol</u>. 30:269 (1966).

35. C.Hackenbrock, Untrastructural basis for metabolically linked mechanical activity in mitochondria. II. Electron transport-linked ultrastructural transformations in mitochondria, <u>J.Cell.Biol</u>. 37: 345 (1968).

36. N.E.Weber and P.V.Blair, Ultrastructural studies of beef mitochondria. II. Adenine nucleotide induced modifications of mitochondrial morphology, <u>Biochem.Biophys.Res.Commun</u>. 41:821 (1970).

37. B.Scherser and M.Klingenberg, Demonstration of the relationship between the adenine nucleotide carrier and the structural changes of mitochondria as induced by adenosine 5'-diphosphate, <u>Biochemistry</u> 13:161 (1974).

38. P.Lundberg, ATP and phosphate induce configurational changes of submitochondrial particles, <u>Biochim.Biophys.Acta</u> 376:458 (1975).

39. S.F.Jakovcic, K.Haddock, G.S.Geta, M.Rabinowitz and H.Swift, Mitochondrial development in liver of foetal and newborn rats, <u>Biochem.J</u>. 121: 341 (1971).

40. G.P.Brierley, The uptake and extrusion of monovalent cations by isolated heart mitochondria, <u>Mol.Cell.Biochem</u>. 10:41 (1976).

41. K.D.Garlid, Unmasking the mitochondrial K^+/H^+ exchanger: tetraethyl-ammonium-induced K^+ loss, <u>Biochem.Biophys.Res.Commun</u>. 87:842 (1979).

42. K.D.Garlid, On the mechanism of regulation of the mitochondrial K^+/H^+ exchanger, <u>J.Biol.Chem</u>. 255:11273 (1980).

43. R.S.Dordick, G.P.Brierley and K.D.Garlid, On the mechanism of A23187-induced potassium efflux in rat liver mitochondria, <u>J.Biol.Chem</u>. 255:10299 (1980).

44. G.P.Brierley, M.S.Jurkowitz, T.Farooqui and D.W.Jung, K^+/H^+ antiport in heart mitochondria, <u>J.Biol.Chem</u>. 259:14672 (1984).

45. A.Panov, S.Flippova and V.Lyakhovich, Adenine nucleotide translocase as a site of regulation by ADP of the rat liver mitochondria permeability to H^+ and K^+ ions, <u>Arch.Biochem.Biophys</u>. 199:420 (1980).

46. A.P.Halestrap, The regulation of the matrix volume of mammalian mitochondria *in vivo* and *in vitro* and its role in the control of mitochondrial metabolism, <u>Biochim.Biophys.Acta</u> 973:355 (1989).

47. E.C.Slater, J.A.Berder and M.A.Herweijer, A hypothesis for the mechanism of respiratory-chain phosphorylation not involving the eletrochemical gradient of protons as obligatory intermediate, <u>Biochim.Biophys.Acta</u> 811:217 (1985).

48. J.M.Cuezva and J.M.Medina, Adenine nucleotide concentrations in liver

of fetal rats. Neonatal changes in the premature newborn, Rev.Esp. Fisiol. 38:161 (1982).

49. J.M.Cuezva, C.I.Chitra and M.S.Patel, The newborn of diabetic rat II. Impaired gluconeogenesis in the postnatal period, Pediatr.Res. 16: 638 (1982).

50. E.Serrano, A.M.Luis, P.Encabo, A.Alconada, L.Ho, M.S.Patel and J.M. Cuezva, Rapid postnatal induction of the pyruvate dehydrogenase complex in rat liver mitochondria, Annals N.Y.Acad.Sci. 573:412 (1989).

51. D.G.Nicholls and R.Locke, Brown adipose tissue in the mammalian neonate, Physiol.Rev. 64:1 (1984).

52. J.Nedergaard, E.Connolly and B.Cannon, Thermogenic mechanisms in brown fat in :"Brown Adipose Tissue", P.Trayhurn and D.G.Nicholls, eds., Edward Arnold Ltd., London, pp. 152 (1986).

53. D.Ricquier, F.Bovillaud, P.Toumelin, G.Mory, R.Bazin, J.Arch and L. Penicaud, Expression of uncoupling protein mRNA in thermogenic or weakly thermogenic brown adipose tissue: Evidence for a rapid β-adrenoreceptor-mediated and transcriptionally regulated step during activation of thermogenesis, J.Biol.Chem. 261:13905 (1986).

54. A.M.Luis and J.M.Cuezva, Rapid postnatal changes in F_1-ATPase proteins and in the uncoupling protein in brown adipose tissue mitochondria of the newborn rat, Biochem.Biophys.Res.Commun. 159:216 (1989).

55. J.M.Cuezva, M.Benito, F.J.Moreno and J.M.Medina, Prematurity in rat. II. Effect of hypothermia, Biol.Neonate. 37:218 (1980).

56. E.D.Saggerson, T.W.J.McAllister and H.Baht, Lipogenesis in rat brown adipocytes, Biochem.J. 251:701 (1988).

57. D.Ricquier, G.Mory, F.Bovillaud, J.Thibault and J.Wissenbach, Rapid increase of mitochondrial uncoupling protein and its mRNA in stimulated brown adipose tissue, FEBS Lett. 178:240 (1984).

58. A.Jacobsson, U.Stadler, M.A.Glotzler and L.P.Kozak, Mitochondrial uncoupling protein from mouse brown fat: Molecular cloning, genetic mapping and mRNA expression, J.Biol.Chem. 260:16250 (1985).

59. J.M.Cuezva, F.J.Moreno and J.M.Medina, Prematurity in the rat. I. Fuels and gluconeogenic enzymes, Biol.Neonate 88 (1980).

60. J.M.Cuezva, F.J.Moreno and J.M.Medina, Blood oxygen concentrations in premature newborn rats during the early neonatal period, IRCS Med. Sci.Biochem. 39:644 (1981).

61. J.M.Cuezva and J.M.Medina, Prematurity in the rat, III. Effect of oxygen supply, Biol.Neonate 39:70 (1981).

62. J.M.Medina, J.M.Cuezva and F.Mayor, Non-gluconeogenic fate of lactate during the early neonatal period in the rat, FEBS Lett. 114:132 (1980)

63. J.M.Cuezva, C.Valcarce and J.M.Medina, Substrate availability for maintenance of energy homeostasis in the inmediate postnatal period of the fasted newborn rat, in: "The Physiological Development of the Fetus and Newborn", C.T.Jones and P.W.Nathanieslz, eds., Academic Press, London, pp. 63 (1985).

64. E.Fernández, C.Valcarce, J.M.Cuezva and J.M.Medina, Postnatal hypo-glycemia and gluconeogenesis in the newborn rat, Biochem.J. 214:525 (1983).

65. J.H.Oppenheimer and H.H.Samuels, "Molecular Basis of Thyroid Hormone Action", Academic Press Inc., New York, (1983).

66. H.H.Samuels, B.M.Forman, Z.D.Horowitz and Z.S.Ye, Regulation of gene expression by thyroid hormone, J.Clin.Invest 81:957 (1988).

67. M.D.Brand and M.P.Murphy, Control of electron flux through the res-piratory chain in mitochondria and cells, Biol.Rev. 62:141 (1987).

68. R.P.Hafner, Thyroid hormone uptake into the cell and its subsequent localization to the mitochondria, FEBS Lett. 224:251-256(1987).

69. K.Sterling, M.A.Brener and T.Sakurada, Rapid effect of triiodo-thyronine on the mitochondrial pathway in rat liver in vivo, Science 210:340 (1980).

70. K. Sterling, Thyroid hormone action at the cell level, N.Engl.J.Med. 300:117 (1979).

71. N.M. Gadaleta, M.Renis, G.R. Minervini, E.Serra, T.Bleve, A.Giovine, G. Zacheo and A.M. Giuffrida, Effect of hypothyroidism on the biogenesis of free mitochondria in the cerebral hemispheres and in cerebellum of rat during postnatal development, Neurochem.Res. 10:163 (1985).

72. C.A. Batie and M.A. Verity, Membrane enzyme development in nerve ending mitochondria during neonatal hypothyroidism, Dev.Neurosci. 2:139 (1979).

73. B.D. Nelson, A.Mutvei and V.Joste, Regulation of biosynthesis of the rat liver inner mitochondrial membrane by thyroid hormone, Arch. Biochem.Biophys. 228:41 (1984).

74. H.J. Seitz, M.J. Muller and S.Soboll, Rapid thyroid hormone effect on mitochondrial and cytosolic ATP/ADP ratios in the intact liver cell, Biochem.J. 227:149 (1985).

75. W.E. Thomas and J.Mowbray, Receptor mediated amplification control of oxidative phosphorylation by tri-iodothyronine, Biochem.Soc.Trans. 15:669 (1987).

76. K. Sterling, Direct thyroid hormone activation of mitochondria: The role of adenine nucleoside translocase, Endocrinology 119:292 (1986).

77. J.E. Silva and P.R. Larsen, Comparison of iodothyronine 5'-deiodinase and other thyroid-hormone-dependent enzyme activities in the cerebral cortex of hypothyroid neonatal rat, J.Clin.Invest. 70:1110 (1982).

78. J.M. Cuezva, E.S. Burkett, D.S. Kerr, H.M. Rodman and M.S. Patel, The newborn of diabetic rat, I. Hormonal and metabolic changes in the postnatal period, Pediatr.Res. 16:632 (1982).

79. J.M. Cuezva and M.S. Patel, Disturbances of fetal liver carbohydrate metabolism and perinatal glucose homeostasis, Biochem.Soc.Trans. 13:83 (1985).

80. J.M. Cuezva and M.S. Patel, Effect of glucose and insulin administration on hepatic adenylate energy charge and the cytosolic redox state in the neonates of normal and insulin-treated diabetic rats, Biol. Neonate 48:221 (1985).

81. J.R. Aprille and M.T. Nosek, Neonatal hypoxia or maternal diabetes delays postnatal development of liver mitochondria, Pediatr.Res. 21:266 (1987).

UNCOUPLING PROTEIN mRNA EXPRESSION

DURING THE PERINATAL PERIOD OF THE RAT

Marta Giralt, Immaculada Martin, Roser Iglesias,
Octavi Viñas, Teresa Mampel and Francesc Villarroya

Departamento de Bioquímica y Fisiología
Universidad de Barcelona
08028 Barcelona, Spain

INTRODUCTION

Brown adipose tissue is specialized in thermogenesis and it is considered as a major site for heat production in newborn mammals as well as in cold-adapted or overfed adult rodents. The thermogenic function of brown fat is the result of the presence of the "uncoupling protein" (UCP), present in the inner mitochondrial membrane. This protein short-circuits the proton electrochemical gradient generated by the respiratory chain activy as a natural uncoupler of oxidative phosphorylation (1). UCP is unique to brown fat and can be considered as a cell marker of brown adipocytes. Recently, a molecular approach to UCP has been developed in several laboratories by isolating UCP cDNAs for rat, mouse, lamb and bovine as well as specific genomic probes for human and ovine UCP. These cDNAs have been used to determine the sequence of UCP which has revealed a striking structural homology with other mitochondrial membrane proteins such as the ADP/ATP carrier or the phosphate carrier. The use of molecular probes as tools to monitor UCP mRNA levels in adult animals has shown that changes in the thermogenic activity of brown fat parallel UCP mRNA levels in the tissue, whereas run-on transcription experiments with isolated brown fat nuclei have established a main transcriptional control of UCP synthesis (for review see 2).

Brown fat thermogenesis has been classically recognized to be specially relevant in the neonatal period not only in small mammals but probably also in larger ones such as ruminants and primates. Brown adipose tissue develops prenatally in most species including rat or humans and UCP is know to be present at birth (3). However, the modulation of UCP gene expression during the pre- and immediately postnatal periods has not been extensively studied and it would be probably more complex than that in adults as both ontogenic and environmental factors may be involved. In the present work we have studied the changes in UCP mRNA expression during the perinatal period using the UCP 36 probe, a full length cDNA for rat UCP (4). We have determined when UCP mRNA first appears in prenatal rat as well as the changes in UCP mRNA levels throughout the pre- and immediately postnatal life. We have also investigated the role of postnatal environmental temperature on UCP mRNA expression as well as the effects of prematurity and postmaturity.

Endocrine and Biochemical Development of the Fetus and Neonate
Edited by J. M. Cuezva *et al.*
Plenum Press, New York, 1990

RESULTS AND DISCUSSION

Caesarian sections of pregnant rats were performed in a thermostatized chamber at 37°C and fetuses were extracted and interscapular brown adipose tissues of each litter were dissected and pooled. RNA was prepared using the hot phenol method. Northern analysis of brown fat RNA using the ^{32}P-UCP cDNA probe showed two bands, a major one of 1.5 kb and a minor one of 1.8 kb, in RNA from fetuses on days 19, 20, 21 and 22 of gestation. No signals were observed in 18 day-old fetuses (Figure 1). The two observed signals correspond to the two UCP mRNAs already characterized in different rodent species (2). The intensity of UCP mRNA signals from day 19 of gestation on ward indicated a progressive increase in the UCP mRNA abundance in the last period of the fetal rat development. In order to unequivocaly assess the absence of UCP mRNA on day 18 of gestation, poly (A)$^+$ RNA was isolated using poly(U)-coated paper-affinity chromatography of RNA samples belonging to 3-4 different litters of either 18 and 19 day-old fetuses. As shown in Figure 1, Northern analysis of poly (A)$^+$ RNA confirmed the absence of UCP mRNA from day 18-old rat fetuses and a sudden appearance on day 19. Overexposure of Northern blot did not show either the 1.5 kb or the 1.8 kb UCP mRNA signals prior to day 19 of gestation. Therefore it is concluded that UCP gene expression is turn on between the day 18 and 19 of fetal life in the rat.

In adult rats UCP gene expression is known to be regulated at the level of transcription mainly via activation of β-adrenoreceptors of plasma membrane by the noradrenaline coming from the sympathetic innervation of the tissue (5). More recently T3 generated "in situ" by iodothyronine 5'-deiodinase activity of brown fat have been claimed to be necessary for the cold-induced increase in UCP mRNA (6). The induction of UCP gene expression and the progressive increase of UCP mRNA during the late fetal life can hardly be attributed to the sympathetic activity upon the tissue as a sympathetic nerve endings are not developed in the fetal rat (3). However, at the stage of prenatal rat development when UCP mRNA initiates its expression, brown adipose tissue shows peak values of iodothyronine 5'-deiodinase activity (7) thus suggesting that T3 may be involved in early stages of prenatal brown fat differentiation. A similar hypothesis has been formulated recently from the comparision of the prenatal iodothyronine 5'-deiodinase activity profile (8) and UCP mRNA expression in perirenal adipose tissue of bovines (9), a species with a very different pattern of brown fat development.

In order to study postnatal changes in UCP mRNA levels, we performed Northern analysis of brown fat RNA extracted from at-term spontaneously delivered neonates at 0 (considered as when all purps are born but they have not yet initiated suckling), 6, 12 and 24 hours after birth (Figure 1). Results showed that UCP mRNA levels increased as soon as 6 hours after birth and remained high at least the first 24 hours of life. This data are essentially in agreement with previous results (5) and they seem to indicate that a rapid postnatal increase in the degree of UCP gene expression is a key event to achieve a high thermogenic activity of brown fat in the neonatal life.

In view of these results, we have tried to study the influence of the different environmental and onthogenic determinants of the postnatal UCP mRNA rise first by establishing the role of postnatal ambient temperature. Caesarian sections of pregnant rats were performed at-term (22 days of gestation) in a thermostatized chamber (37°C). Two- three pups were sacrificed immediately (0 hours) and half the remaining litter was placed at 21°C whereas the other half was kept at 37°C. Northern analysis of brown adipose tissue RNA extracted from 0 hour-old, 6 hour-old or 12 hour-old pups kept at different temperatures were performed. Results (data not

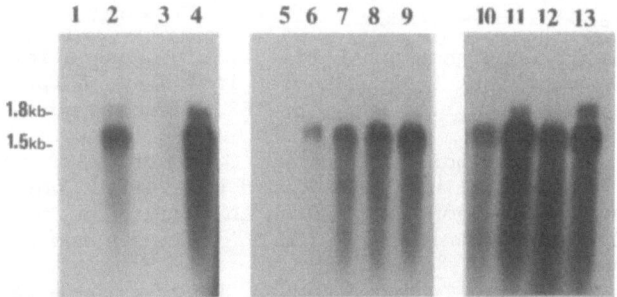

Fig. 1. Uncoupling protein mRNA levels during the perinatal
development of the rat. Northern analysis of brown
adipose tissue RNA hybridized to the ^{32}P-labeled UCP
36 cDNA. Lanes 1 and 2 correspond to 1 µg of poly(A)$^{+}$
RNA from fetuses on day 18 and 19 of gestation,
respectively. Lanes 3 and 4 are an overexposure of
lanes 1 and 2. Lanes 5 to 9 correspond to 10 µg of
total RNA from fetuses on day 18 (lane 5), 19 (lane
6), 20 (lane 7), 21 (lane 8) and 22 (lane 9) of
gestation. Lanes 10 to 13 correspond to 10 µg of
total RNA from newborns 0 (lane 10), 6 (lane 11), 12
(lane 12) and 24 (lane 13) hours after birth.

shown) indicated that the maintenance of newborn pups at 37°C abolishes
the postnatal rise in UCP mRNAs whereas there is a substantial increase
in the intensity of UCP mRNA bands in pups kept at 21°C. Therefore it was
concluded that UCP mRNA expression is sensitive to the cold stimulus at
birth. In order to assess if this capacity of responding to the thermic
stress is already present in early stages of development, we studied the
postnatal environment temperature sensitivity in premature neonates by
means of performing caesarian sections of pregnant rats on day 21 of
pregnancy and keeping pups in the same situations mentioned above for
at-term fetuses. Results of Northern blot analysis (data not shown)
indicated the absence of a postnatal rise in UCP mRNA levels both in
pups kept at 37°C and in those kept at 21°C. It was concluded that the
capacity of UCP mRNA expression to respond to the cold-stress stimulus
appears in the last day of fetal life.

In order to clarify the role of birth itself in UCP mRNA expression,
Northern analysis were performed comparing brown fat RNA from at-term
fetuses, 24 hours-old pups and postmature fetuses obtained by caesarian
sections of progesterone-treated pregnant rats on day 23 of gestation.
Results (data not shown) indicated that there was a substantial increase in
UCP mRNA levels in post-mature fetuses when compared with fetuses at-term,
although they did not achieve the levels found in 24 hours-old pups born
at term. Therefore we concluded that probably thermic stress can not be
claimed as the exclusive factor responsible for UCP mRNA rise occuring
inmediately after birth and that developmental aspects independent of birth
are involved in the increased UCP mRNA expression in the postnatal life.

ACKNOWLEDGEMENTS

Thanks are given to Dr. D.Ricquier for the kind gift of UCP 36 probe.
This work has been supported in part by Dirección General de la Investigación Científica y Técnica (Grant PB-577/86) Spain.

REFERENCES

1.   D. Nicholls, S. A. Cunnigham and E. Rial, The bioenergetic mechanisms of brown adipose tissue thermogenesis, in: "Brown Adipose Tissue", (P. Trayhurn and D. Nicholls, eds), pp. 52, Edward Arnold, London, (1986).

2.   D. Ricquier, F. Bouillaud, F. Villaroya, O. Champigny, E. Hentz, S. Raimbault, L. Casteilla, G. Mory and J. P. Revelli, Control of uncoupling protein gene expression in rodents and humans, in: "Highlights on Endocrinology", (C. Christiansen and B. J. Riins, eds), Copenhaguen, (1987).

3.   J. Nedergaard, E. Connolly and B. Cannon, Brown adipose tissue in the mammalian neonate, in: "Brown Adipose Tissue", (P. Trayhurn and D. G. Nicholls, eds), pp. 153, Edward Arnold, London, (1986).

4.   F. Bouillaud, D. Ricquier, J. Thibault and J. Weissenbach, Molecular approach to thermogenesis in brown adipose tissue: cDNA cloning of the mitochondrial uncloupling protein, Proc. Nat. Acad. Sci. USA., 82: 445 (1985).

5.   D. Ricquier, F. Bouillaud, P. Toumelin, G. Mory, R. Bazin, J. Arch and L. Penicaud, Expression of uncoupling protein mRNA in thermogenic or weakly thermogenic brown adipose tissue, J. Biol. Chem., 261: 13905 (1986).

6.   A. C. Bianco and J. E. Silva, Intracellular conversion of thyroxine to triiodothyronine is required for the optimal thermogenic function of brown adipose tissue, J. Clin. Invest., 249: 5421 (1987).

7.   M. Giralt, I. Martín, T. Mampel, F. Villaroya, R. Iglesias and O. Viñas, Evidence for a differential physiological modulation of brown fat iodothyronine 5'-deiodinase activity in the perinatal period, Biochem. Biophys. Res. Commun., 156: 493 (1988).

8.   M. Giralt, L. Casteilla, O. Viñas, T. Mampel, J. Robelin and F. Villarroya, Iodothyronine 5'-deiododinase activity as an early event of prenatal brown-fat differentiation in bovine development, Biochem. J., 259: 555 (1989).

9.   L. Casteilla, O. Champigny, F. Bouillard, J. Robelin and D. Ricquier, Sequential changes in the expression of mitochondrial protein mRNA during the development of brown adipose tissue in bovine and ovine species, Biochem. J., 257: 665 (1989).

IODOTHYRONINE 5'-DEIODINASE ACTIVITY IN BROWN

ADIPOSE TISSUE DURING PERINATAL DEVELOPMENT

Marta Giralt, Octavi Viñas, Immaculada Martin,
Louis Casteilla[*], Francesc Villarroya,
Teresa Mampel and Roser Iglesias

Departamento de Bioquímica y Fisiología
Universidad de Barcelona
08028 Barcelona, Spain

[*]Laboratoire de Production de Viande, INRA
Theix, and Centre de Recherches sur la Nutricion
CNRS, Meudon-Bellevue, France

INTRODUCTION

An important type II iodothyronine 5'-deiodinase activity has been reported to be present in rat brown adipose tissue, whereas it is absent from white adipose tissue. The iodothyronine 5'-deiodinase activity is responsible for the T_4 deiodination to T_3 and the available data, based mainly in rodents studies, point to the existence of at least two different enzymatic forms. Type I enzyme is mainly present in liver and kidney, has Km values for T_4 and rT_3 in the micromolar range and is inhibited uncompetitively by propylthiouracil. Type II enzyme has been found in the brain, pituitary and brown fat, it has a Km for T_4 and rT_3 in the nanomolar range and it is not inhibited by propylthiouracil (1,2).

The functional role of iodothyronine 5'-deiodinase in rat brown adipose tissue seems to have two different aspects. First, it is a substantial source of systemic T_3 in the rat, at least in situations of high thermogenic activity (3). Secondly, the correlation noted between the levels of brown fat thermogenesis and iodothyronine 5'-deiodinase activity in different physiological or experimental situations of adult rodents have suggested that T_3 produced in the tissue may be involved in its thermogenic activity (4).

Brown adipose tissue is known to be especially important during neonatal life, when brown fat thermogenesis is considered to be necessary to overcome the neonatal thermic stress. However, there are important differences in the pattern of brown fat development among different species. In small rodents such as the rat or mouse, brown fat develops prenatally, becomes especially relevant in the first day of life and remains as a differenciated tissue throughout adult life. In other larger species such as dogs, primates or ruminants brown fat is especially relevant during fetal life, and "disappears" some days after birth, when it is replaced by white adipose tissue, unequivocally differentiated brown adipose tissue being absent in adult life (5).

Endocrine and Biochemical Development of the Fetus and Neonate
Edited by J. M. Cuezva *et al.*
Plenum Press, New York, 1990

141

We have studied the pattern of brown adipose tissue iodothyronine 5'-deiodinase activity during development in two species with clearly different brown adipose tissue ontogeny: rat and bovine. We were interested to check whether the close parallelism between brown fat thermogenesis and 5'-deiodinating activity occuring in adult rats is present during development, and to determine if the consideration of iodothyronine 5'-deiodinase activity as a cell marker of brown adipose tissue can be extended to species different from small rodents.

RESULTS AND DISCUSSION

Kinetic Characterization

Outer ring (5') deiodinating activity was assayed by quantifying the ^{125}I-liberated from L-(5'-^{125}I)-rT3. Standard conditions for rat and bovine brown adipose tissue have been described previously (6,7). The reaction kinetics of bovine brown adipose tissue iodothyronine 5'-deiodinase in the presence of different DTT and rT3 concentrations, were completely different from those obtained in rat brown fat.

In the study concerning iodothyronine 5'-deiodinase in rat brown fat, increased DTT concentrations altered the intercepts of double-reciprocal plots, yielding a pattern of intersecting lines, as it is characteristic of sequential-type kinetics. When bovine brown fat 5'-deiodinase kinetics were studied, increased DTT concentrations altered the intercepts of double-reciprocal plots without changing the slops, thus resulting in a group of essentially parallel lines, indicating ping-pong type reaction kinetics. On the other hand, the apparent Michaelis constant (Km) for rT3 was 6nM in rat brown fat while in bovine perirenal tissue it was 2.6μM.

When the effect of PTU on the iodothyronine 5'-deiodinase activity was studied, a strong inhibitory effect was found for bovine brown fat, ranging from 90 to 99% inhibition, depending on the rT3 and DTT concentration in the assay. In contrast, less than 1% inhibition was observed in rat brown adipose tissue.

The pattern of characterics of bovine brown fat iodothyronine 5'-deiodinase found in this study is essentially different from that found in rat brown adipose tissue, whereas it is coincident with that previously reported for rat liver (1). Taking into account the iodothyronine 5'-deiodinase classification based on rat's studies (1) present data lead us to conclude that in the bovine species, differently from rodents, brown fat 5'-deiodinase is essentially a Type I enzyme.

Developmental Changes in Rat and Bovine Brown Adipose Tissue Iodothyronine 5'-deiodinase Activity

Changes in rat interscapular brown adipose tissue and bovine perirenal brown adipose tissue iodothyronine 5'-deiodinase activities as a function of the pre and postnatal age are shown in Figure 1. Despite the difference in the duration of pregnancy and in the 5'-deiodinase type in both species, the profile of prenatal 5'-deiodinase activity was surprisingly similar. Its most remarkable feature was the presence of peak values of the enzyme activity around the last third of gestation, which were followed by a substantial decrease before birth. After parturition there was also a similar profile of 5'-deiodinase activity in both species: a slight postnatal increase was followed by a progressive decay. However, whereas rat brown fat iodothyronine 5'-deiodinase activity is always present in adult life, bovine perirenal adipose tissue iodothyronine 5'-deiodinase activity became undetectable from around day 100 of life. In fact, the

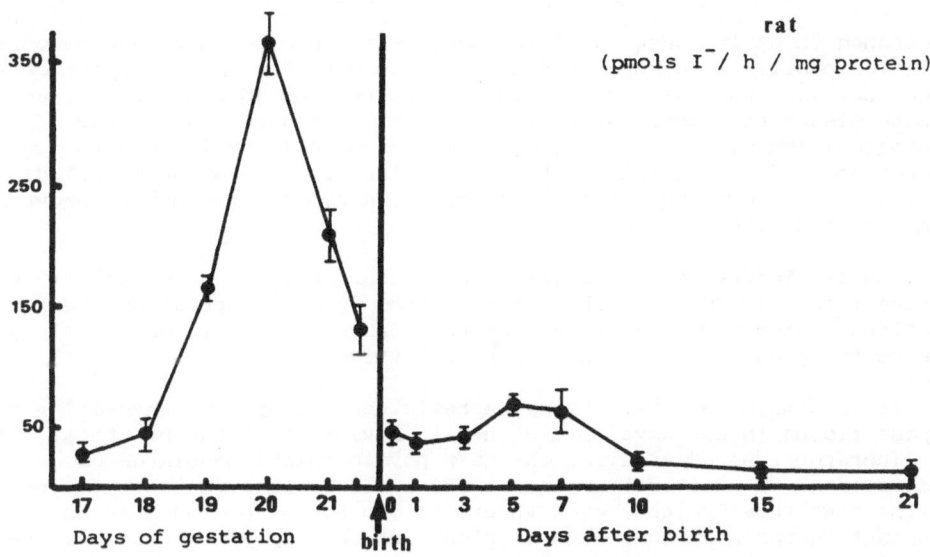

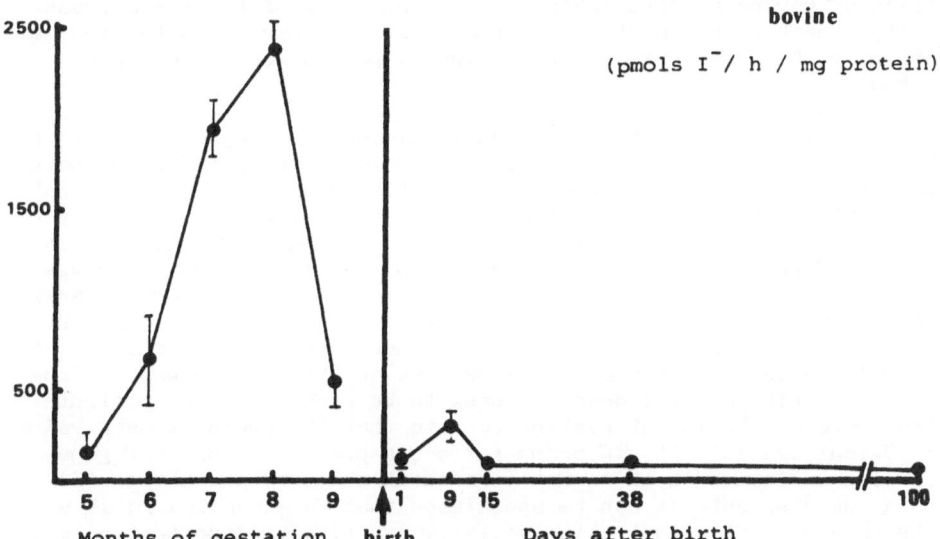

Fig. 1. Iodothyronine 5'-deiodinase activity in rat interscapular
brown adipose tissue and bovine perirenal brown adipose
tissue during development. Points are means ± SEM of
samples corresponding to pooled rat interscapular brown
adipose tissue of 5-8 independent litters for rat fetuses
and neonates, and of samples corresponding to 3-7
perirenal adipose tissue of different bovine individuals.

progressive decrease in 5'-deiodinase activity in bovine perirenal adipose
tissue correlates with the replacement from a brown to a white appearance
of this adipose depot in bovine postnatal life. When we studied
5'-deiodinase activity in other bovine adipose depots having a brown fat

appearence in early stages of life (pericardic, peritoneal, intermuscular) a similar profile of changes and correlation with the brown appearance was noted. However, when we studied bovine subcutaneous adipose tissue, an adipose tissue that does not present either the biochemical or the cytological aspect of brown fat, 5'-deiodinase activity was always undetectable. It was concluded that in bovines iodothyronine 5'-deiodinase activity, despite being a type I enzyme, is a typical feature of brown fat when compared with white fat.

The similarity between developmental changes of 5'-deiodinase in so different species suggests that common mechanisms of regulation and common functional roles may be expected for this enzyme activity, especially as regards the presence of the high prenatal values.

It is considered that the sympathetic nervous system innervating brown adipose tissue is not developed either in bovine or in the rat fetal life (8). Therefore, noradrenaline, the main physiological modulator of 5'-deiodinase in adult rat brown fat (2) would probably not be responsible for the high fetal values, even though it can not be excluded as an important factor for the postnatal rise. Interestingly, insulin has been recently reported to be a powerful stimulator of rat brown fat 5'-deiodinase activity (9). Taking into account the similarity between the profiles of perinatal circulating insulin and brown fat 5'-deiodinase activity, the hypothesis of an important role of insulin on the modulation of brown fat 5'-deiodinase in the perinatal period would need future attention.

Concerning the possible functional role of the enzyme activity in the perinatal period it must be said first that the known parallelism between brown fat 5'-deiodinase and thermogenic activity occuring in adult rats is not present in the prenatal period of rats and bovines. Thus, a very high 5'-deiodinase activity occurs during fetal life when thermogenesis is not activated. Furthermore, the key event of postnatal brown fat thermogenic activation, namely the increase in UCP gene expression, occurs clearly before the slight postnatal peak of 5'-deiodinase activity is present in rats (preceding chapter) or bovines (5). Therefore, the close link between 5'-deiodinase increase and UCP gene expression stimulus suggested from studies in adult rats (4) does not seem to be present in the perinatal period. However, it is interesting to note that the prenatal peak values of 5'-deiodinase activity in brown fat occur precisely when UCP gene expression is induced both in rat (preceding chapter) and bovine (5) prenatal development. It can be speculated that T3 generated in situ thanks to a high 5'-deiodinating activity may play an important role in the prenatal differentiation of brown fat and perharps in the induction of the UCP gene, whereas noradrenaline would be the main modulator of the UCP gene expression later in development.

ACKNOWLEGMENTS

This work has been supported in part by Dirección General de la Investigación Científica y Técnica, Ministerio de Educación y Ciencia (Grant PB-577-86) and CIRIT, Generalitat de Catalunya. Spain.

REFERENCES

1.   P. R. Larsen, J. E. Silva and M. M. Kaplan, Relationships between circulating and intracellular thyroid hormones. Physiological and clinical implications, Endocr. Rev. 2:87 (1981).

2. J. E. Silva and P. R. Larsen, Adrenergic activation of T3 production in brown adipose tissue, <u>Nature</u> 305:712 (1983).

3. J. A. Fernández, T. Mampel, F. Villarroya and R. Iglesias, Direct assessment of brown adipose tissue as a site of systemic triiodothyronine production in the rat, <u>Biochem. J.</u> 243:281 (1987).

4. A. C. Bianco and J. E. Silva, Intracellular conversion of thyroxine to triiodothyronine is required for the optimal thermogenic function of brown adipose tissue, <u>J. Clin. Invest.</u> 79:295 (1987).

5. L. Casteilla, O. Champigny, F. Bouillaud, J. Robelin and D. Riquier, Sequential changes in the expression of mitochondrial protein mRNA during the development of brown adipose tissue in bovine and ovine species, <u>Biochem. J.</u> 257, 665 (1989).

6. R. Iglesias, J. A. Fernández, T. Mampel, M. J. Obregon and F. Villarroya, Iodothyronine 5'-deiodinase activity in rat brown adipose tissue during development, <u>Biochim. Biophys. Acta</u> 923:233 (1987).

7. M. Giralt, L. Casteilla, O. Viñas, T. Mampel, R. Iglesias, J. Robelin and F. Villarroya, Iodothyronine 5'-deiodinase activity as an early event of prenatal brown-fat differentiation in bovine development, <u>Biochem. J.</u> 259: 555 (1989).

8. J. Nedergaard, E. Connolly and B. Cannon, Brown adipose tissue in the mammalian neonate, <u>in</u>: "Brown Adipose Tissue", P. Trayhurn and D. G. Nicholls, eds., pp. 153, Edward Arnold, London (1986).

9. J. E. Silva and P. R. Larsen, Hormonal regulation of iodothyronine 5'-deiodinase in rat brown adipose tissue, <u>Amer. J. Physiol.</u> 251: E639 (1986).

CHARACTERIZATION OF A SYSTEM TO STUDY THE UNCOUPLING

PROTEIN EXPRESSION IN BROWN ADIPOCYTE PRIMARY CULTURES

Almudena Porras, Matilde Peñas, Margarita Fernández
and Manuel Benito

Departamento de Bioquímica y Biología Molecular
Centro Mixto C.S.I.C./U.C.M.
Facultad de Farmacia
28040 Madrid, Spain

INTRODUCTION

Brown adipose tissue at birth is the main tissue involved in the neonatal thermogenesis. The brown fat mitochondria have a highly specialized protein called "uncoupling protein" which is the major component of the inner membrane and only present in this tissue. This protein is a carrier for protons and allows the re-entry of protons extruded by the respiratory chain, thus dissipating the proton-electrochemical gradient as heat (1). Synthesis of the uncoupling protein in vivo at birth is essentially controlled by the sympathetic fibers, releasing noradrenaline, stimulated by neonatal hypothermia. This acute effect is mimiced by the injection of isoproterenol (2). The possibility of the use of primary cultures of brown adipose tissue as a model to study the short-term and long-term regulation of uncoupling protein expression remains to be established. Thus, we report here the successful culture of brown adipocytes from rat fetuses in the last day of gestation to investigate the genetic expression of the uncoupling protein under several experimental conditions.

MATERIAL AND METHODS

Isolation and Culture of Brown Adipocytes

Interscapular brown adipose tissue was taken from 22-day fetuses of Wistar rats. After removal under sterile conditions, the tissue was finally chopped with scissor and placed in 20 ml sterile bottles containing each one 3 ml of 100 mM HEPES/4% (w/v) albumin isolation buffer (3) with 2 mg of collagenase/ml and incubated at 37°C for 45-60 min in a shaking water bath at 100 strokes/min. Every five minutes, bottles were removed and shaken vigorously for 10 s in an automatic mixer to facilitate tissue digestion. The collagenase action was stopped by addition of the isolation buffer without calcium and 1.2 mM EDTA. Then, tissue was filtered through a 100 μm-pore-size nylon mesh to remove any undigested material. The free cells from the filtrate were collected at the bottom of a sterile plastic tube by centrifugation at 80g for 5 min, an washed once with the isolation buffer. The number of cells obtained by this method was approx. 25-35 x 10^6 cells/g

Endocrine and Biochemical Development of the Fetus and Neonate
Edited by J. M. Cuezva *et al.*
Plenum Press, New York, 1990

147

of fetal brown adipose tissue. Cell viability, determined by trypan blue exclusion, was over 95%.

The isolated cells were plated in 6 cm-diam. plastic dishes and were cultured under different conditions depending on the type of experiment. For the experiment under proliferative conditions the culture method was essentially as described previously (4), inoculating 0.8-1.2 x 10^6 cells/ dish and being maintained in culture during 7 days. For the experiments under non-proliferative conditions, to study the regulation of uncoupling protein expression, 2 x 10^6 cells/dish were inoculated and the same culture medium, supplemented with 10% fetal-calf serum, was used during the first 6 h of culture to provide good cellular attachment. After that, the medium was changed every 12 h, containing a reduced fetal-calf serum concentration (0.5%), 100 μM ascorbate, and + 10 μM isoproterenol. Cells were removed after 24, 48 and 72 h of culture, they were pelleted by centrifugation at 12500 r.p.m. for 5 min and the supernatant was discarded using the cell pellet subsequently. The DNA content was determined by a spectrofluori-metric method (5). For this determination, cell pellet was resuspended in a buffer containing 50 mM Na_2 HPO4, 2 M NaCl and 2 mM EDTA pH 7.4 and sonicated at 1.5 mA during 30 s. The cytochrome-c-oxidase activity (6) and protein content (7) were determined.

Polyacrylamide Gel Electrophoresis and Immunoblotting

Cell pellet was resuspended in a buffer containing 10 mM Tris-HCl and was sonicated during 30 s at 1.5 mA and adequately diluted in 0.094 M Tris-HCl, 30% glycerol, 6% (w/v) SDS, 20 mM DTT and 30 μg/ml bromophenol blue pH 6.8. The samples prepared as described above were submitted to vertical slab-gel electrophoresis in an SDS-11% polyacrylamide gel. The electrophoresis buffer containing 25 mM Tris-HCl, 192 mM glycine and 1% (w/v) SDS, pH 8.3. Proteins were transferred to nitrocellulose overnight (16 h at 250 mA). The transfer buffer contained 25 mM Tris-HCl, 192 mM glycine, 20% (v/v) methanol, pH 8.3. After transfer, nitrocellulose filters were stained reversibly with Ponceau to check the protein transfer and then blocked by incubation during 30 min with 10 mM Tris-HCl, 150 mM NaCl (TBS) and 3% albumin pH 7.3. After washing with TBS-0.05% Tween pH 10.5, filters were incubated during 2 h with the rabbit anti-(uncoupling protein) serum (generously given by Dr. D.G. Nicholls and Dr. E. Rial, Dundee, U.K.) diluted 1/160 in TBS-3% albumin. Then, filters were washed with the medium described above. Bound antibody was detected by sheep-anti-(rabbit IgG) conjugated with peroxidase, developed with 4-chloronaphtol, scanned in a densitometer and photographed.

RESULTS AND DISCUSSION

Development of the Uncoupling Protein in Fetal Brown Adipocytes Cultured under Proliferative Conditions

Brown adipocytes cultured in a medium supplemented with 10% (v/v) fetal-calf serum proliferate and total protein and DNA contents increased with the time of culture reaching the confluence in 5-7 days under our conditions (results not shown). These cells gradually acquire spindle shape. These results are in agreement with those ultrastructural observations made with cultured precursor cells from rat mature brown fat (9).

A characteristic of thermogenic active brown adipocyte is the high content in mitochondria. *In vivo* stimulation of the tissue by cold or noradrenaline, is characterized by an increase in the mitochondrial mass and in the total amount of uncoupling protein (2). Therefore, both

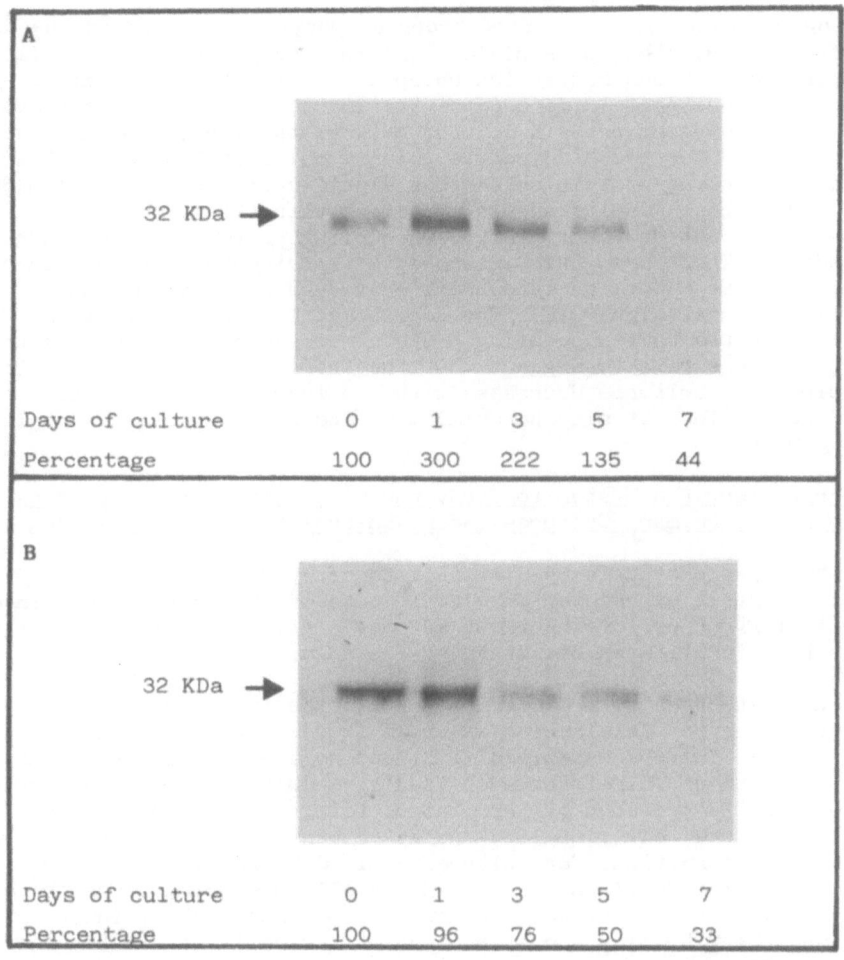

Fig. 1. Representative immunoblotting experiments of the uncoupling
protein in fetal rat brown adipocytes in primary culture
after 0, 1, 3, 5 and 7 days of culture under proliferative
conditions. Figure 1A: the protein content layered on each
lane was 12 μg. Figure 1B: the cytochrome-c-oxidase
activity layered on each lane was 5.5 mU.

parameters were considered in order to follow the thermogenic capacity of
these cells in culture. Cytochrome-c-oxidase activity was measured as an
index of mitochondrial mass and the uncoupling protein was detected and
relatively quantified by immunoblot. The cytochrome-c-oxidase activity,
either expressed per mg of protein or per μg DNA, decreased with the time
of culture, being 50% of the initial content after 7 days of culture; in
agreement with results described previously (10) for brown adipocytes from
adult mice and reveal a decrease in mitochondrial content per cell. The
uncoupling protein detected by immunoblotting experiments, using a well
characterized immune serum, gradually decreased with the time of culture
(Figures 1 A, B).

The uncoupling protein is clearly detected in freshly isolated
adipocytes in suspension at time zero, prior to inoculating the dishes for
culture. After one day of culture, this content seems higher since the

uncoupling protein only comes from brown adipocytes attached to the dishes and is free of any other possible contaminants (e.g. red blood cells) present in the previous suspension which could affect the amount of protein. The densitometric measure of the immunoblots indicated a 85% decrease in uncoupling protein content between the first and the seventh day of culture. Mice brown adipocyte primary cultures have also shown that levels of uncoupling protein suffered a significative decrease with days of culture (10). These results are in agreement with the mitochondrial mass decrease but do not show whether the uncoupling protein is also reduced per mitochondrion. Therefore other immunoblots experiments were carried out layering the same amount of cytochrome-c-oxidase activity on the polyacrylamide gel (Figure 1B). The densitometric scanning of these immunoblots showed that uncoupling protein decreased by 70% during the time of culture. These results suggest that the thermogenic capacity of fetal brown adipocytes cultures decreases with the increased culture time parallel to the loss of mitochondrial mass and mitochondrial uncoupling protein content.

Characterization of a System to Study the Uncoupling Protein Expression in Brown Adipocyte Primary Cultures under Non-Proliferative Conditions

It has been proved to be difficult to study the regulation of uncoupling protein expression *in vivo* because of the many factors involved. The possibility of performing brown adipocyte primary cultures simplifies and therefore facilitates the study of its regulation.

Brown adipocytes do proliferate in cultures in the presence of sufficient amount of fetal serum as shown above. However, fetal serum may contain certain factors capable of influencing the expression of the uncoupling protein. Therefore, such studies should be done under defined conditions in the presence of the minimum fetal serum concentration required to assure cell viability and attachment in culture. We assayed four different conditions for culture: 2, 1, 0.5% fetal-calf serum or 1% albumin instead of fetal-calf serum. Under all these conditions the cells did not proliferate, as revealed by protein and DNA content (data not shown), and the viability and attachment of the cells were maintained for 72 h, specially when serum was present. Medium containing 0.5% fetal-calf serum was chosen for further experiments to assess the cell response in terms of uncoupling protein expression. Cells were maintained in culture for 24, 48 and 72 h. in the medium containing 0.5% fetal-calf serum in the presence or absence of 10 μM isoproterenol (β-adrenergic agonist) and collected after those periods of time to measure DNA and protein content and cytochrome-c-oxidase activity, as well as to determine the presence and relative amounts of uncoupling protein by immunoblotting.

DNA and protein contents confirmed no cell proliferation. Figures 2A and 2B show two representative immunoblots with the same amount of total protein (Figure 2A) or the same amount of cytochrome-c-oxidase (Figure 2B) layered on the gel in each lane for the three periods studied, in the presence or in the absence of isoproterenol.

The densitometric measures of the immunoblots clearly indicate an inductory effect of isoproterenol, increasing the uncoupling protein content in the mitochondria of treated cells as compared with their controls. In addition, the inductory effect of isoproterenol is better observed after 48 h of treatment under our conditions of culture. The stimulatory effect of β-adrenergic agonists on brown adipose tissue thermogenic capacity has been well established in studies *in vivo* (11,12, 13). However, during the last few years several attempts to maintain brown adipocytes in primary cultures have been reported (3,9,10,14,15) but remained unsuccesful to reproduce all the *in vivo* effects of catecholamines

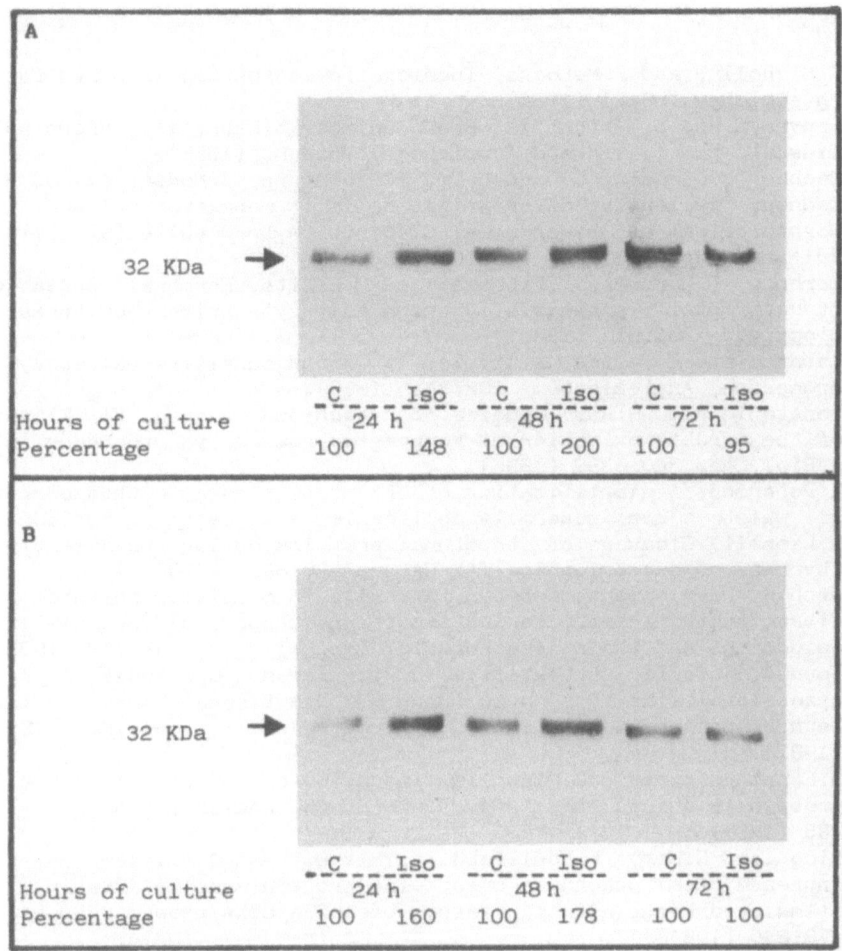

Fig. 2. Representative immunoblotting experiments of the
uncoupling protein in fetal rat brown adipocytes
in primary cultures after 24, 48 and 72 hours of
culture under non-proliferative conditions (0.5%
fetal-calf serum present) in the presence (Iso)
and absence (C) of 10 μM isoproterenol. Figure
2A: the protein content layered on each lane was
30 μg. Figure 2B: the cytochrome-c-oxidase
activity layered on each lane was 8 mU.

and β-adrenergic agonists. In this paper, we report some preliminary
results with fetal brown adipocyte primary cultures, in a system which
has proved to reproduce the *in vivo* effects of β-adrenergic agents on the
uncoupling protein content. However, further investigations are required
to elucidate the regulation of the expression of the uncoupling protein.

ACKNOWLEDGMENTS

This work has been totally supported by a grant from C.A.Y.C.I.T /
C.S.I.C (PR 84-0023), Spain.

REFERENCES

1. D. G. Nicholls, and R. M. Locke, Thermogenic mechanisms in brown fat, Physiol. Rev., 64:1 (1984).
2. P. Trayhurn and D. G. Nicholls, Brown adipose tissue, in: "Brown adipose tissue", pp. 1, (Edward Arnold ed.) London, (1986).
3. M. Néchad, P. Kunsela, C. Carneheim, P. Björntorp, J. Nedergard and B. Cannon, Development of brown fat cells in monolayer culture, I: Morphological and biochemical distinction from white fat cells in culture, Exp. Cell. Res. 149:105 (1983).
4. M. Lorenzo, C. Roncero, I. Fabregat and M. Benito, Hormonal regulation of rat foetal lipogenesis in brown adipocyte primary cultures, Biochem. J. 251:617 (1988).
5. C. Labarca and K. Paigen, A simple, rapid and sensitive DNA assay procedure, Anal. Biochem. 102:344 (1980).
6. T. Yonetani and G. S. Ray, Studies on cytochrome oxidase, VI: Kinetics of the aerobic oxidation of ferrocytochrome c by cytochrome oxidase, J. Biol. Chem 240:3392 (1965).
7. G. L. Peterson, A simplification of the protein assay method of Lowry et al. which is more generally applicable, Anal. Biochem. 83:346 (1977).
8. U. K. Laemmli, Cleavage of structural proteins during the assembly of the head of bacteriophage T4, Nature 227:680 (1970).
9. M. Néchad, Development of brown fat cells in monolayer culture, II: Ultrastructural characterization of precursors, differentiating adipocytes and their mitochondria, Exp. Cell. Res. 149:119 (1983).
10. C. Forest, A. Doglio, L. Casteilla, D. Ricquier and G. Ailhaud, A preadipocyte clonal line from mouse brown adipose tissue. Short and long-term responses to insulin and β-adrenergics, Exp. Cell. Res. 168:233 (1987).
11. S. A. Cunningham and D. G. Nicholls, Induction of functional uncoupling protein in guinea pigs infused with noradrenaline, Biochem. J. 245:485 (1987).
12. D. Ricquier, G. Mory, F. Bouillaud, J. Thibault and J. Weissenbach, Rapid increase of mitochondria uncoupling protein and its mRNA in stimulated brown adipose tissue. Use of a cDNA probe, FEBS Lett. 178:240 (1984).
13. D. Ricquier, F. Bouilland, P. Toumelin, G. Mory, R. Bazin, J. Arch and L. Penicaud, Expression of uncoupling protein mRNA in thermogenic or weakly thermogenic brown adipose tissue, J. Biol. Chem. 261:13905 (1986).
14. M. Cigolini, S. Cinti, O. Bosello, L. Brunetti and P. Björntrop, Human brown adipose cells in culture, Exp. Cell. Res. 159:261 (1985).
15. M. Cigolini, S. Cinti, O. Bosello, L. Brunetti and P. Björntrop, Isolation and ultrastructural features of brown adipocytes in culture, J. Anat. 145:207 (1986).

DEVELOPMENT AND MOLECULAR BIOLOGY OF MAMMALIAN

PYRUVATE DEHYDROGENASE COMPLEX

Mulchand S. Patel, Lap Ho and Donna J. Carothers

Departments of Biochemistry and Nutrition and
Pew Center for Molecular Nutrition
Case Western Reserve University School of Medicine
Cleveland, Ohio 44106, U.S.A.

INTRODUCTION

Glucose is the major fuel for oxidative metabolism in the developing fetus. A continuous supply of glucose and other nutrients from the maternal circulation to the fetus via the placenta is essential for normal fetal development. The initial metabolism of glucose to pyruvate occurs by the glycolytic pathway; further oxidation of pyruvate to CO_2 takes place through the tricarboxylic acid cycle in mitochondria. These two pathways are functionally linked by the pyruvate dehydrogenase complex (PDC) in the mitochondria. The level of oxidative metabolism of pyruvate continually increases in different tissues beginning in the early postnatal period, and gradually reaches the maximal levels observed in tissues from adults. In this chapter, we will discuss the structure, function and regulation of mammalian PDC with special emphasis on recent advances in the cloning of several components of the complex. Additionally, the expression of PDC activity in differentiating cultured cells and developing animals will be discussed. The importance of normal pyruvate oxidation during early mammalian development will be underlined by discussing the effects of inborn errors of PDC.

STRUCTURE AND FUNCTION OF THE COMPLEX

The mammalian PDC catalyzes the irreversible oxidative decarboxylation of pyruvate to acetyl-CoA according to the overall reaction:

$$\text{Pyruvate} + \text{CoA} + \text{NAD}^+ \longrightarrow \text{Acetyl-CoA} + CO_2 + \text{NADH} + \text{H}^+$$

The enzyme complex contains multiple copies of three catalytic and three regulatory components (Table 1). Five coenzymes or cofactors (thiamin pyrophosphate, lipoic acid, coenzyme A, FAD and NAD) are also involved in the component reactions (for review see refs. 1-5). The catalytic components are pyruvate dehydrogenase (E1), dihydrolipoamide acetyltransferase (E2), and dihydrolipoamide dehydrogenase (E3). The E1 component is a tetramer ($\alpha_2\beta_2$) composed of two non-identical subunits with molecular masses of 41,000 (α) and 36.000 (β) daltons. E1 catalyzes the thiamin pyrophosphate-dependent decarboxylation of pyruvate to 2-(1-hydroxyethylidene)-thiamin pyrophosphate (Figure 1). E1 also catalyzes the

Endocrine and Biochemical Development of the Fetus and Neonate
Edited by J. M. Cuezva *et al.*
Plenum Press, New York, 1990

153

Table 1. Subunit composition of mammalian heart pyruvate dehydrogenase complex

Component	Subunit/Component			Component/Complex
	Subunit	No.	Mass[*]	
E1	α	2	41,000	30
	β	2	36,000[***]	60
E2		1	52,000	6
E3		2	55,000	6
X		1	50,000	
Kinase	C[**]	1	48,000	
	R[**]	1	45,000	
Phosphatase	C	1	50.000	
	R	1	97,000	

[*]Molecular mass in daltons.
[**]C: catalytic subunit and R: regulatory subunit.
[***]70,000 daltons in SDS-PAGE.

transfer of the resultant acetyl group to lipoyl moieties which are covalently attached to the E2 component. E2 catalyzes the trasfer of acetyl groups from the acetyl-lipoyl moiety to Coenzyme A (CoA) to form acetyl-CoA, with the resulting donation of electrons from the dihydrolipoyl moieties of E2 to the E3 component. Structurally, sixty E2 monomers interact noncovalently to form the central core of the complex (1-3). E3, a homodimer, contains one noncovalently bound FAD per subunit, and utilizes NAD^+ as the final electron acceptor.

The regulatory components of the complex are pyruvate dehydrogenase kinase (E1 kinase), phospho-pyruvate dehydrogenase phosphatase (phospho-E1 phosphatase) (1-5) and a recently identified component referred to as protein X (6,7). Three specific serine residues (phosphorylation sites 1, 2, and 3) on the E1α subunit are phosphorylated by the E1 kinase (1,4,5,7). Phosphorylation of the E1α subunit inactivates the E1 component, which is rate limiting for the entire complex. The major inactivation of E1 results

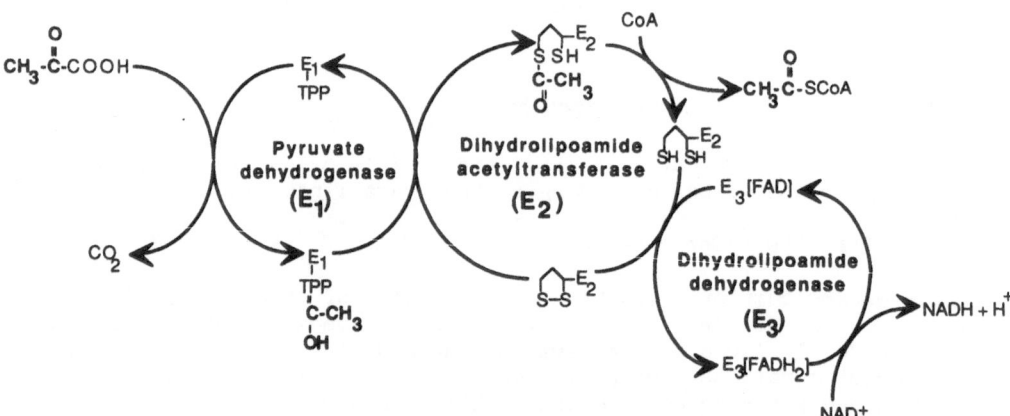

Fig. 1. Reaction sequence of the three catalytic components in the pyruvate dehydrogenase complex. The abbreviations used are: TPP, thiamin pyrophosphate, S-S and SH SH oxidized and reduced forms of lipoyl moiety covalently linked to E2(1,2).

154

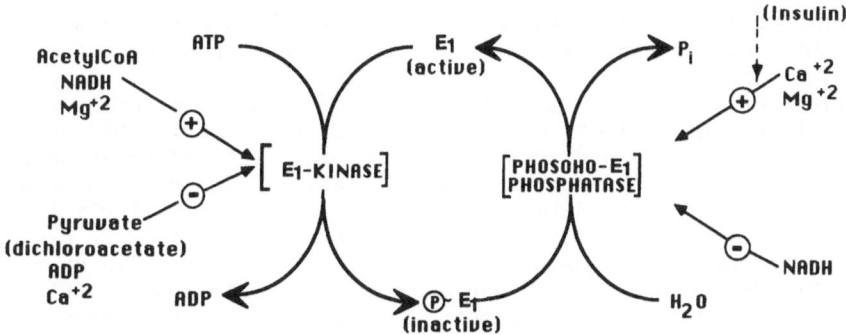

Fig. 2. Phosphorylation-dephosphorylation cycle of the E1 component
of mammalian pyruvate dehydrogenase complex. Abbreviations
used are: E1, dephosphorylated E1; (P)-E1, phosphorylated
E1; activator (+) or inhibitor (-), of both E1 kinase and
phospho-E1 phosphatase (1,4).

from the phosphorylation of site 1. Phosphorylation of site 2 is slightly
inactivating and phosphorylation of the third site has no inactivating
effect (8). Dephosphorylation of the phospho-E1α subunit by phospho-E1
phosphatase restores the catalytic activity of the E1 component (Figure 2).
Phospho-E1 phosphatase requires both Ca^{2+} and Mg^{2+} for activity. Protein X
contains a covalently bound lipoyl moiety which participates in an
acetylation reaction similar to that observed in the E2 component, and may
play an important role in the interactions of specific catalytic and
regulatory components of PDC (6,7,9).

The PDC reaction is essentially irreversible; there is no mechanism
allowing acetyl-CoA to be converted to pyruvate and subsequently to
glucose. It is a "point of no return" in the metabolism of carbohydrate
(4,10) and control of the complex is therefore crucial. The activities of
E1-kinase and phospho-E1 phosphatase are both regulated by ATP/ADP, acetyl-
CoA/CoA and NAD^+/NADH, and high pyruvate levels (1,4,10). Such modulators
may alter active PDC levels without concomitant alteration in total (active
plus inactive) PDC levels. The complex activity is regulated by end-product
inhibition (NADH, acetyl-CoA), and this inhibition is reversed by NAD^+ and
CoA. Acetyl-CoA inhibition is probably at the level of the E2 component,
while NADH levels affect the activity of E3 (1,4).

CLONING OF THE PDC COMPONENTS

In the past 18 months, several laboratories have independently
isolated and characterized cDNA clones for the E1 (11-15), E2 (16-20), and
E3 (21,22) components of mammalian PDC. Most of these cDNAs were isolated
from λgt11 libraries using component- or complex-specific antibodies. The
identity of each cDNA clone was established by similarity between the amino
acid sequence deduced from the cDNA nucleotide sequence and the known
partial amino acid sequence of the corresponding mammalian PDC component
protein. Using this experimental approach, we have isolated and partially
characterized cDNA clones for the E1 (14,15), E2 (17,18), and E3 (22) of
human liver PDC (Figure 3).

Our human liver E1α cDNA clone (1423 base pairs) has an open reading
frame of 1170 nucleotides and encodes an E1α prepeptide of 390 amino acid
residues (calculated molecular mass of 43,246 daltons) (unpublished data).

The first 29 amino acids of the prepeptide are consistent with a leader peptide which shares many features commonly present in other mitochondrial matrix signal sequences (11). The mature (mitochondrial form) human E1α polypeptide contains 361 amino acids with a calculated molecular mass of 40,183 daltons. The nucleotide sequence of three recently published E1α cDNAs are significantly different from each other (11-13) and from our human liver E1α cDNA sequence (unpublished data). Discrepancies range from single base insertions or deletions to major deletions causing shifts in the reading frame of up to 64 amino acids. This has resulted in marked variations in the deduced amino acid sequence of the human E1α mature peptide. The deduced amino acid sequence of mature E1α from our E1α cDNA clone (unpublished data) closely matches that published for human hepatoma E1α (13) except for two amino acids. In generating different cDNA libraries, cloning artifacts or peculiarities of the cells used may have contributed to the observed discrepancies among these various reported E1α cDNA sequences.

In Northern analysis of human tissue RNA, two hybridizable E1α mRNA species have been identified (a more abundant 1.6 kilobase species and a less abundant species of 3.3. kilobases) (11,13,15) (Figure 4). These two E1α mRNA species appear to have identical 5' untranslated and coding regions, but differ in the length of the 3' untranslated region (11).

The amino acid sequences around the three phosphorylation sites (sites 1-3) in the deduced sequence of human E1α are identical to those previously

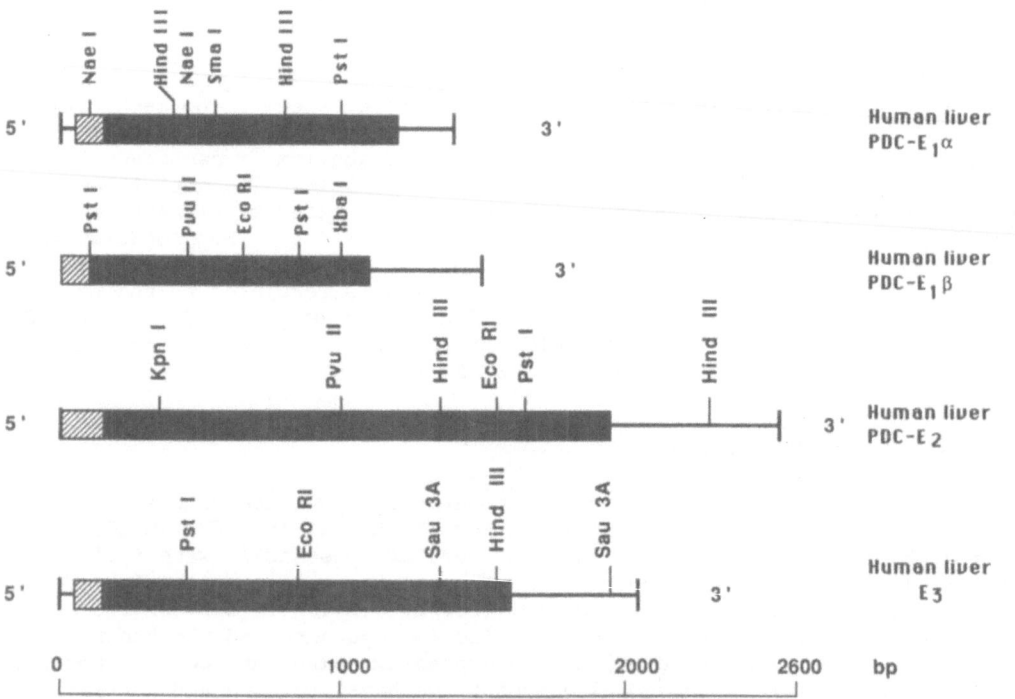

Fig. 3. Size and partial endonuclease restriction maps of the human PDC E1α, E1β, E2 and E3 cDNA clones. The hatched and solid boxes represent coding regions for the leader peptide and mature polypeptide respectively, and solid lines represent the 5'- and 3'-noncoding regions of the clones. bp, base pairs (25; unpublished data).

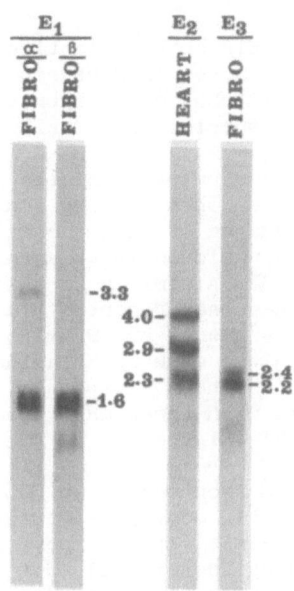

Fig. 4. Northern blot analyses of RNA from human heart and
cultured skin fibroblasts. Total RNA was fractionated
by electrophoresis on a formaldehyde-agarose gel,
transferred to Gene Screen membrane, and probed with
labelled cDNA probes (14,15,18,22). E1α, 20 μg total
RNA; E1β, 20 μg total RNA; E2, 25 μg of total RNA;
E3, 15 μg of poly (A$^+$) RNA. The size of the RNA in
kilobases was determined by comparing its electrophoretic mobility with those of ribosomal RNAs (14,
15,18,22).

determined by the amino acid sequence analysis of the phosphorylated
peptides of porcine E1α (23). In bovine E1α, only one residue between sites
1 and 2 is changed from aspartic acid to asparagine (8,23). Based on
indirect evidence, the adjacent lysine (214) and tryptophan (215) residues
located in a hydrophobic region of the human E1α subunit are a proposed
candidate for the thiamin pyrophosphate binding site (11). Presently, very
little is known about the structure of E1α and its interaction with the
other components of the complex.

A cDNA clone (1.69 kilobases) for the human foreskin E1β subunit has
recently been reported (12). The deduced E1β precursor peptide contains 359
amino acid residues (30 amino acid residues in the leader sequence and 329
in the mature polypeptide) (12). We have isolated and characterized two
cDNA clones for the E1β subunit from a human liver λgt11 cDNA library (14;
Figure 3, unpublished data). The composite human liver E1β cDNA sequence
encodes the entire mature E1β peptide and most of the E1β precursor peptide
(26 of the 30 amino acid residues reported for the deduced human foreskin
E1β leader sequence (12)) (Figure 3). Although the deoxynucleotide sequence
of human liver E1β cDNA is in good agreement with that derived from the
cultured human foreskin fibroblast E1β cDNA, the two nucleotide sequences
differ at 20 positions over the entire length of the cDNA. These differences are the results of nucleotide additions, deletions or substitutions in
both non-coding and coding regions. Six base changes involving single-
nucleotide insertions or substitutions in the coding region have resulted
in reading-frame shifts affecting stretches of 3 to 8 deduced amino acids
(unpublished data). The cause of these discrepancies between the human

liver and foreskin fibroblast E1β cDNA nucleotide sequences is not clear, but may be due to cloning artifacts. In Northern analysis of total RNA from human skin fibroblasts, the E1β cDNA clone hybridized to an abundant 1.6 kilobases species and a minor species of 5.2 kilobases (Figure 4) (14).

An E2-cDNA clone (2583 base pairs) isolated from a human liver λgt11 library in our laboratory has an open reading frame of 1848 base pairs encoding a partial leader sequence of 54 amino acids and a mature polypeptide of 561 amino acid residues (with a calculated molecular mass of 59,551 daltons) (18) (Figure 3). A similar E2 cDNA clone from a human placental cDNA library also has been reported (19). However, approximately the first half of the leader sequence and 8 amino acid residues in the mature peptide differ between the deduced E2 amino acid sequences from these two human E2 cDNAs. In Northern analysis of RNA from human heart, we found that E2 cDNA hybridizes with three mRNA species which are 2.7, 2.9 and 3.2 kilobases in size (18) (Figure 4).

The E2 component isolated from mammalian PDC is composed of three distinct domains, (i) a catalytic domain (also referred to as the subunit-binding domain or compact domain), (ii) a lipoyl-bearing domain, and (iii) an E3-binding (or E3/E1-binding) domain (24,25). The locations of these three domains in the deduced amino acid sequence of the mature E2 component of human PDC are as follows; residues 1 to 254 as the amino-terminal lipoyl-bearing domain; residues 255 to 330 as the E3/E1 binding domain; and residues 331 to 561 as the carboxyl-terminal catalytic domain (Figure 5). Although there are variations in the size of the lipoyl-binding domain among the PDC-E2 components derived from different species, the relative sizes of the other E2 domains are largely conserved among the different species (Figure 5) (for a review see 25). The carboxyl-terminal catalytic domain confers specificity for subunit-subunit interactions and substrate specificity. Sixty E2 monomers interact non-covalently with each other forming the core subcomplex (icosahedral 532 symmetry) of the PDC complex (2,3). Approximately 24% of the amino acid residues in this domain are conserved among PDC-E2 from four different species (human, *Saccharomyces cerevisiae*, *Escherichia coli* and *Azotobacter vinelandii* (for review see 25). Recently, we have identified three separate regions with considerable homology (ranging from 36 to 62 %) within this domain (25). Although it is difficult to assign any specific functions to these regions, it is suggested that a histidine residue (amino acid 534) present in a stretch of 14 amino acids may be involved in interaction with CoA (18,25).

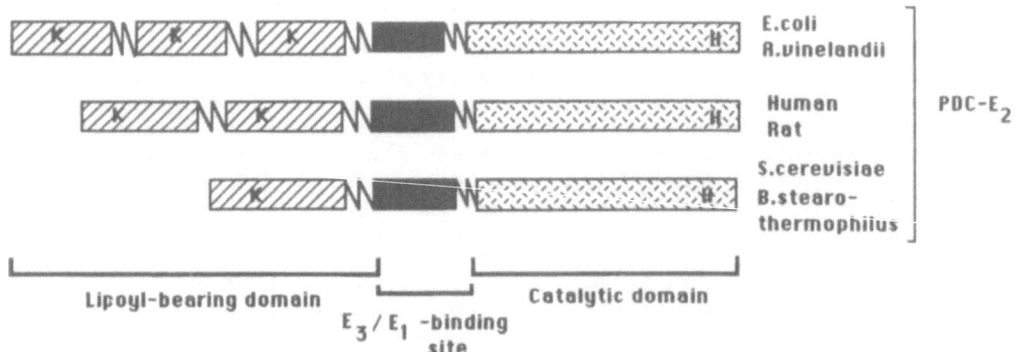

Fig. 5. Comparison of the domain structures of the E2 component of PDC from prokaryotes and eukaryotes. The boundaries of the domains are approximate. K indicates the lipoyl-binding lysine residue; H indicates a histidine residue (25).

Since E3 is a common component among the three α-ketoacid dehydrogenase complexes (PDC, α-ketoglutarate dehydrogenase complex and the branched-chain α-ketoacid dehydrogenase complex), it is not surprising that a conserved stretch of 32 amino acids has been identified in the E3/E1 binding domain of the E2 components from the α-ketoacid dehydrogenase complexes isolated from different species (for review see 25). There is a high degree of homology ranging from 50 to 91% in this region among the different PDC-E2s which have been compared. However, identification of specific amino acid residues involved in the interaction between E2 and E3 remains to be established.

The lipoyl-bearing domain of human PDC-E2 shows different structural characteristics when compared to the same domain present in E2s from prokaryotes and lower eukaryotes. Two similar repeating units of 127 amino acids each in length (with approximately 55 % homology) have been identified in the lipoyl-bearing domain of human E2 vs three repeating units (about 100-110 amino acids in length) in E2 from *Escherichia coli* and *Azotobacter vinelandii* and only one repeating unit in the E2 from *Bacillus stearothermophilus* and *Saccharomyces cerevisiae* (Figure 5) (25). Each repeating unit shows the following characteristics: (i) one lysine residue which participates in covalent linkage of lipoic acid is located approximately 40 to 47 amino acids from the amino terminus of each unit, (ii) a highly conserved region of 38 amino acids surrounds the lipoyl-bearing lysyl residue, and (iii) the carboxyl-terminal stretch of each repeating unit in PDC-E2 is rich in alanine and proline residues, imparting some flexibility in movement to each repeating unit. In *Escherichia coli*, removal of two of the three repeating units in the lipoyl-bearing domain in E2 has no apparent effect on overall PDC catalytic activity or active site coupling of the component (26). The evolutionary and functional significance of having varying numbers of repeating units in the E2 remains unclear.

The E3 component is shared among the three α-ketoacid dehydrogenase complexes. This conclusion is based on the following observations: (i) reconstitution experiments, (ii) immunological cross-reactivity, and (iii) genetic disorders of E3 affecting the three complexes (27). Two laboratories have simultaneously reported isolation of full-length cDNA clones for human E3 (21,22) and pig heart (21). The nucleotide sequence of our human liver E3 cDNA (2082 base pairs) (Figure 3) has an open-reading frame of 1527 base pairs encoding the precursor E3 polypeptide of 509 amino acids. Based on the structural characteristics of a mitochondrial leader sequence and identification of the mature E3 amino-terminus amino acid in the deduced peptide based on the pig heart E3 sequence, the first 35 amino acids of the deduced precursor E3 peptide are identified as the signal sequence (22). The calculated molecular mass of the deduced mature E3 (474 amino acids) is 50,216 daltons. The primary amino acid sequences of the human mature E3 peptide derived independently by two laboratories are identical except in two positions (amino acids 69 and 119) (21,22). There is a very high degree of homology (96%) between the primary amino acid sequence of mature E3s of human liver and pig heart (21). In Northern analysis, two hybridizable E3 mRNA species of approximately 2.2 and 2.4 kilobases have been observed in human tissues (Figure 4), although only one mRNA species of about 2.4 kilobases has been observed in rat tissues (22). The presence of two polyadenylation signals located approximately 0.2 kilobase apart in human liver E3 cDNA clones may be responsible for the two E3 mRNA species observed in human tissues (25).

Based on the known tertiary structure of human erythrocyte glutathione reductase (28) and the considerable homology (33%) between human glutathione reductase and human E3, four domains (FAD-binding, NAD-binding, central and interface domains) have been identified in human E3, as well as

E3 derived from other species (for review see 25,27). Comparisons of the secondary structures (both β-sheets and α-helices) of human E3 and glutathione reductase reveals considerable conservation, especially for β-sheets. This comparison has been further extended to identify likely active site residues in these domains, particularly for their coenzymes, which are identical or closely related. In addition, subunit-subunit interactions within the dimeric protein are found to be conserved in analogous domains for both human E3 and glutathione reductase, indicating a great degree of overall structural and functional similarity between these two flavoproteins (25,27).

REGULATION OF THE PDC

During carbohydrate feeding, the PDC is activated to catalyze increased formation of acetyl-CoA for use in either the tricarboxylic acid cycle or synthesis of fatty acids (for review see 4,10). During fasting, PDC activity is markedly reduced by phosphorylation to spare carbohydrate oxidation in most tissues. When glucose is the carbon source, acetyl-CoA formed by PDC acts as a substrate for fatty acid synthesis. Regulation of fatty acid synthesis, both acute and long term, is due to changes in a variety of plasma hormones, including insulin, glucocorticoids, and catecholamines. Insulin is the most well-characterized of the effectors (29). The proportion of PDC in its active form is increased after insulin treatment. Insulin may cause an increase in polyamines which lower the phospho-E1 phosphatase's requirements for Mg^{2+} ions. There seems to be no evidence to suggest that insulin stimulates phospho-E1 phosphatase by altering the intramitochondrial Ca^{2+} level (29) or that cAMP is involved in regulation of PDC phosphorylation. E1 kinase is not inhibited under this condition (31) and no significant changes in NAD^+/NADH, acetyl-CoA/CoA, or ATP/ADP levels have been observed (32,33). It is likely that insulin treatment increases or decreases the concentration of some effector of phospho-E1 phosphatase, other than Ca^{2+} or Mg^{2+} (29). A "second messenger" for insulin has been postulated (34,35), and partially purified. The mediator is said to stimulate phospho-E1 phosphatase activity, and thus to activate E1 (36). More recently, a phosphatidylinositol-specific phospholipase C has been implicated in mediating insulin stimulation of PDC (37).

Active E1 levels in the heart and kidneys decrease with fasting or with diabetes (38). This reflects the high degree of fat oxidation in these tissues, causing the accumulation of acetyl-CoA which stimulates phosphorylation and inactivation of PDC by E1-kinase. Regulation of the levels of "active" and inactive E1 in the liver is more complicated, since the liver is the site of gluconeogenesis from pyruvate while it oxidizes fatty acids for energy (4,10,39). Insulin treatment increases the level of active E1 in the liver, but this is attributed to lowered concentrations of plasma free fatty acids (40). Glucagon does not have an effect on E1 activity in perfused rat liver or isolated liver cells (41,42). Vasopressin increases E1 activity in perfused rat liver (43), possibly due to an increased intracellular Ca^{2+} concentration which may activate phospho-E1 phosphatase. Epinephrine is known to cause a rise in cytoplasmic Ca^{2+} (44). Dependence of PDC activity in rat heart on intramitochondrial free Ca^{2+} concentration has recently been shown (45). The "total" PDC activity level is not affected in rat liver in the hyperthyroid state but is reduced by a third in the hypothyroid state (46). Decreased "total" PDC activity in the hypothyroid state is accompanied by reduced levels of E1 proteins due to alterations in rates of synthesis of these proteins without significant changes in the degradation rates (46). It is likely that the other PDC component proteins are coordinately regulated in a similar fashion in the hypothyroid state. Detailed mechanisms for hormonal regulation of PDC activity in liver have not been established.

Pyruvate oxidation in muscle is controlled by the amount of "active" PDC. In aerobic conditions pyruvate oxidation parallels ATP utilization by muscle. When lipids are mobilized as fuels (diabetes, starvation), they are oxidized preferentially over pyruvate (39). This is because acetyl-CoA produced by β-oxidation inhibits PDC by stimulating E1-kinase activity (47). The E1-kinase may also be activated, especially during starvation or diabetes, by a thermolabile, non-dialyzable factor in heart mitochondria (48). Both E1-kinase activation and phospho-E1 phosphatase inhibition seem to contribute to the low PDC activity in muscle when fatty acids are oxidized. In the heart, PDC activity can be modulated by the action of epinephrine. This modulation is mediated by the activation of the phospho-E1 phosphatase which in this case seems to correlate with increased intra-mitochondrial Ca^{2+}. Thus, acute hormonal regulation of PDC seems to be mediated through the specific kinase and phosphatase.

DIFFERENTIATION

Differentiation of 3T3-L1 preadipocytes (derived from fetuses of the Swiss albino mice) to 3T3-L1 adipocytes is characterized by the acquisition of several morphological and biochemical characteristics of adipocytes. Differentiation of 3T3-L1 preadipocytes is accompanied by marked increases in both the capacity to synthesize long chain fatty acids and triglycerides from glucose and acetate and in the activities of several lipogenic enzymes. Since PDC occupies a central position in the conversion of glucose to acetyl-CoA during lipogenesis, changes in PDC activity have also been monitored during differentiation of 3T3-L1 cells. The activity of the complex, as well as of the E1 component, increases several fold with differentiation (49) (Figure 6). In differentiating 3T3-L1 cells, E1 activity increases in parallel with increases in E1 protein, reflecting increased rates of synthesis for the two non-identical E1 subunits (49). Activities of the other two catalytic components (E2 and E3) also increase (50,51). The increased E3 activity is also associated with a rise in E3 protein content, which results from an increase in the rate of E3 synthesis in differentiating 3T3-L1 cells (51). These results suggest that the expression of PDC proteins are coordinately regulated in differentiating 3T3-L1 cells, most likely at a transcriptional level.

Just as the "total" PDC activity in differentiating 3T3-L1 cells is hormonally determined, the "active" PDC levels are modulated by the regimen of drugs or hormones used to promote cellular differentiation (52). Chronic exposure to dexamethasone plus 1-methyl-3-isobutyl xanthine, with co-treatment with insulin, causes a significant increase in the level of "active" PDC in 3T3- L1 cells, as compared to the levels in cells exposed chronically to insulin, or to acute treatment with dexamethasone and 1-methyl- 3-isobutyl xanthine. The time course of PDC induction varies with the drug regimen, as does the magnitude of the induced PDC activity (52). Molecular mechanisms for these differences remain to be determined, but an alteration in activities of E1 kinase or phospho-E1 phosphatase seems likely.

DEVELOPMENTAL ASPECTS

PDC activity, like that of other mitochondrial enzymes, is low in most fetal tissues and increases gradually during the postnatal period. For instance, "total" PDC activity is low in fetal rat liver (53,54) and increases slowly to the adult level over 3-4 postnatal weeks. Moreover,

most of the PDC in developing rat liver is present in the inactivated form
because of the presence of low levels of plasma insulin which, in turn, is
a consequence of the consumption of a high fat-low carbohydrate milk during
the suckling period. However, the "active" and "total" PDC activities in
rat liver increase significantly during the first 2 and 6 postnatal hours,
respectively (54) (Figure 7). Using immunological techniques, it has
recently been shown that the rise in PDC activity in the immediate
postnatal period is due to increased content of all PDC catalytic component
proteins (55). The increase in proportion of PDC activity in its "active"
form in 2 hour-old rat liver plays an important role in clearance of
pyruvate derived from a high circulating lactate pool at the time of birth
(56). It is possible that the postnatal increase in hepatic pyruvate
concentration is responsible for the increased fraction of "active" PDC in
newborn rat livers because high pyruvate concentration inhibits E1-kinase
activity. Plasma insulin rapidly declines in newborn rats during the first
few hours after birth (57), and hence it may not have a significant effect
on the postnatal increase of "active" PDC in the newborn rat. Administration of glucose (0.111 mmol/pup) to the newborn rat at birth causes a
significant increase in both "active" and "total" PDC activity and in the
proportion of "active" PDC in "total" PDC activity in liver (54) (Figure
7). It is possible that the administration of glucose enhances the
secretion of insulin and increases the concentration of hepatic pyruvate
derived from glucose. The transient increase in "active" PDC in livers of
newborn rats during the first 2 postnatal hours plays an important role in
rapid clearance of the lactate pool through oxidative metabolism (56).

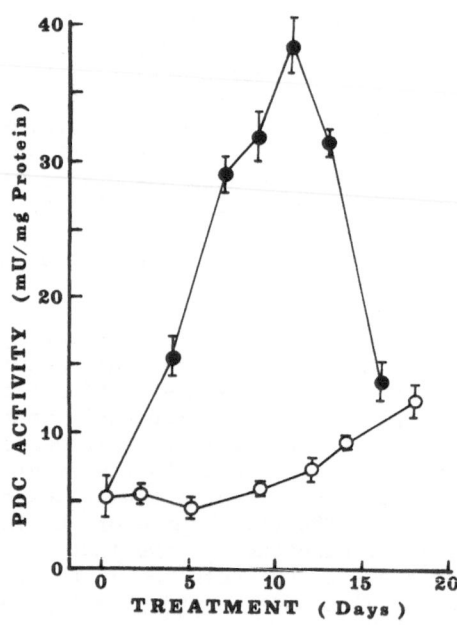

Fig. 6. Differentiation-induced changes in the specific activity
of "total" PDC in cultured 3T3-L1 preadipocytes and
adipocytes. Confluent cultures of 3T3-L1 preadipocytes
were treated as follows: open circle, no addition to
medium, closed circle, medium containing insulin (10
μg/ml), dexamethasone (0.25 μM), 1-methyl-3-isobutyl
xanthine (0.5 mM) and biotin (8 μg/ml) for 48 hours
followed by the medium containing insulin and biotin
only (49).

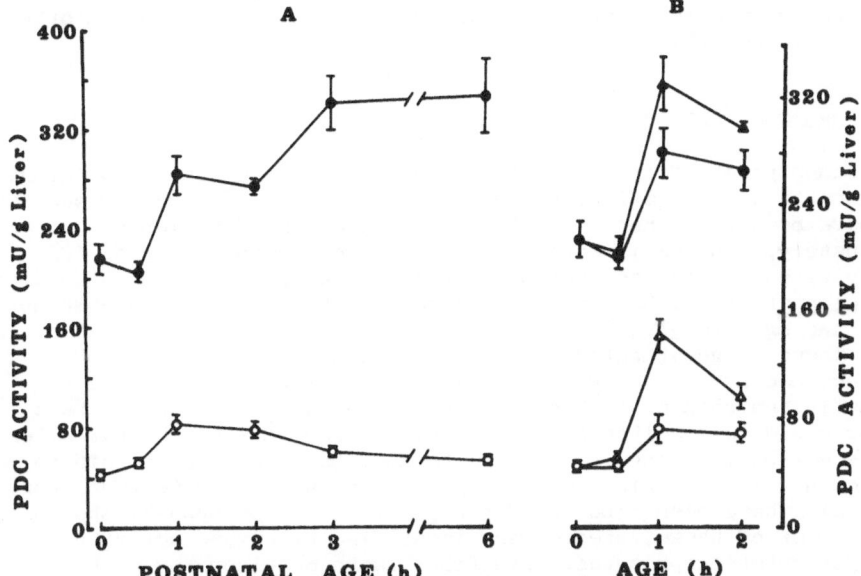

Fig. 7. Immediate postnatal changes (A) and the effect of
glucose administration at birth (B) on "active"
and "total" (dichloroacetate-activated) PDC
activity in livers of rat neonates. A: open
circle, "active" PDC; closed circle, "total" PDC;
B: saline-injected pups - open circle "active"
PDC and closed circle "total" PDC; glucose (0.111
mmol/pup)-injected pups - open triangle, "active"
PDC and closed triangle, "total" PDC (54).

The postnatal development of PDC activity in rat brain has been
extensively investigated because of the importance of oxidative metabolism
in cerebral maturation. "Total" PDC activity is very low in rat brain at
birth but increases rapidly after the tenth postnatal day and reaches adult
levels by the end of the fourth postnatal week (58,59). The postnatal
increase in "total" PDC activity in the rat brain has been correlated with
a coordinated increase in the content of PDC component proteins measured by
immunoblotting (60). The postnatal increase in PDC activity in the rat
brain reflects both an increase in mitochondrial number and in the amount
of PDC protein per mitochondrion (60). Ontogenic changes in PDC activity in
various regions of rat brain show significant differential developmental
patterns. For instance, PDC activity in medulla/pons attains adult levels
by postnatal day 21, whereas the development of this activity in cerebral
cortex, hippocampus, striatum, and cerebellum progresses more slowly (61).
Region-specific maturation profiles for PDC may play an important role in
development of metabolic compartmentation in rat brain. Immunocytochemical
localization of PDC in rat brain has revealed that PDC is present
throughout the neuropil but is enriched in selective neuronal perikarya
which are especially abundant in regions containing cholinergic neurons
(62). Training-induced alteration in the phosphorylation status of the α
subunit of PDC-E1 in rat brain has been observed (63) but the significance
of this observation remains obscure.

Both "total" PDC activity and the fraction of "active" form in rat
mammary gland increase steadily from near term to mid-lactation and remain
at high levels until the end of lactation (64). PDC activity declines
precipitously within 3 days after weaning (64). Rat mammary PDC possesses

many of the regulatory properties discussed above for the PDC in other mammalian tissues (64).

INBORN ERRORS OF PDC

Another facet of the role of pyruvate metabolism comes from studying human genetic defects affecting pyruvate metabolism (for review see 65-69). There have been more than 100 reported cases of PDC deficiency. Unfortunately, many early reports were based on measurements of PDC activity without prior activation of the complex by treatment with either dichloroacetate (an inhibitor of E1-kinase), phosphatase or a combination of Ca^{2+} and Mg^{2+} to give "total" PDC activity as opposed to "active" PDC levels. There is great potential for genetic heterogeneity in PDC deficiency because there are in total nine catalytic and regulatory proteins in the complex (Table 1). Presumably, any mutation in these proteins could cause altered pyruvate metabolism. Mutations that alter pyruvate metabolism range from complete absence of enzyme protein to more subtle changes in the kinetic binding of substrates or effectors. Enzyme deficiencies have been reported for E1, E2, E3 and phospho-E1 phosphatase (65-69). Most of these defects were identified by enzyme activity analyses of the patients' lymphocytes, skin fibroblasts or tissue extracts. To further characterize the defect, immunological and Northern analyses of the complex proteins and mRNAs have been used. Perturbations of the E1 component appear to be the most prevalent cause of PDC deficiency (65,69,70). The next most prevalent cause of PDC deficiency involves the E3 component, which is common to all the α-ketoacid dehydrogenase complexes, namely PDC, α-ketoglutarate dehydrogenase complex, and branched-chain α-ketoacid dehydrogenase complex (71-73). Deficiency of E3 activity impairs the catalytic activity of all three of these complexes. Deficiencies in E2 or phospho-E1 phosphatase activity are less common.

In most cases the localization of the defect in the complex was provided by activity measurements of PDC (both "active" and "total") as well as E1, E2 and E3 components measured using partial reactions (74-77). Findings of low levels of E1 activity require an assessment of phospho-E1 phosphatase activity. Immunological analysis of PDC deficiency by the Western blot technique is now routinely used in several laboratories to identify a change in the content or the catalytic efficiency of one of the components of the complex (15,75-82). This technique also detects significant alterations in the size of the affected component protein (15,79). Western blot analyses of several E1- deficient patients have demonstrated simultaneous reduction of both E1α and E1β proteins (Figure 8) (75-77). Since the E1 component exists as an α2β2 tetramer in the mitochondria, it is likely that in the absence of one peptide the other peptide is rapidly degraded (75). Additionally, tissue-specific expression of E1 deficiency has been observed using Western blot analysis (77). Availability of cDNA clones for the PDC components has aided in further characterization of mutations in E1-deficient subjects (15,69). The presence, absence, or reduced level of specific E1α mRNA levels has been described in several E1 deficient patients (Figure 9) (15), although the nature of genetic defects in these patients remains to be further investigated.

The most common clinical manifestations of PDC deficiency are lactic acidosis and varying degrees of central nervous system dysfunction (65-69). Patients often show progressive neurological dysfunction, including ataxia, seizures, hypotonia and psychomotor retardation. No correlation has been documented between the clinical severity of patient condition and the level of residual PDC activity measured in skin fibroblasts or other readily

164

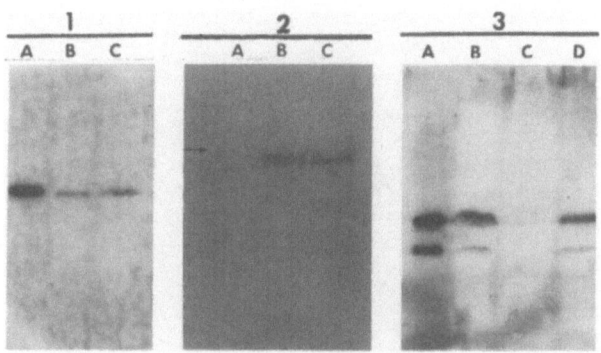

Fig. 8. Immunoblot analysis of E1, E2 and E3 components in
cultured skin fibroblasts from a control and a PDC-
deficient patient. For each panel, crude mitochondrial
pellets prepared from fibroblasts containing similar
quantities of cellular proteins were separated by
electrophoresis, electrotransferred to a nylon-based
membrane, and treated with a specific antibody followed
by ^{125}I-protein A. Panel 1, E3 immunoblot; Panel 2, E2
immunoblot; Panel 3, E1 immunoblot (upper band E1α and
lower band E1β). In all panels: lane A, purified PDC
component, lane B, control; lane C, patient. In panel
3, lane D is an additional control (75).

available tissue specimens. Dietary management of PDC-deficient patients is
primarily based upon provision of acetyl-CoA by feeding a ketogenic diet.
This may slow, but does not stop, the progression of neurological
deterioration (65). Pharmacological treatments include administration of
dichloroacetate, thiamin, and lipoic acid; however, the benefits of these
treatments have not been established (61,65,72,83). The efficacy of
therapeutic interventions is likely to depend on early and accurate
diagnosis of the defects.

CLINICAL CONDITIONS AFFECTING PDC ACTIVITY

Reduction in PDC activity has been observed in affected areas of brain
from patients with Huntington's disease (Caudate, putamen, hippocampus) and
Alzheimer's disease (frontal, cortex) (84,85). Interestingly, however, no
abnormality in PDC activity was detected in skin fibroblasts from
Huntington and Alzheimer patients (84). Reduction in cerebral PDC activity
in patients with Alzheimer's disease is correlated with decreased amounts
of normal PDC antigens (85). It is suggested that PDC deficency in
Alzheimer's disease may be related to mitochondrial damage and to impaired
calcium homeostasis in patients' nerve cells.

The major serologic feature of primary biliary cirrhosis, a chronic
idiopathic autoimmune liver disease, is the presence of anti-mitochondrial
antibodies in the majority of affected patients (86,87). One of the
mitochondrial proteins reacting strongly with the antiserum is the 70,000
dalton protein (20,88,89) identified as the E2 component of mammalian PDC
(88,89). The region of E2 containing the lipoic acid binding lysine residue
appears to be the target site for interaction with the antibodies (90). The
significance of these specific anti-mitochondrial E2 antibodies in patients
with primary biliary cirrhosis remains elusive at the present time.

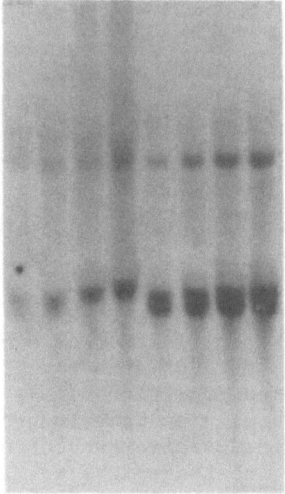

Fig. 9. Northern blot analysis of total RNA from fibroblasts of a
control and an E1-deficient patient. Varying amounts of
total RNA as identified for each lane was fractionated
by electrophoresis, transferred to a nylon membrane, and
probed with a labelled E1α cDNA. Two E1α mRNA species of
1.6 kb (lower band) and 3.3 kb (upper band) were
identified. Densitometric analysis showed a 78% reduction
in the content of E1α mRNA species in fibroblasts from
this patient (unpublished data).

SUMMARY

Mammalian PDC, a multi-component mitochondrial enzyme complex, is
comprised of three catalytic and three regulatory components. It is
regulated by covalent modification involving phosphorylation (inactivation)
/dephosphorylation (activation) by a specific E1-kinase and phospho-E1
phosphatase, respectively. Acute regulation is modulated by changes in the
intramitochondrial NAD^+/NADH, Acetyl-CoA/CoA, and ATP/ADP ratios as well as
the concentrations of Mg^{2+} and Ca^{2+} caused by dietary and/or hormonal
manipulations. Developmental patterns of PDC in rat tissues show tissue-
specificity and hence appear to play a role in fuel economy in the
developing organism. Furthermore, the simultaneous increase of various PDC
component proteins in rat organs during postnatal development and in
differentiating 3T3-L1 cells suggests that the PDC genes are coordinately
regulated during cellular differentiation and development.

Isolation and characterization of cDNA clones specific for human PDC
component proteins in the past 18 months has provided, for the first time,
the complete primary amino acid sequences for the E1α, E1β, E2 and E3
components of human PDC. This has led to (i) comparison of the primary
amino acid sequences of human PDC components with those of other species,
(ii) prediction of structural features, and (iii) in some cases tentative
identification of active site residues. Availability of the PDC cDNA clones
has opened up a new dimension in characterization of genetic mutations in
human PDC components. Furthermore, characterization of structural and
regulatory aspects of the PDC component genes would enhance our
understanding of regulation of the complex at a molecular level.

ACKNOWLEDGMENTS

The work performed in this laboratory and reported in this review was supported by U.S. Public Health Service Grants AM 20478, HD 12643 and Metabolism Training Grant AM 07319. We are indebted to Dr. Thomas J. Thekkumkara for the preparation of illustrations. We would like to acknowledge the contributions over the years made by our associates, collaborators and colleagues on the subject matter reviewed in this article. We thank Drs. Douglas Kerr, Isaiah Wexler, Gary Johanning and Deborah Crawford for critical reading of this manuscript.

REFERENCES

1. L. J. Reed, Multienzyme complexes, Acc. Chem. Res., 7:40 (1974).
2. R. M. Oliver and L. J. Reed, Multienzyme complexes, in: "Electron Microscopy of Proteins", J. R. Harris, ed., Academic Press, London, pp. 1 (1982).
3. L. J. Reed and R. O. Oliver, Structure-function relationships in pyruvate and α-ketoglutarate dehydrogenase complexes, Adv. Exp. Med. Biol., 148: 231 (1982).
4. O. H. Wieland, The mammalian pyruvate dehydrogenase complex: structure and regulation, Rev. Physiol. Biochem. Pharmacol. 96:123 (1983).
5. S. J. Yeaman, The mammalian 2-oxoacid dehydrogenases: a complex family, TIBS, 11:293 (1986).
6. O. de Marcucci and J. G. Lindsay, Component X: an immunologically distinct polypeptide associated with mammalian pyruvate dehydrogenase multi-enzyme complex, Eur. J. Biochem., 149:641 (1985).
7. J. M. Jilka, M. Rahmatullah, M. Kazemi and T. E. Roche, Properties of a newly characterized protein of the bovine kidney pyruvate dehydrogenase complex, J. Biol. Chem., 261:1858 (1986).
8. S. J. Yeaman, E. T. Hutcheson, T. E. Roche, F. H. Pettit, J. R. Brown, L. J. Reed, D. C. Watson and G. H. Dixon, Sites of phosphorylation on pyruvate dehydrogenase from bovine kidney and heart, Biochem., 17:2364 (1978).
9. M. Rahmatullah, J. M. Jilka, G. A. Radke and T. E. Roche, Properties of the pyruvate dehydrogenase kinase bound to and separated from the dihydrolipoyl transacetylase-protein X subcomplex and evidence for binding of the kinase to protein X, J. Biol. Chem., 261:6515 (1986).
10. R. M. Denton and A. P. Halestrap, Regulation of pyruvate metabolism in mammalian tissues, in: "Essays in Biochemistry", P. N. Campbell and R. D. Marshall, eds., 15:37 (1979).
11. H. H. M. Dahl, S. M. Hunt, W. M. Hutchison and G. K. Brown, The human pyruvate dehydrogenase complex. Isolation of cDNA clones for the E1α subunit, sequence analysis, and characterization of the mRNA, J. Biol. Chem., 262:7398 (1987).
12. K. Koike, S. Ohta, Y. Urata, Y. Kagawa and M. Koike, Cloning and sequencing of cDNAs encoding the α and β subunits of human pyruvate dehydrogenase, Proc. Natl. Acad. Sci. USA, 85:41 (1988)
13. L. DeMeirleir, N. MacKay, A. M. L. H. Wah and B. H. Robinson, Isolation of a full-length complementary DNA coding for human E1α subunit of the pyruvate dehydrogenase complex, J. Biol. Chem. 263:1991 (1988).
14. L. Ho, A. A. Javed, R. A. Pepin, T. J. Thekkumkara, C. Raefsky, J. E. Mole, A. M. Caliendo, M. S. Kwon, D. S. Kerr and M. S. Patel, Identification of a cDNA clone for the β-subunit of the pyruvate dehydrogenase component of human pyruvate dehydrogenase complex, Biochem. Biophys. Res. Comm., 150:904 (1988).
15. I. D. Wexler, D. S. Kerr, L. Ho, M. M. Lusk, R. A. Pepin, A. A. Javed, J. E. Mole, B. W. Jesse, T. J. Thekkumkara, G. Pons and M. S. Patel, Heterogeneous expression of protein and mRNA in pyruvate dehydrogenase deficiency, Proc. Natl. Acad. Sci. USA, 85:7336 (1988).

16. S.Matuda, S.Matuo, K.Nakano and T.Saheki, Molecular cloning ofcDNA for rat liver lipoate acetyltransferase, a component of pyruvate dehydrogenase complex, <u>Biochem.Biophys.Res.Comm.</u>, 142:953 (1987).

17. T.J.Thekkumkara, B.W.Jesse, L.Ho, C.Raefsky, R.A.Pepin, A.A.Javed, G. Pons and M.S.Patel, Isolation of a cDNA clone for the dihydrolipo-amide acetyltransferase component of the human liver pyruvate dehydrogenase complex, <u>Biochem.Biophys.Res.Comm.</u>, 145:903 (1987).

18. T.J.Thekkumkara, L.Ho, I.D.Wexler, G.Pons, T.C.Liu and M.S.Patel, Nucleotide sequence of a cDNA for the dihydrolipoamide acetyl-transferase component of human pyruvate dehydrogenase complex, <u>FEBS Letters</u>, 240:45 (1988).

19. R.L.Coppel, L.J.McNeilage, C.D.Surh, J.Van de Water, T.W.Spithill, S.Whittingham and M.E.Gershwin, Primary structure of human M2 mitochondrial autoantigen of primary biliary cirrhosis: dihydro-lipoamide acetyltransferase, <u>Proc.Natl.Acad.Sci.USA</u> 85:7317 (1988).

20. M.E.Gershwin, I.R.Mackay, A.Sturgess and R.L.Coppel, Identification and specificity of a cDNA encoding the 70-KD mitochondrial antigen recognized in primary biliary cirrhosis, <u>J.Immun.</u>, 138:3525 (1987).

21. G.Otulakowski and B.H.Robinson, Isolation and sequence determination of cDNA clones for porcine and human lipoamide dehydrogenase, <u>J.Biol.Chem.</u> 262:17313 (1987).

22. G.Pons, C.Raefsky-Estrin, D.J.Carothers, R.A.Pepin, A.A.Javed, B.W. Jesse, M.K.Ganapathi, D.Samols and M.S.Patel, Cloning and cDNA sequence of the dihydrolipoamide dehydrogenase component ofhuman α-ketoacid dehydrogenase complexes, <u>Proc.Natl.Acad.Sci.USA</u> 85:1422 (1988).

23. P.H.Sugden, A.L.Kerbey, P.J.Randle, C.A.Waller and K.B.M.Reid, Amino acid sequences around the sites of phosphorylation in the pig heart pyruvate dehydrogenase complex, <u>Biochem.J.</u>, 181:419 (1979).

24. D.M.Bleile, M.L.Hackert, F.H.Pettit and L.J.Reed, Subunit structure of dihydrolipoyl transacetylase component of pyruvate dehydrogenase complex from bovine heart, <u>J.Biol.Chem.</u> 256:514 (1981).

25. T.J.Thekkumkara, G.Pons, S.Mitroo, J.E.Jentoft and M.S.Patel, Molecular biology of human pyruvate dehydrogenase complex: structural aspects in the E2 and E3 components, <u>Annals.NY.Acad.Sci.</u> 573:113 (1989).

26. J.R.Guest, H.M.Lewis, L.D.Graham, L.C.Packman and R.N.Perham, Genetic reconstruction and functional analysis of the repeating lipoyl domains in the pyruvate dehydrogenase multienzyme complex of *Escherichia coli*, <u>J.Mol.Biol.</u>, 185:743 (1985).

27. D.J.Carothers, G.Pons and M.S.Patel, Dihydrolipoamide dehydrogenase: functional similarities and divergent evolution of the pyridine nucleotide-disulfide oxidoreductases, <u>Arch.Biochem.Biophys.</u>, 268: 409 (1989).

28. G.E.Schulz, R.H.Schirmer, W.Sachsenheimer and E.F.Pai, The structure of the flavoenzyme glutathione reductase, <u>Nature</u>, 273:120 (1978).

29. S.E.Marshall, J.G.McCormack and R.M.Denton, Role of Ca^{2+} ions in the regulation of intramitochondrial metabolism in rat epidydimal adipose tissue, <u>Biochem.J.</u>, 218:249 (1984).

30. Z.Damuni, J.S.Humphreys and L.J.Reed, Stimulation of pyruvate dehydrogenase activity by polyamines, <u>Biochem.Biophys.Res.Commun.</u>, 124:95 (1984).

31. W.A.Hughes and R.M.Denton, Incorporation of Pi into pyruvate dehydrogenase phosphate in mitochondria from control and insulin-treated adipose tissue, <u>Nature</u> 264:471 (1976).

32. R.M.Denton, W.A.Hughes, B.J.Bridges, R.W.Brownsey, J.G.McCormack and D.Stansbie, Regulation of mammalian pyruvate dehydrogenase by hormone, <u>in</u>: "Hormone and Cell Regulation", vol. 2, J.Dumont and J.Nunez, eds., Elsevier/North Holland Biomedical Press, Amsterdam, pp. 191 (1978).

33. I. Paetzke-Brunner, H. Schon and O. H. Wieland, Insulin activates pyruvate dehydrogenase by lowering the mitochondrial acetyl-CoA/CoA ratio as evidenced by digitonin fractionation of isolated fat cells, FEBS Lett. 93:307 (1978).

34. L. Jarret and J. R. Seals, Pyruvate dehydrogenase activation in adipocyte mitochondria by an insulin-generated mediator from muscle, Science, 206:1407 (1979).

35. S. Suzuki, T. Toyota, H. Suzuki and Y. Goto, Partial purification from human mononuclear cells and placenta plasma membrane of an insulin mediator which stimulates pyruvate dehydrogenase and supresses glucose-6-phosphatase, Arch. Biochem. Biophys. 235:418 (1984).

36. S. L. Macaulay and L. Jarret, Insulin mediator causes dephosphorylation of the α subunit of pyruvate dehydrogenase by stimulating phosphatase activity, Arch. Biochem. Biophys. 237:142 (1985).

37. A. R. Saltiel, Insulin generates an enzyme modulator from hepatic plasma membranes: regulation of adenosine 3',5'-monophosphate phosphodiesterase, pyruvate dehydrogenase, and adenylate cyclase, Endocrinology, 120:967 (1987).

38. O. Wieland, E. Siess, F. H. Schulze-Wethmar, H. G. V. Funcke and B. Winton, Active and inactive forms of pyruvate dehydrogenase in rat heart and kidney: effect of diabetes, fasting and refeeding on pyruvate dehydrogenase interconversion, Arch. Biochem. Biophys., 143:593 (1971).

39. P. J. Randle, P. H. Sudgen, A. L. Kerbey, P. M. Radcliffe and N. J. Hutson, Regulation of pyruvate oxidation and the conservation of glucose, Biochem. Soc. Sympos. 43:47 (1978).

40. O. H. Wieland, C. Patzelt and G. Loffler, Active and inactive forms of pyruvate dehydrogenase in rat liver, Eur. J. Biochem., 26:426 (1972).

41. C. Patzelt, G. Loffler and O. Wieland, Interconversion of pyruvate dehydrogenase in the isolated perfused rat liver, Eur. J. Biochem., 33:117 (1973).

42. T. H. Claus and S. J. Pilkis, Effect of dichloroacetate and glucagon on the incorporation of labeled substrates into glucose and on pyruvate dehydrogenase in hepatocytes from fed and starved rats. Arch. Biochem. Biophys., 182:52 (1977).

43. D. A. Hems, J. G. McCormack and R. M. Denton, Activation of pyruvate dehydrogenase in the perfused rat liver by vasopressin, Biochem. J., 176:627 (1978).

44. R. M. Denton and J. G. McCormack, Ca^{2+} transport by mammalian mitochondria and its role in hormone action, Am. J. Physiol., 249:E543 (1985).

45. R. Moreno-Sanchez and R. G. Hansford, Dependence of cardiac mitochondrial pyruvate dehydrogenase activity on intramitochondrial free Ca^{2+} concentration. Biochem. J., 256:403 (1988).

46. M. B. Weinberg and M. F. Utter, Effect of thyroid hormone on the turnover of rat liver pyruvate carboxylase and pyruvate dehydrogenase, J. Biol. Chem., 254:9492 (1979).

47. P. J. Randle, P. J. England and R. M. Denton, Control of the tricarboxylate cycle and its interaction with glycolysis during acetate utilization in rat heart, Biochem. J., 117:677 (1970).

48. A. L. Kerbey and P. J. Randle, Thermolabile factor accelerates pyruvate dehydrogenase kinase reaction in heart mitochondria of starved or alloxan diabetic rats, FEBS Letters. 127:188 (1981).

49. C-W. C. Hu, M. F. Utter and M. S. Patel, Induction of pyruvate dehydrogenase in 3T3-L1 during differentiation, J. Biol. Chem., 258:2315 (1983).

50. D. T. Chuang, C.-W. C. Hu and M. S. Patel, Induction of the branched-chain 2-oxo acid dehydrogenase complex in 3T3-L1 adipocytes during differentiation, Biochem. J., 214:177 (1983).

51. D. J. Carothers, G. Pons and M. S. Patel, Induction of dihydrolipoamide dehydrogenase in 3T3-L1 cells during differentiation, Biochem. J., 249:897 (1988).

52. M. S. Patel, C. Raefsky, C.-W. C. Hu and L. Ho, Modulation by dexamethasone

of the pyruvate dehydrogenase complex activity in 3T3-L1 adipocytes, Biochem.J., 226:607 (1985).

53. S. E. Knowles and F. J. Ballard, Pyruvate dehydrogenase activity in rat liver during development. Biol. Neonate, 24:41 (1974).

54. C. I. Chitra, J. M. Cuezva and M. S. Patel, Changes in the activity of 'active' pyruvate dehydrogenase complex in the newborn of normal and diabetic rats, Diabetologia, 28:148 (1985).

55. E. Serrano, A. M. Luis, P. Encabo, A. Alconada, L. Ho, M. S. Patel and J. M. Cuezva, Rapid postnatal induction of the pyruvate dehydrogenase complex in rat liver mitochondria, Annals.NY.Acad.Sci., 573:412 (1989).

56. J. M. Medina, J. M. Cuezva and F. Mayor. Non-gluconeogenic fate of lactate during the early neonatal period in the rat, FEBS Lett., 114:132 (1980).

57. J. M. Cuezva, E. S. Burkett, D. S. Kerr, H. M. Rodman and M. S. Patel, The newborn of diabetic rats, I. Hormonal and metabolic changes in the postnatal period, Pediatr.Res., 16:632 (1982).

58. J. E. Cremer and H. M. Teal, The activity of pyruvate dehydrogenase in rat brain during postnatal development, FEBS Lett. 39:17 (1974).

59. J. M. Land, R. F. G. Booth, R. Berger and J. B. Clark, Development of mitochondrial energy metabolism in rat brain, Biochem.J. 164:339 (1977).

60. G. D. A. Malloch, L. A. Munday, M. S. Olsen and J. B. Clark, Comparative development of the pyruvate dehydrogenase complex and citrate synthase in rat brain mitochondria, Biochem.J., 238:729 (1986).

61. R. F. Butterworth and J. -F. Giguere, Pyruvate dehydrogenase activity in regions of the rat brain during postnatal development, J.Neurochem. 43:280 (1984).

62. T. A. Milner, C. Aoki, K. -F. R. Sheu, J. P. Blass and V. M. Pikel, Light microscopic immunocytochemical localization of pyruvate dehydrogenase complex in rat brain: topological distribution and relation to cholinergic and catecholaminergic nuclei, J.Neurosci., 7:3171 (1987).

63. D. G. Morgan and A. Routtenberg, Brain pyruvate dehydrogenase: phosphorylation and enzyme activity altered by a training experience, Science, 214:470 (1981).

64. H. G. Coore and B. Field, Properties of pyruvate dehydrogenase of rat mammary tissue and its changes during pregnancy, lactation and weaning, Biochem.J. 142:87 (1974).

65. J. P. Blass, Inborn errors of pyruvate metabolism, in: "The Metabolic Basis of Inherited Disease", Fifth Edition, J. B. Stanbury, J. B. Wyngaarden, D. S. Fredickson, J. L. Goldstein, and M. S. Brown, eds., McGraw-Hill Book Co., New York, pp. 193 (1983).

66. B. H. Robinson, Inborn errors of metabolism leading to lactic acidemia, TIBS 151 (1982).

67. R. F. Butterworth, Pyruvate dehydrogenase deficiency disorders, in: "Cerebral Energy Metabolism and Metabolic Encephalopathy", D. W. McCandless, ed., Plenum Pub. Corp. pp. 121 (1985).

68. B. H. Robinson, Cell culture studies on patients with mitochondrial diseases: molecular defects in pyruvate dehydrogenase. J.Bioenergetics Biomembranes, 20:313-323(1988).

69. L. Ho, I. D. Wexler, D. S. Kerr and M. S. Patel, Genetic defects in human pyruvate dehydrogenase, Annals NY Acad.Sci. 573:347 (1989).

70. B. H. Robinson, H. MacMilan, R. Petrova-Benedict and W. G. Sherwood. Variable clinical presentation in patients with defective E1 component of pyruvate dehydrogenase complex. J.Pediatr. 111:525 (1987).

71. B. H. Robinson, J. Tailor and W. G. Sherwood, Deficiency of dihydrolipoyl dehydrogenase (a component of the pyruvate and α-ketoglutarate dehydrogenase complexes): a cause of congenital chronic lactic acidosis, Pediatr.Res., 11:1198 (1977).

72. R. Matalon, D. A. Stumpf, K. Michels, R. D. Hart, J. K. Parks and S. I. Goodman, Lipoamide dehydrogenase deficiency with primary lactic acidosis: favorable response to treatment with oral lipoic acid, J. Pediat., 104:65 (1984).

73. Y. Sakaguchi, M. Yoshino, S. Aramaki, I. Yoshido, F. Yamashita, T. Kuhara, I. Matsumoto, and T. Hayashi, Dihydrolipoyl dehydrogenase deficiency: a therapeutic trial with branched-chain amino acid restriction, Eur. J. Pediatr., 145:271 (1986).

74. K.-F. R. Sheu, C.-W. C. Hu and M. F. Utter, Pyruvate dehydrogenase complex activity in normal and deficient fibroblasts, J. Clin. Invest., 67:1463 (1981).

75. L. Ho, C.-W. C. Hu, S. Packman and M. S. Patel, Deficiency of the pyruvate dehydrogenase component in pyruvate dehydrogenase complex-deficient human fibroblasts, J. Clin. Invest., 78:844 (1986).

76. D. S. Kerr, L. Ho, C. M. Berlin, K. F. Lanoue, J. Towfighi, C. L. Hoppel, M. M. Lusk, C. M. Gondek and M. S. Patel, Systemic deficiency of the first component of the pyruvate dehydrogenase complex, Pediatr. Res., 22:312 (1987).

77. D. S. Kerr, S. A. Berry, M. M. Lusk, L. Ho and M. S. Patel, A deficiency of both subunits of pyruvate dehydrogenase which is not expressed in fibroblasts, Pediatr. Res., 24:95 (1988).

78. C. A. Wicking, R. D. Scholem, S. M. Hunt and G. K. Brown, Immunochemical analysis of normal and mutant forms of human pyruvate dehydrogenase, Biochem. J., 239:89 (1986).

79. N. McKay, R. Petrova-Benedict, J. Thoene, B. Bergen, W. Wilson and B. Robinson, Lacticacidaemia due to pyruvate dehydrogenase deficiency, with evidence of protein polymorphism in the α-subunit of the enzyme, Eur. J. Pediatr., 144:445 (1986).

80. G. K. Brown, R. D. Scholem, S. M. Hunt, J. R. Harrison and A. C. Pollard, Hyperammonaemia and lactic acidosis in a patient with pyruvate dehydrogenase deficiency, J. Inher. Metab. Dis., 10:359 (1987).

81. A. Kitano, I. Akaboshi, F. Endo, I. Matsuda, Y. Okano, Y. Hase, Y. Nagao, S. Kamoshita, S. Miyabayashi and K. Narisawa, Immunochemical evidence of pyruvate dehydrogenase (E1) deficiency, J. Inher. Metab. Dis., 11:329 (1988).

82. M. A. Birch-Machin, I. M. Shepherd, M. Solomon, S. J. Yeaman, D. Gardner-Medwin, H. S. A. Sherratt, J. G. Lindsay, A. Ansley-Green and D. M. Turnbull, Fatal lactic acidosis due to deficiency of E1 component of the pyruvate dehydrogenase complex, J. Inher. Metab. Dis., 11:207 (1988).

83. Y. Kuroda, M. Ito, K. Toshima, E. Takeda, E. Naito, T. J. Hwang, T. Hashimoto, M. Miyao, M. Masuda, K. Yamashita, T. Adachi, Y. Suzuki and K. Nishiyama, Treatment of chronic congenital lactic acidosis by oral administration of dichloroacetate, J. Inher. Metab. Dis., 9:244 (1986).

84. S. Sorbi, E. D. Bird and J. P. Blass, Decreased pyruvate dehydrogenase complex activity in Huntington and Alzheimer brain, Ann. Neurol., 13:72 (1983).

85. K.-F. R. Sheu, Y.-T. Kim, J. P. Blass and M. E. Weksler, An immunological study of the pyruvate dehydrogenase deficit in Alzheimer's disease brain, Ann. Neurol., 17:444 (1985).

86. I. Ghadiminejad and H. Baum, Evidence for the cell-surface localization of antigens cross-reacting with the "mitochondrial antibodies" of primary biliary cirrhosis, Hepatology, 7:743 (1987).

87. M. E. Gershwin, R. L. Coppel and I. R. Mackay, Primary biliary cirrhosis, and mitochondrial autoantigens - Insights from molecular biology, Hepatology, 8:147 (1988).

88. S. J. Yeaman, S. P. M. Fussey, D. J. Danner, O. F. W. James, D. J. Mutimer and M. F. Bassendine, Primary biliary cirrhosis: Identification of two major M2 mitochondrial autoantigens, Lancet i:1067 (1988).

89. S. P. M. Fussey, J. R. Guest, O. F. W. James, M. F. Bassendine and S. J. Yeaman, Identification and analysis of the major M2 autoantigens in primary

biliary cirrhosis, Proc. Natl. Acad. Sci. USA. 85:8654 (1988).

90. J. Van de Water, M. E. Gershwin, P. Leung, A. Ansari and R. L. Coppel, The autoepitope of the 74-KD mitochondrial autoantigen of primary biliary cirrhosis corresponds to the functional site of dihydrolipoamide acetyltransferase, J. Exp. Med., 167:1791 (1988).

METABOLIC ASPECTS OF DEVELOPMENT

FUEL SUPPLY TO THE BRAIN DURING THE

EARLY POSTNATAL PERIOD

José M. Medina, Emilio Fernández, Juan P. Bolaños,
Carlos Vicario and Carmen Arizmendi

Departamento de Bioquímica y Biología Molecular
Facultad de Farmacia
Universidad de Salamanca, Spain

INTRODUCTION

Perinatal time spans over three periods so called the gestational, suckling and weaning period. Gestation ends in the labor which interrupts maternal supply of metabolic substrates giving way to an autonomous metabolism. The fuel supply is rapidly regained by the milk nutrients which supply the newborn with the energy and carbon skeletons required for its development. However, immediately after birth and before the onset of suckling takes place there is a time lapse during in which the newborn has to withstand a unique starvation. This period, herewith referred to the presucking period, is a consequence of incompatibility of the first strokes of ventilation and suckling itself, although more probably it is due to the time elapsed to the compulsory change in the metabolic substrates from the womb to the mammary glands. Nevertheless, during the presuckling period, the newborn has to survive from its own reserves during an unique period of stress and vulnerability. To get through this short but difficult period, the fetus accumulates important energy reserves and adapts its metabolic machinery to the expected changes in its surroundings; the period devoted to this preparation can be called "prepartum" (Figure 1).

MAIN FUELS DURING THE EARLY NEONATAL PERIOD

During late gestation the fetus accumulates a substantial amount of glycogen in different tissues (1). Although muscle glycogen reserves are mobilized inmediately after delivery, the liver is the only tissue which significantly contributes to supplying glucose to the different neonatal tissues. Thus, the accumulation of glycogen in liver is substantial (2,3), reaching concentrations about twice those found in the livers of well-fed adult rats. However, most of the liver glycogen is rapidly depleted within the first 6 h (4-6). Despite the high rate of liver glycogenolysis, normal glycemia is only transiently regained (Figure 2), showing that glucose is actively utilized during the early neonatal period. However, the decrease in availability of glucose is balanced by the rise of plasma free fatty acids (Figures 3,4) which supply energy to most of the neonatal tissues. In addition, the occurrence of an active ketogenesis at term (7) provides the brain with ketone bodies (Figures 3,4) as alternative substrates for

Endocrine and Biochemical Development of the Fetus and Neonate
Edited by J. M. Cuezva *et al.*
Plenum Press, New York, 1990

175

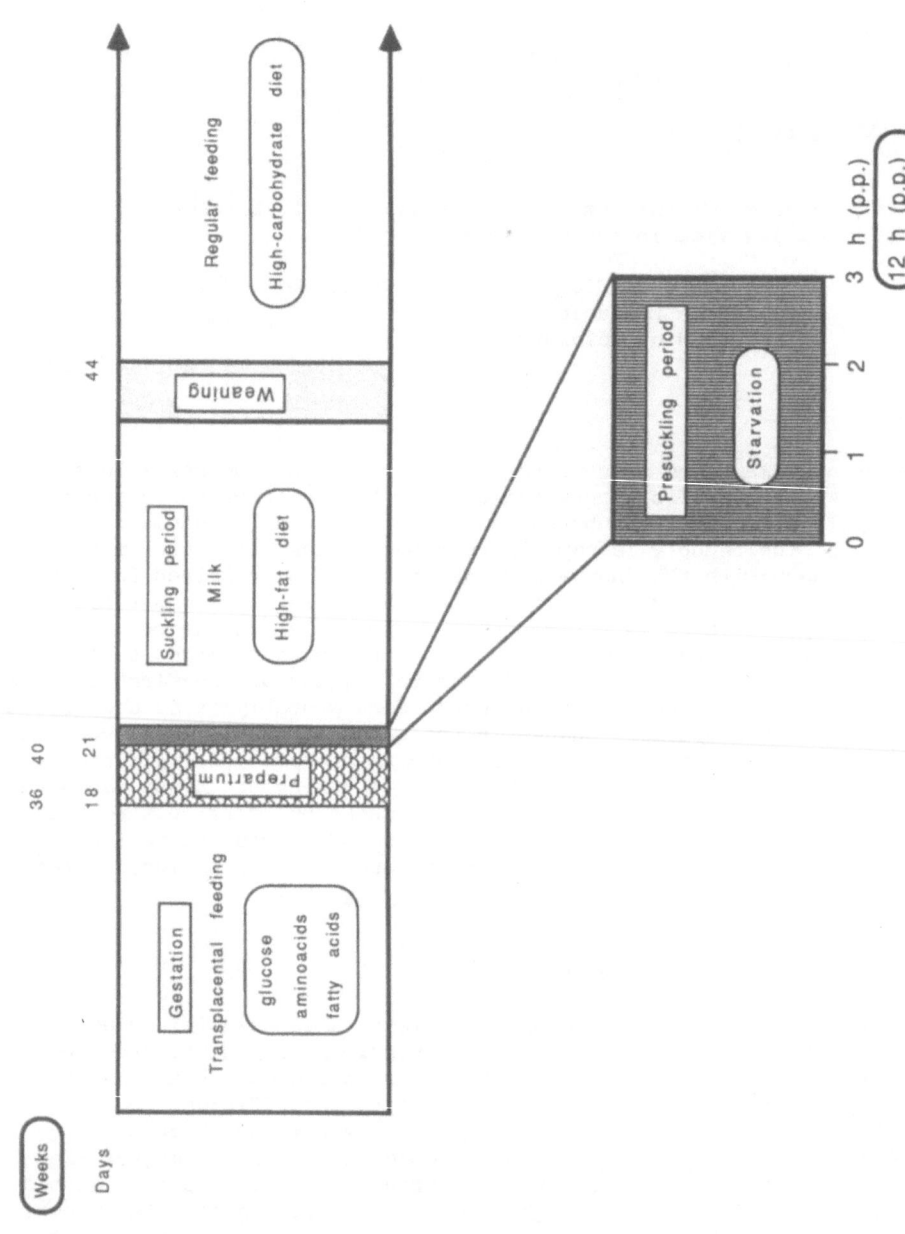

Fig. 1. Perinatal period in man and rat. Times indicated refer to post-conception. Time indicated in vound lables (weeks and hours) are for humans, p.p.: post-partum.

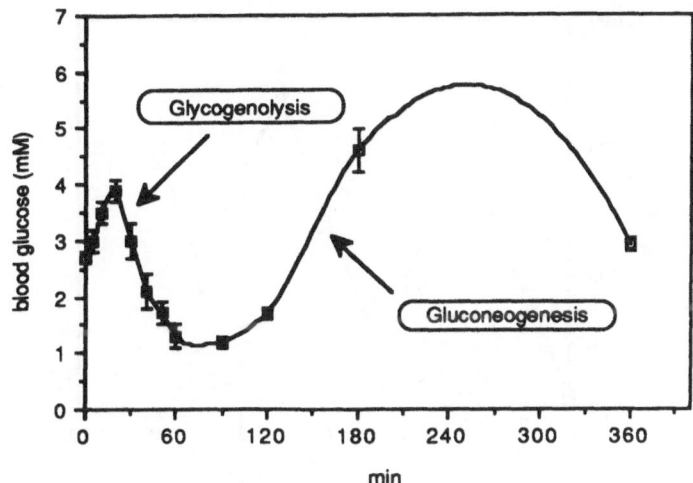

Fig. 2. Plasma glucose concentrations in the rat during
the postnatal period. Arrows show when the
processes are probably initiated. Fetuses were
delivered by cesarian section after cervical
dislocation of the mother. Newborns were
maintained with their mother up to the time of
the experiments. Fetuses and newborns were
decapited, exsanguinated and the livers freeze-
clamped. Metabolites were determined as
described elsewhere (12). Results are means ±
SEM (n=8-10).

glucose (8). However, the substrate supply is interrupted during the
presucking period. Thus, in human baby (Figure 4) free-fatty acid release
from the adipose tissue is delayed (9,10) together with the inhibition of
ketone-body synthesis due to a lack of carnitine which is supplied with the
milk (11); in the case of the rat the lack of adipose tissue at birth
prevents free-fattty acid supply until the onset of lactation (12) because
they come from milk triacylglycerols. During this lag period, the newborn
has to depend from its own reserves, increasing the efficiency of the
metabolic machinery in order to survive this untimely period of starvation.
Indeed, mortality is significant (3%) during the first 2 h of extrauterine
life, but negligible 1 h later (13). Such evidence points to the presucking
period as a high risk phase in adaptation to extrauterine life.

Consequently, the scarce glucose available during the presuckling
period has to be supplemented with other metabolic substrates to account
for the energy expenditures. Thus, glycerol (Figure 5) might play a role as
an energy substrate in these circumstances because its concentrations
increase after birth, mantaining high levels throughout the suckling period
(2). However, blood glycerol concentrations are very low during the
presuckling period (Figure 5, inset) because it is derived from the
hydrolysis of milk triacylglycerols (12). In addition, alanine
concentrations in the blood of newborns at delivery are 2-fold higher than
those found in adult animals (14,15). Moreover, alanine is rapidly removed
within the presuckling period. However, energy and carbon skeletons yielded
by alanine metabolism under these circumstances may be insignificant
because its concentrations are very low as compared to other relevant
substrates (14,15).

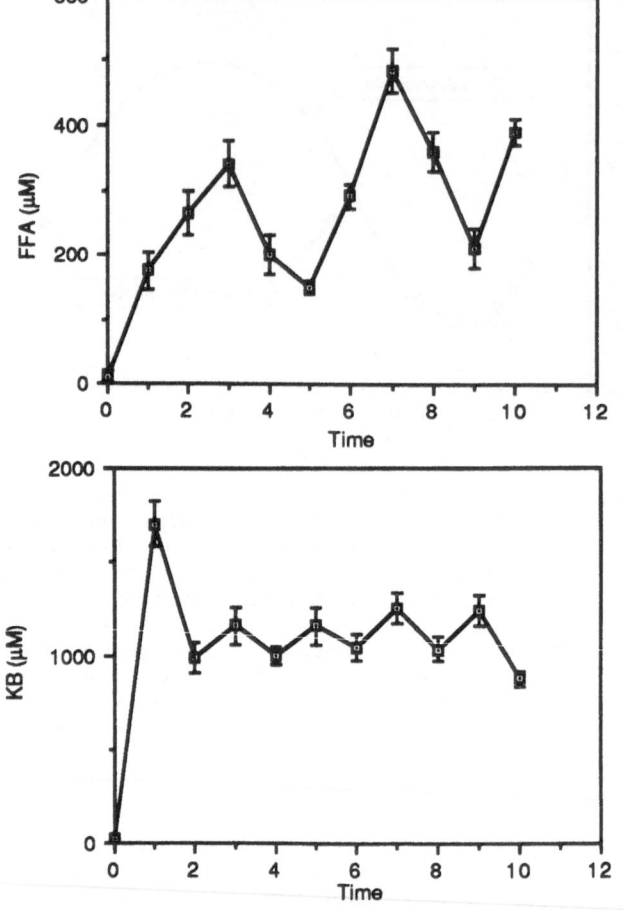

Fig. 3. Plasma free fatty acids and ketone bodies concentrations during the postnatal period. See legend
of Figure 2 for experimental details (2).

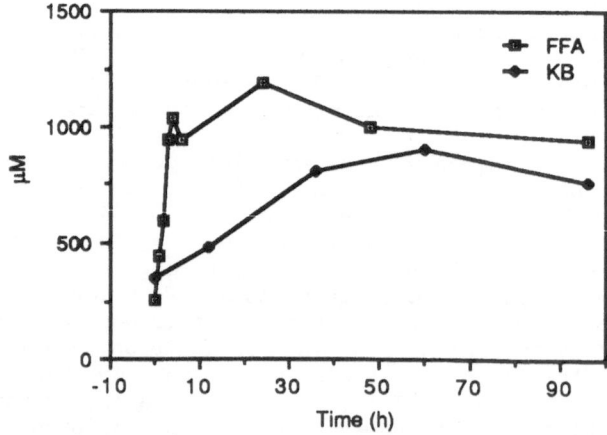

Fig. 4. Plasma free fattty acids and ketone bodies
concentrations in human baby during the postnatal period. Data collected by Exton (11).

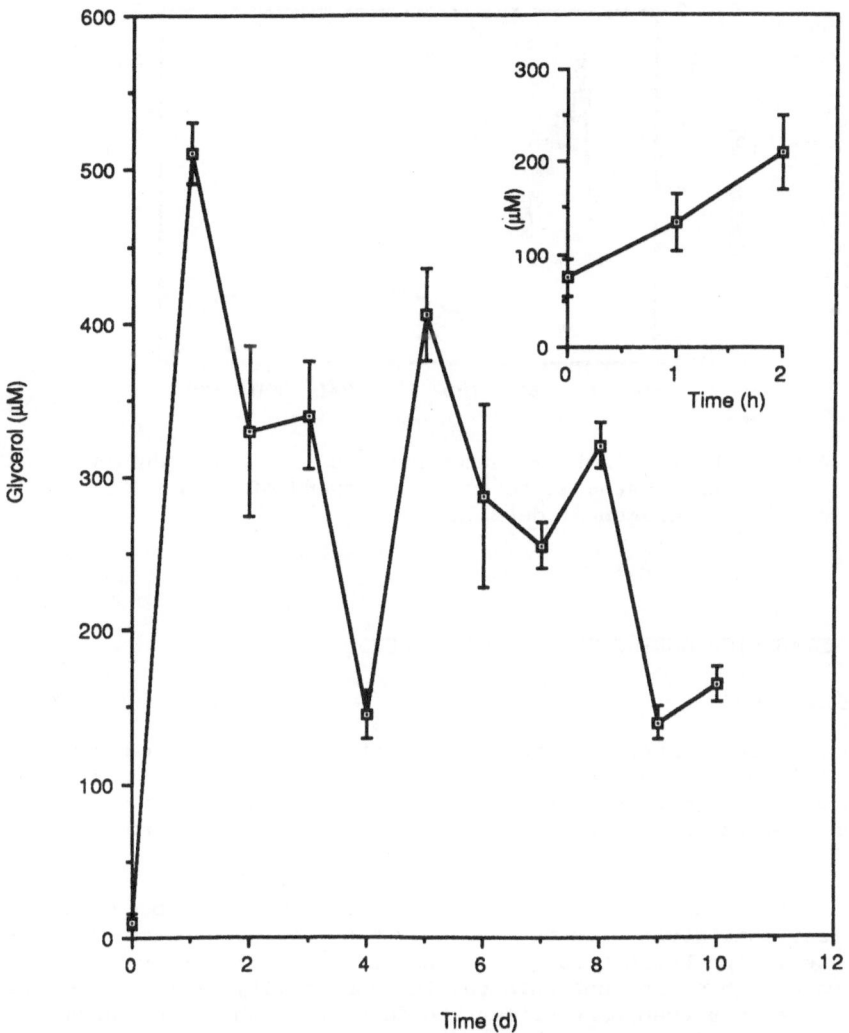

Fig. 5. Plasma glycerol concentrations in the rat during
the postnatal period. See legend of Figure 2 for
experimental details (2).

On the other hand, lactate may play a relevant role in the energy
homeostasis of the newborn. Figure 6 shows the time course of plasma
lactate concentrations during the early neonatal period (2,14,16). The high
plasma lactate concentration observed at birth is further increased
immediately after delivery. However, once lactate concentrations peak, a
sharp reduction is observed which results in the removal of most of blood
lactate within the first two h of extrauterine life (Figure 6). These
results show that most blood lactate is removed before the onset of
suckling takes place (i.e., during the presuckling period). This finding
emphasises the importance of lactate as a metabolic fuel for the newborn
immediately after delivery. Indeed, during the first 2 postnatal h, liver
glycogenolysis is not initiated (4) or its rate is quite low (10,12);
glycemia (Figure 2) is very low (4,12,14), and the plasma concentrations of
other relevant substrates are negligible (2,4,12,14,15). Consequently,
lactate may play an important role as a source of energy and carbon
skeletons for neonatal tissues during the presuckling period.

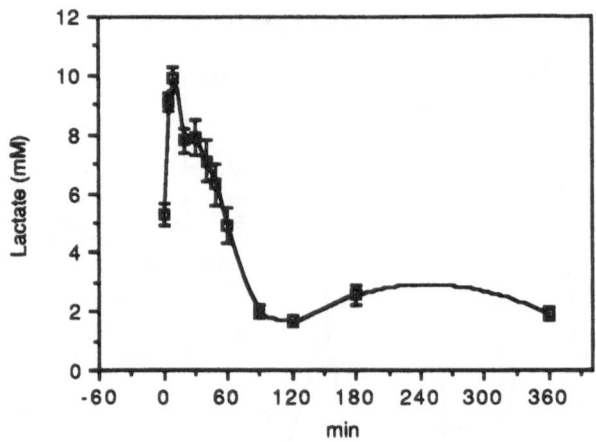

Fig. 6. Plasma lactate concentrations in the rat during
the postnatal period. See legend of Figure 2 for
experimental details.

LACTATE METABOLISM DURING THE PERINATAL PERIOD

Origin of Lactate

During late gestation fetal blood lactate can come from several
sources: i) from the mother transferred through placenta, ii) from the
placenta where it is formed from glucose and presumably from alanine, and
iii) formed by the fetus itself from glucose and probably from alanine
(Figure 7).

Thus, transfer of maternal lactate to the fetus has been reported
(17), a fact consistent with the recently reported occurrence of a carrier
for lactate in the brush border membrane of human (18) and rat placenta
(19). However, this does not rule out the possibility of the occurrence of
a reported reverse transport of lactate fetus-to-mother (20) which may be
active when lactate accumulates into the fetal compartment; lactate, once
in the maternal compartment, can be converted into glucose through
gluconeogenic pathway. Nonetheless, the high activities of the main
gluconeogenic enzymes during late pregnancy (21), in particular liver and
kidney phosphoenolpyruvate carboxykinase, together with the observed
increase in the maternal rate of gluconeogenesis from lactate during late
gestation (Figure 8) suggest that maternal lactate is mostly converted to
glucose, a fact that can prevent its transfer to the fetal compartment. If
so, most of lactate transferred to the fetus may be of placental origin.

The formation of lactate from glucose in human and rat placenta is
well documented *in vivo* (22-24) and *in vitro* (25-27). However, synthesis
of lactate from alanine in placenta needs futher clarification. Thus,
Palacin et al. (28) has reported the conversion of alanine into lactate in
isolated perfused womb with their rat fetuses. However, in such a
preparation it is not possible to decide in wich compartment (placenta or
fetus) alanine is converted into lactate. Moreover, we have not been able
to demonstrate lactate synthesis from alanine in rat placenta *in vitro*
(27), a fact consistent with the idea that alanine conversion into lactate
is probably occurring in the fetal compartment. The occurrence of low
activity of alanine amminotransferase in rat placenta (29) supports this
idea.

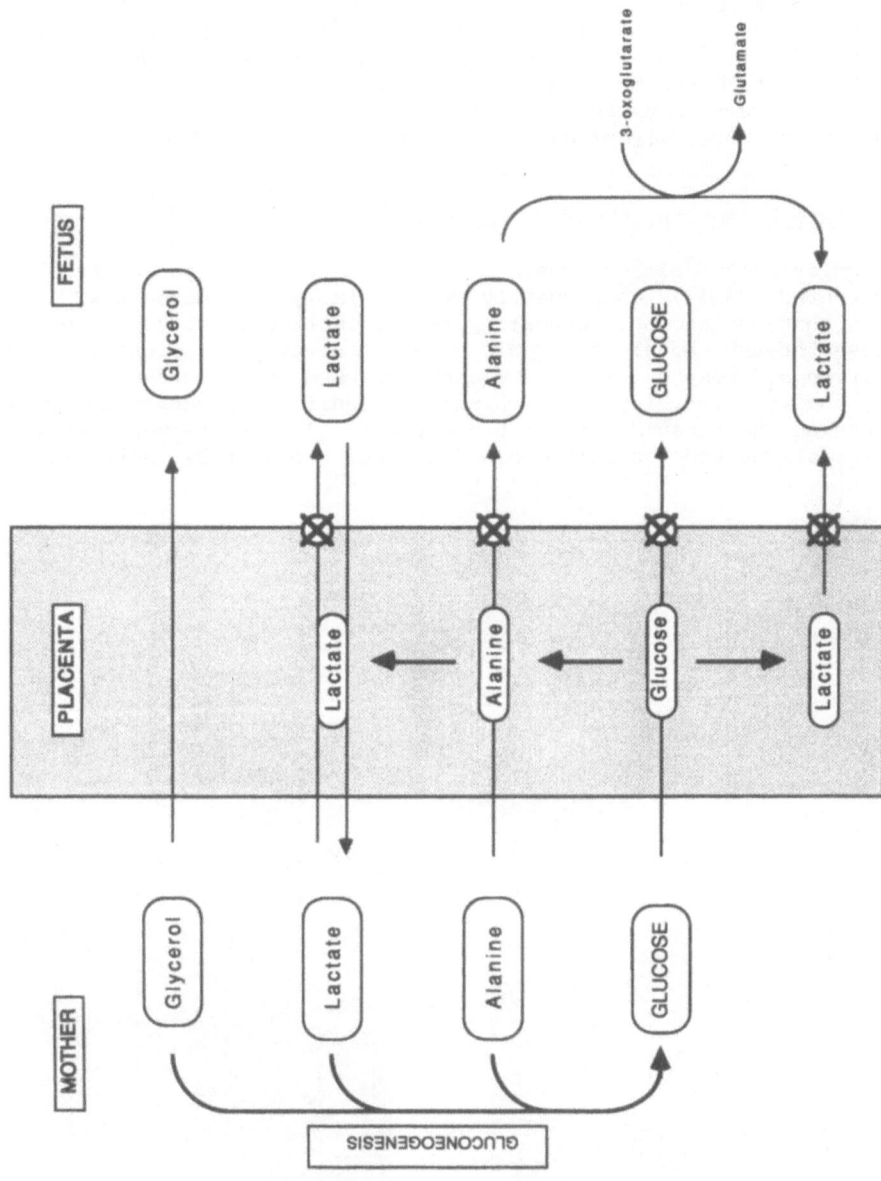

Fig. 7. Mother-conceptus relationships for the main metabolic substrates during late gestation.

181

Glucose is probably the main source of lactate in the fetal compartment because fetal tissues have an active anaerobic metabolism. Therefore, as mentioned above, some lactate may come from alanine, which would be converted to lactate in the fetal liver by alanine aminotransferase and lactate dehydrogenase-catalyzed reactions. However, lactate synthesis from alanine may be limited by the low activity of liver alanine aminotransferase during this period.

As a consequence of its synthesis in placenta and/or fetal tissues lactate accumulates in fetal blood during late gestation, reaching levels ten times higher than those observed in adult animals (Figure 6). The origin of the lactate accumulated in the newborn blood immediately after delivery is difficult to assess. However, it may be suggested that lactate is formed via anaerobic glycolysis from glucose released by muscle glycogen.

Fate of Lactate During the Presuckling Period

The relevance of gluconeogenesis during the neonatal period has been well demostrated (30-32). Consequently, we initially considered the possibility that during the presuckling period lactate might be converted into glucose through the gluconeogenic pathway. However, we found (10) that the activities of phosphoenolpyruvate carboxykinase and fructose-1,6-bisphosphatase were very low at birth and no significant changes could be observed during the first 2 h after birth (Table 1); glucose-6-phosphatase activity was almost undetectable during the same period (12). Actually,

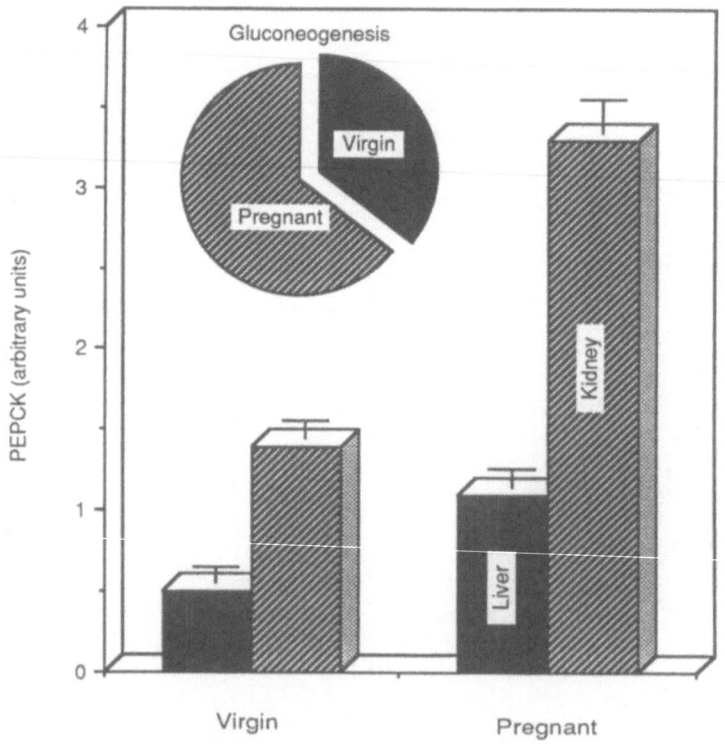

Fig. 8. Liver and kidney phosphoenolpyruvate carboxykinase (PEPCK) and the rate of gluconeogenesis from lactate in virgin and pregnant (last day of gestation) rats (21).

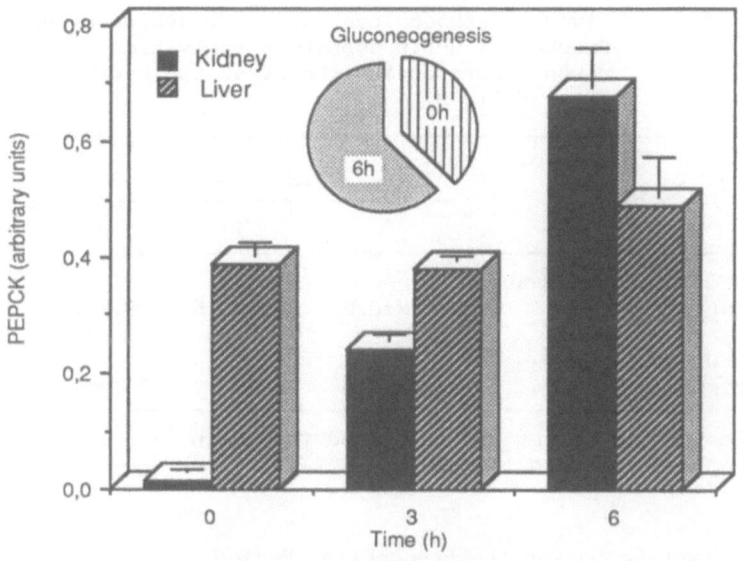

Fig. 9. Liver and kidney phosphoenolpyruvate carboxykinase
(PEPCK) and the rate of gluconoegenesis from
lactate in newborn rats during the first hours
after delivery (6).

very low activities of liver and kidney phosphoenolpyruvate carboxykinase
are observed (Figure 9) until the 6 h after birth (6,33), suggesting that
during the presuckling period there is not enough activity of key gluconeogenic enzymes (see also 34) to support an active gluconeogenesis.
Futhermore, very low radioactivity was incorporated in liver glycogen and
plasma glucose 2h after the injection of U-^{14}C-lactate, suggesting that
gluconeogenesis is negligible during the presuckling period (35). We have
also been able to evaluate the actual rates of gluconeogenesis from lactate
under these circumstances (6) which were very low during the first 3 h
after birth (Figure 9), remaining below that observed in well-fed adult
rats (6,36). Consequently, it may be inferred that gluconeogenesis is not
qualitatively important during the early neonatal period, and that lactate
is being utilized through an alternative pathway.

Dependence of Lactate Utilization on Oxygen Supply

It has been observed that mild premature newborns (12) and severe
premature newborns (37) show a resistance to plasma lactate utilization
during the first 1-2 h of extrauterine life. The delay observed in lactate
utilization coincides with the occurrence of hypoxemia (38), suggesting
that postnatal lactate utilization depends on blood oxygen concentrations. A
statistically significant negative correlation is observed between
naturally occurring lactate and oxygen concentrations in the blood of
normal and prematurely delivered newborns rats during the early neonatal
period (38). Futhermore, hypoxia brought about by decreasing oxygen
concentrations in the gas breathed by the newborns (15), induces the
accumulation of lactate in plasma, whereas hyperoxia increases the rate of
plasma lactate removal (39). These results clearly show that blood lactate
removal depends on oxygen availability, suggesting that lactate may undergo
oxidation during this period. In addition, the dependence of lactate
utilization on the oxygen supply is consistent with the idea that the
increase in the oxygen supply occurring with the onset of ventilation may
be the sole signal for lactate utilization after birth.

Table 1. Liver phosphoenolpyruvate carboxykinase and
fructose-1,6-biphosphatase activities in
newborn rats during the first two hours
after delivery

	TIME (hours)		
	0	1	2
Phosphoenolpyruvate carboxykinase	7.7±0.9	7.0±0.5	8.7±0.7
Fructose-1,6-bisphosphatase	43.3±3.6	35.3±2.7	31.1±2.8

Enzyme activities are expressed in $mU \cdot min^{-1} \cdot g^{-1}$
Results are mean ± S.E.M. (n=5-7) (12).

Oxidation of Lactate During the Presuckling Period

Since gluconeogenesis is insignificant during the presuckling period
and plasma lactate removal largely depends on oxygen availability, we
considered the possibility that plasma lactate might be oxidized to CO_2
during this period. We therefore trapped the CO_2 exhaled by the newbors
over 2 h period after postnatal injection of radioactively labelled lactate
(35). The results of these experiments (Table 2) showed that about 65% of
the newborn's total lactate (lactate body pool size) is respired during the
first 2 h of extrauterine life. The lactate respired by prematurely
delivered newborns was substantially lower (about 32%), in agreement with
the time course of plasma lactate observed under these circumstances (12).
These results clearly suggest that lactate is oxidized during the
presuckling period through the tricarboxylic acid cycle. This provides a
system for the complete utilization of muscle glycogen once it has been
converted into lactate by anaerobic glycolysis. In addition, the lactate
accumulated as a reserve by the fetus during late gestation can also be
utilized by terminal oxidation, increasing the energy yield.

Table 2. The fate of lactate in term and preterm newborn rats during
the early neonatal period

Time from delivery (h)	Radioactive CO_2 evolved $dpm \times 10^{-3}$/h	Lactate body pool size (bps)(μmol)	Lactate respired (μmol/h)	Lactate produced (μmol/h)
0 term	–	46.7	–	–
1	2096±27	48.0	13.3	+14.6
2	1758±50	27.9	16.9	-3.2
0 preterm	–	22.2	–	–
1	423±26	27.1	1.3	+6.2
2	1118±44	26.1	5.7	+4.7

Results are means ± SEM. Data without SEM are calculated (35).

TISSUE DISTRIBUTION OF LACTATE UTILIZATION

Since lactate removal takes place at a very high rate, several neonatal tissues have been suggested as site for lactate utilization. Thus, it has been shown that neonatal lung (40), heart (41) and liver (42) utilizes lactate for energy production and/or lipogenic purposes. Recently, it has been shown that brown adipose tissue from adult rats consumes lactate as a metabolic substrate (43). Accordingly, it is tempting to speculate that brown adipose tissue may utilize lactate during the perinatal period. Thus, during late gestation, the fetus may utilize lactate as a source of carbon skeletons for the synthesis of fat reserves of brown adipose tissue. However, special attention has been paid to lactate utilization by the brain, probably because this organ has to continue its development even under the starvation condition occurring during the presuckling period. Lactate utilization by the brain has been reported in suckling newborn rats (44), in fetal rats (45), in newborn dogs (46), in fetal sheep liver (47), porcine placenta (48) and in glucose-6-phosphatase-deficient human babies (49). Likewise, we have shown that lactate plays an important role as a metabolic substrate for the brain during the early neonatal period (50). We have also shown that lactate is preferentially used than glucose and ketone bodies which are the main substrates for the brain during the suckling period (51). It may therefore be suggested that lactate is utilized by a wide range of tissues playing, a singular role during the perinatal period.

LACTATE UTILIZATION BY THE NEONATAL BRAIN

Because lactate is not a gluconeogenic substrate during the presuckling period but rather is oxidized through the tricarboxylic acid cycle, we were prompted to investigate the possibility that in these circumstances it might utilized by the neonatal brain. Accordingly, neonatal brain slices were incubated with increasing concentrations of radioactively labelled lactate and the CO_2 evolved was measured to calculate lactate oxidation (50). As shown in Figure 10, the rate of oxidation approached saturation at lactate concentrations of about 10 mM. It is noteworthy that this concentration is the highest found in newborns under physiological circumstances (Figure 6), a fact that strongly suggests that the lactate oxidation observed in our experiments may be of physiological relevance. Additionally, lactate oxidation strictly depends on oxygen concentration since the radioactive CO_2 evolved was drastically reduced when the oxygen was substituted by N_2 in the gas phase (50). This is in agreement with the previously observed dependence of plasma lactate removal on oxygen availability (38). In addition, lactate oxidation in neonatal brain slices (Figure 10) was 10 times higher than maximal glucose oxidation (5 mM glucose) and 3 times higher than maximal D-3-hydroxy-butyrate oxidation (2 mM DL-3-hydroxybutyrate). These results (51) point to the importance of lactate as an energy substrate for the brain during the presuckling period. In addition, the rate of lipogenesis from lactate in the same experimental conditions (Figure 10) was twice that of glucose and 5-fold higher than that from 3-hydroxybutyrate, suggesting that lactate is also an excellent lipogenic substrate for the perinatal brain. Consequently, it could be suggested that lactate is the main substrate for the brain during the early neonatal period, being utilized not only as a source of energy but also supplying the brain with carbon skeletons for the synthesis of cerebral structures (51). The metabolic significance of lactate utilization by the neonatal brain and its dependence on the oxygen supply must be stressed taking into account their possible relevance in premature and anoxic babies.

Lactate Competition With Glucose and 3-Hydroxybutyrate

During gestation glucose is the main metabolic substrate for the brain although it is occasionally supplemented with ketone bodies (45). However, during the suckling period since glucose is scarce in the milk of most of species it has to be largely supplemented with ketone bodies. Therefore, during the presuckling period the glucose supply is very scarce because the transplacental supply has been interrupted. Since ketone bodies are not yet available due to the delay in the initiation of ketogenesis, lactate accounts for the most of energy expenses under these circumstances. This prompted us to investigate the possible competition of lactate with the main metabolic substrates, namely glucose and 3-hydroxybutyrate (51).

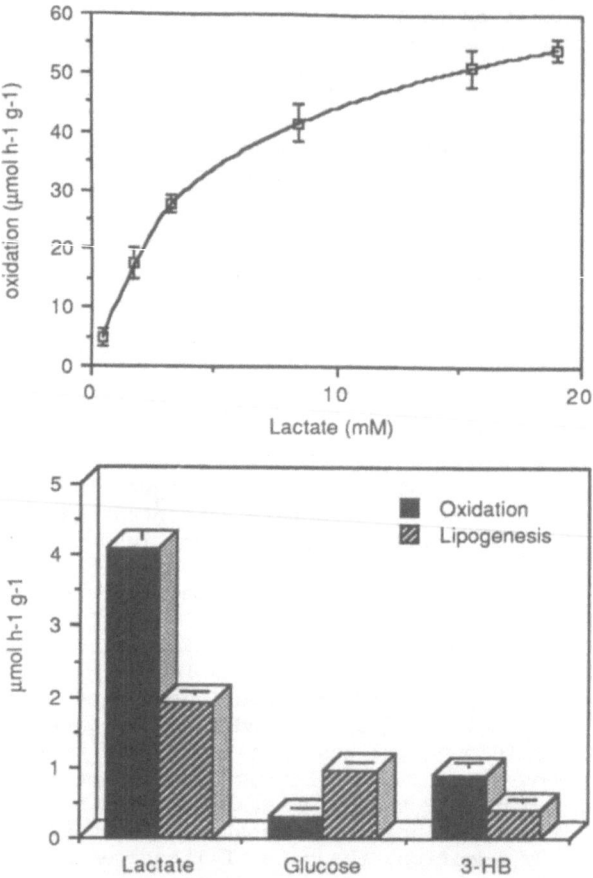

Fig. 10. Dependence of lactate oxidation on substrate concentration (upper panel) and utilization (lower panel) of lactate (10 mM), glucose (5 mM) and D-3-hydroxybutyrate (2 mM), by slices from newborn rat brain. Newborn (21.5 days) brain slices (70-80 mg) were incubated at 37°C in phosphate-physiological saline containing 0.5 μCi of labeled substrate and radioactive CO_2 was trapped in KOH. Incorporation into lipids was measured by the radiactivity counted in the washed chloroform: methanol extracts of the slices previously incubated. Results are means ± SEM (n=20-30) (50,51).

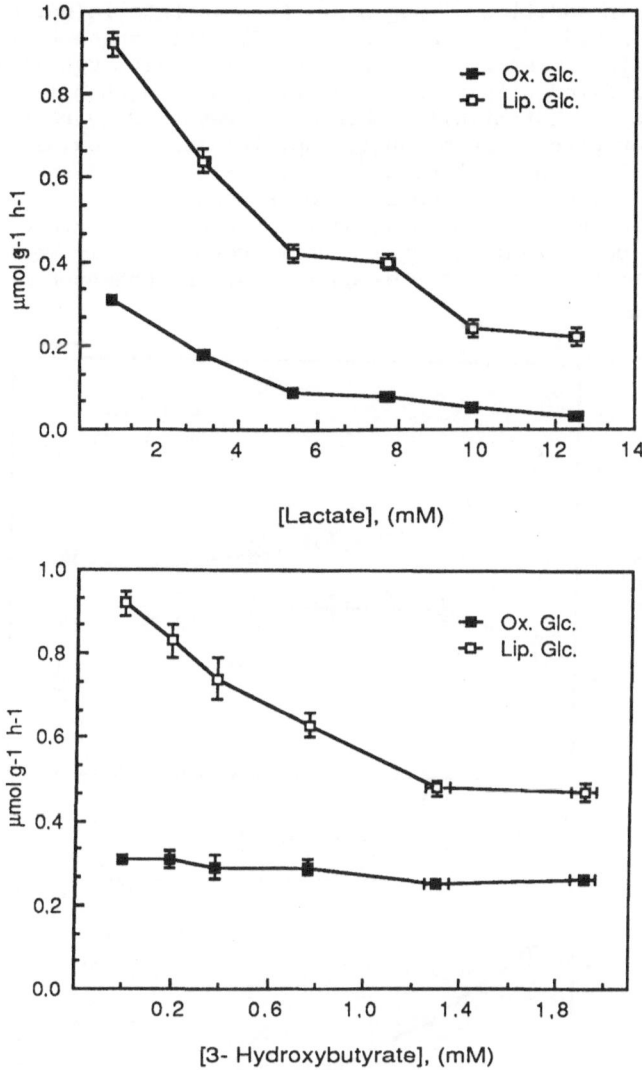

Fig. 11. Inhibition of glucose utilization in presence of
lactate (upper panel) and D-3-hydroxybutyrate (lower
panel) by neonatal brain slices incubated *in vitro*.
Early neonatal (1-2 h) brain slices were incubated
with 5 mM glucose and 0.5 μCi of D-(6-^{14}C)glucose.
See legend of Figure 10 for experimental details
(51).

Actually, lactate inhibits glucose utilization, decreasing its oxidation
and incorporation into lipids (Figure 11); the inhibition is similar to
that exerted by 3-hydroxybutyrate on glucose utilization (Figure 11). This
suggests that in these circumstances the brain prefers lactate to glucose,
presumably diverting glucose to other neonatal tissues that use it as the
sole metabolic substrate. In addition, lactate utilization can relieve
newborns of the metabolic acidosis brought about by lacticacidemia. On the
other hand, glucose slightly inhibits lactate oxidation without showing any
effect on lactate incorporation into lipids (Figure 12). This suggest that
lactate can be utilized as a lipogenic precursor independently of the

glucose supply. In addition, lipogenesis from lactate is not affected by the presence of 3-hydroxybutyrate (Figure 12). This suggests that the main physiological fate of lactate is lipid synthesis, although it is also significantly oxidized within the presuckling period when lactate concentrations are substantially enhanced. However, 3-hydroxybutyrate at physiological concentrations strongly inhibits lactate oxidation, a fact consistent with idea that ketone-body availability may divert lactate for gluconeogenesis. Since the surge of blood ketone bodies (Figure 3) is concurrent with an active gluconeogenesis (Figure 9), lactate may be utilized as a gluconeogenic substrate while ketone bodies would be accounted for the bulk of energy expenditures. The enhancement of

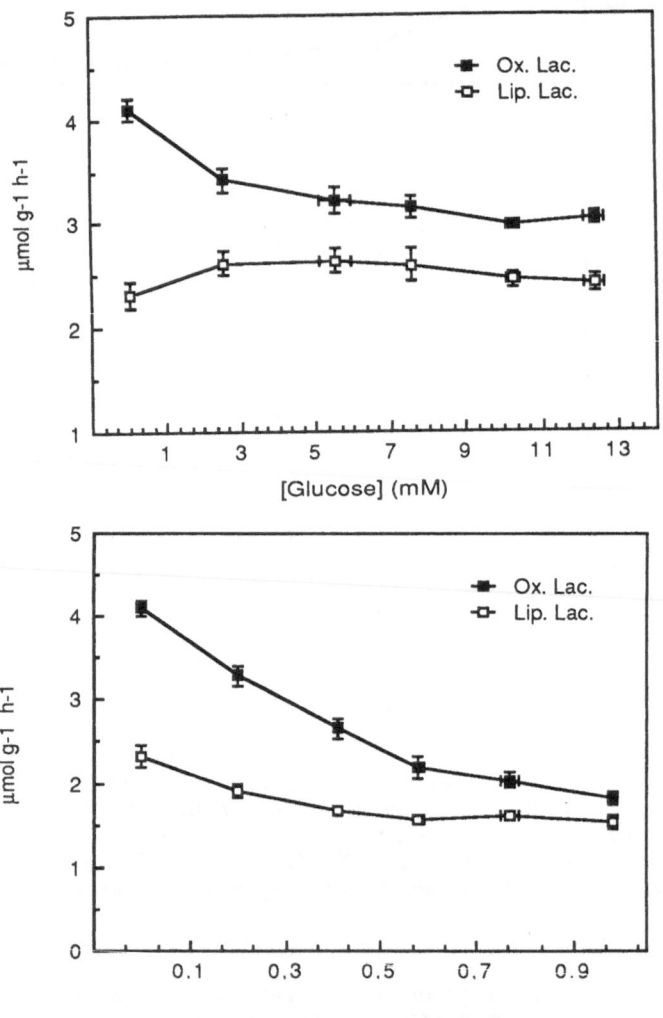

Fig. 12. Inhibition of lactate utilization in presence of glucose (upper panel) and D-3-hydroxybutyrate (lower panel) by neonatal brain slices incubated *in vitro*. Early neonatal (1-2 h) brain slices were incubated with 12 mM lactate and 0.5 µCi of L-(U-^{14}C)lactate. See legend of Figure 10 for experimental details (51).

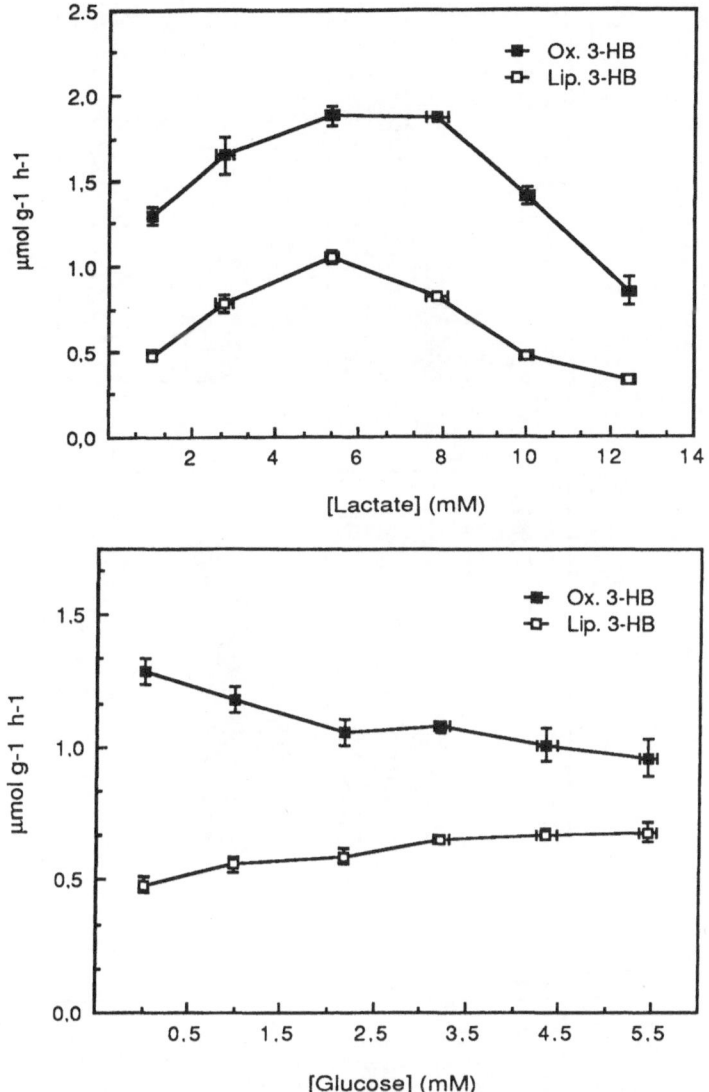

Fig. 13. Inhibition of D-3-hydroxybutyrate utilization in
presence of lactate (upper panel) and glucose
(lower panel) by neonatal brain slices. Early
neonatal (1-2 h) brain slices were incubated
with 2 mM D-3-hydroxybutyrate and 0.5 μCi of
D-3-hydroxy-(3-^{14}C)butyrate. See legend of
Figure 10 for experimental details (51).

3-hydroxybutyrate utilization by physiological concentrations of lactate
but not of glucose (Figure 13) seems to support this idea.

In conclusion, these results are consistent with the idea that
lactate is utilized during the presuckling period supplementing glucose for
energy and synthetic purposes. However, once ketone bodies are available
lactate is mainly utilized as a gluconeogenic substrate.

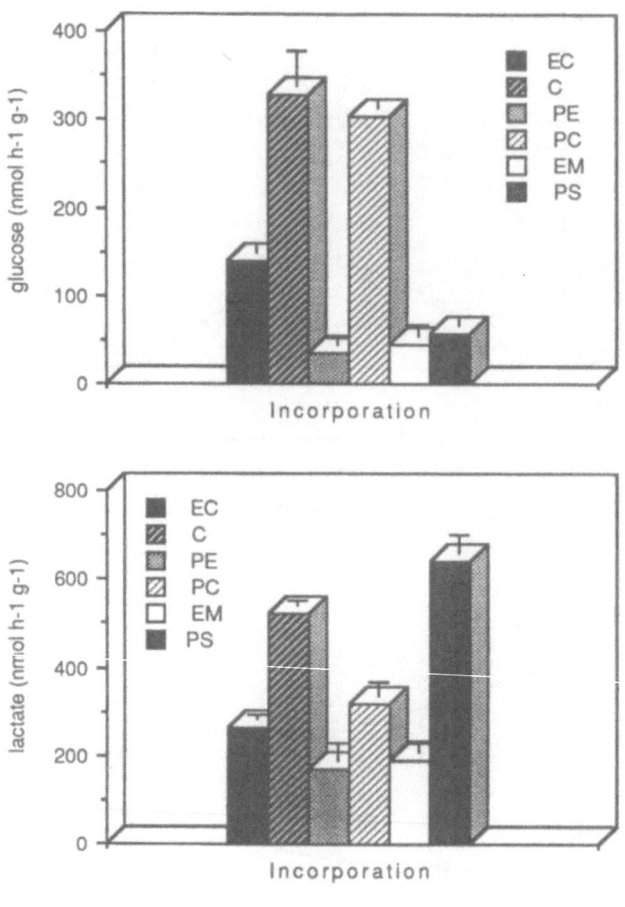

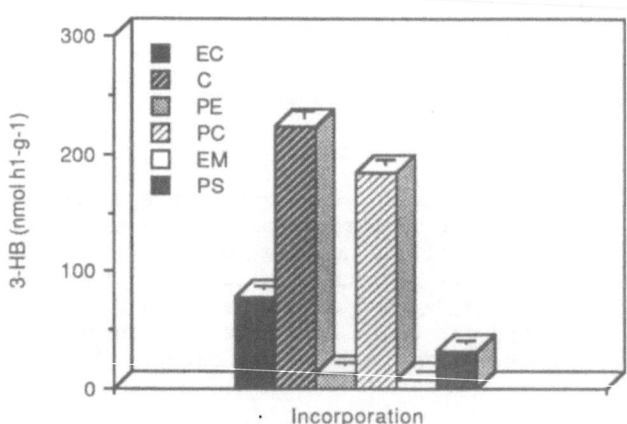

Fig. 14. Incorporation of lactate (upper panel), glucose (middle) and D-3-hydroxybutyrate (lower) into brain lipids. Early neonatal (1-2 h) brain slices incubated with 12 mM L-lactate, 5 mM glucose or 2 mM D-3-hydroxybutyrate and 0.5 µCi of L-(U-^{14}C)lactate, D-(6-^{14}C)glucose or D-3-hydroxy-(3-^{14}C) butyrate, respectively. EC: cholesterol ester; C: cholesterol; PE: phosphatidyl ethanolamine; PC: phosphatidyl choline; EM: sphyngomyelines; PS: phosphatidyl serine. See legend of Figure 10 (52).

Lactate Incorporation Into Brain Lipids

The importance of lactate as a lipogenic substrate prompted us to investigate the incorporation of lactate to brain lipids. Thus, the washed chloroform: methanol extracts of brain slices previously incubated with unlabeled and labeled substrates were chomatographed on TCL plates or HPLC columms and the identified fractions counted for radioactivity. The results are shown in Figure 14 (52). Lactate is preferentially incorporated in sterols and phosphatidyl serine and to lower extent phosphatidyl choline, sphingomyelines and phosphatidyl ethanolamine. This contrasts with glucose and 3-hydroxybutyrate incorporation, which was preferentially observed in sterols and phosphatidyl choline. These results suggest that lactate is specifically incorporated in phosphatidyl serine by a specific pathway. However, it has been reported that the "base-exchange pathway" is the only one that is active during the perinatal period (53), suggesting that phosphatidyl serine is synthesized from phosphatidyl choline by base exchange. Accordingly, the occurrence of some regulation of this pathway which specifically enhances phosphatidyl serine synthesis in contrast with other phospholipid species could be suggested. The physiological role played by the specific incorporation of lactate into phosphatidyl serine is dificult to assess. However, it is tempting to speculate that phosphatidyl serine is required for the assembly of neural structures. Thus, it has been reported that phosphatidyl serine is an absolute requirement for the Na$^+$/K$^+$ ATPase activity (54,55). If so, the specific incorporation of lactate to phosphatidyl serine under these circumstances would be directed in supplying neural membranes with the time-to-time required components. This suggestion may be important because lactate utilization takes place over a short period of time, a fact consistent with idea that an important period of neural development occurs during the presuckling period.

SUMMARY AND CONCLUSIONS

Inmediately after delivery newborn rats undergoes starvation due to the delay in the onset of suckling. During this period glucose is scarce and ketone bodies are not available due to a lack of ketogenesis. Under these circumstances, newborn rats is supplied with another metabolic fuel, lactate, which is utilized as a source of energy and carbon skeletons. Lactate inhibits glucose metabolism but enhances 3-hydroxybutyrate utilization. In addition, lactate utilization is inhibited by 3-hydroxy-butyrate but not by glucose, suggesting that lactate is mainly utilized immediately after birth, when ketone bodies are not yet available. Lipo-genesis from lactate is higher than from glucose and 3-hydroxybutyrate, suggesting that lipid synthesis may be an important fate of lactate *in vivo*. Lactate is specifically incorporated in phosphatidyl serine in contrast to glucose and 3-hydroxybutyrate are used for phosphatidyl choline synthesis. The specific incorporation of lactate in phosphatidyl serine is explained on the basis of the role played for this compound on the activation of Na$^+$/K$^+$ATPase.

ACKNOWLEDGEMENTS

The technical assistance by J.Villorica is gratefully acknowledge. C.Vicario was recipient of a fellowship from the P.F.P.I.-M.E.C., Spain. J.P.Bolaños was recipient of a fellowship from the F.I.S.S.S. This work was supported by a grant from the C.I.C.Y.T., Spain.

REFERENCES

1. G.S.Dawes, and H.J.Shelly, Physiological aspects of carbohydrate metabolism in the fetus and newborn, in: "Carbohydrate metabolism and its disorders", vol. 2, pp. 87-121, Dickens, Randle and Whelan eds. Academic Press, New York (1968).

2. J.M.Cuezva, Efecto de la hipomadurez fetal en el mantenimiento de la homeostasis calorica neonatal; phil.thesis, Madrid (1978).

3. P.Devos, and H.G.Hers, Glycogen metabolism in the liver of the fetal rat. Biochem.J. 140:331 (1974).

4. J.R.Girard, G.S.Cuendet, E.B.Marliss, A.Kevran, M.Rieutort, and R.Assan, Fuels, hormones, and liver metabolism at term and during the early postnatal period in the rat, J.Clin.Inves. 53:3190 (1973).

5. A.Martin, T.Caldes, M.Benito and J.M.Medina, Regulation of glycogenolysis in the liver of the newborn rat in vivo. Inhibitory effect of glucose, Biochim.Biophys.Acta 672:262(1981).

6. E.Fernández, C.Valcarce, J.M.Cuezva, and J.M.Medina, Postnatal hypoglycaemia and gluconeogenesis in the newborn rat. Delayed onset of gluconeogenesis in prematurely delivered newborns, Biochem.J. 214:525 (1983).

7. P.Ferré, P.Sabatien, J.F.Decaux, F.Escriva and J.Girard, Development and regulation of ketogenesis in hepatocytes isolated from newborn rats, Biochem.J 214:937 (1983).

8. D.H.Williamson and B.M.Buckley, The role of ketone bodies in brain development; in: "In born errors of metabolism", pp. 81-96 Homes and Van der Berg eds. Academic Press, London (1973).

9. C.M.Van Duyne and R.J.Havel, Plasma unesterified fatty acid concentration in fetal and neonatal life, Proc.Soc.exp.Biol.Med. 102:599 (1959).

10. C.H.Chen, P.A.J.Adam, D.E.Laskowski, M.L.McCamm and R.Schwart, The plasma free fatty acid composition and blood glucose of normal and diabetic pregnant women and their newborn, Pediatrics 36:843 (1965).

11. J.H.Exton, Gluconeogenesis, Metabolism, 21:945 (1972).

12. J.M.Cuezva F.J.Moreno, J.M.Medina and F.Mayor, Prematurity in the rat.I. Fuels and gluconeogenic enzymes, Biol.Neonate 37:88 (1980).

13. C.Arizmendi, Metabolismo energetico del neonato prematuro de rata. Influencia de las disponibilidades de oxigeno; phil.thesis Madrid (1983)

14. C.Arizmendi, M.Maties, M.Benito and J.M.Medina, Effect of postnatal hypoxia on the energy homeostasis of the newborn rat during the early neonatal period, Biol.Neonate 44:36 (1983).

15. M.C.Rodriguez-Figueroa, M.V.Garcia, J.Martin-Barrientos and J.M.Medina, Hyperammonemia in the premature newborn rat: its association with lactic acidosis, Medical Science Research 16:701 (1988)

16. M.J.Juanes, C.Arizmendi and J.M.Medina, Attenuation of postnatal hypoxia in the premature newborn rat by maternal treatment with dexamethasone: Its relatioship with lung phospholipid content, Biol.Neonate 50:337 (1986).

17. R.Deron, Anaerobic metabolism in the human fetus, Am.J.Obset.Gynecol. 89:241 (1964)

18. D.F.Balkovetz, F.H.Leibach, V.B.Nahesh and V.Ganapathy, J.Biol.Chem. 263:13823 (1988)

19. S.R.Alonso, M.A.Serrano and J.M.Medina, Umpublished results.

20. G.P.B.Kraan and T.Dias, Size of L-lactate transport from the fetal rat to the mother animal, Biol.Neonate 26:9 (1975)

21. C.Valcarce, J.M.Cuezva and J.M.Medina, Increased gluconeogenesis in the rat at term gestation, Life.Sci. 37:553 (1985).

22. L.I.Burd, M.D.Jones, M.A.Simmons, E.L.Makowski, G.Meschis and F.C.Battaglia, Placental production and fetal utilization of lactate and pyruvate, Nature, 253:710 (1975).

23. E. Herrera, M. Palacin, M. Martin and M. A. Lasuncion, Relationships between maternal and fetal fuels and placental glucose transfer in rats with maternal diabetes of varying severity, _Diabetes_ 34 (sup. 2): 42 (1985).

24. M. Palacin, M. A. Lasuncion and E. Herrera, Lactate production and absence of gluconeogenesis from placental trasferred substrates in fetuses from fed and 48-h starved rats, _Pediatr. Res_ 22:6 (1987).

25. C. A. Villee, The metabolism of human placenta _in vitro_, _J. Biol. Chem._ 205:113 (1953).

26. I. R. Holzman. A. F. Philipps and F. C. Battaglia, Glucose metabolism, lactate and ammonia production by the human placenta _in vitro_, _Pediatr. Res._ 13:117 (1979).

27. M. M. Calonge, J. Martin-Barrientos and J. M. Medina, Unpublished results.

28. M. Palacin, M. A. Lasuncion, R. Martin del Rio and E. Herrera, Placental formation of lactate from trasferred L-alanine and its impairment by aminooxyacetate in late-pregnant rat, _Biochim. Biophys. Acta_ 841:90 (1985).

29. Y. Z. Diamant and E. Shafrir, Placental enzymes of glycolysis, gluconeogenesis and lipogenesis in diabetic rat and in starvation, _Diabetologia_ 15:481 (1978).

30. R. G. Vernon and D. G. Walker, Glucose metabolism in the developing rat. Studies _in vivo_, _Biochem. J._ 127:521 (1972).

31. P. Ferré, J. P. Pegorier and J. R. Girard, The effects on inhibition of gluconeogenesis on suckling newborn rats, _Biochem. J._ 162:209 (1977).

32. F. J. Ballard and L. T. Oliver, Glycogen metabolism in embryonic chick and neonatal rat liver, _Biochim. Biophys. Acta_ 71:578 (1963).

33. J. M. Cuezva, C. Valcarce and J. M. Medina, Substrates availability for maintenance of energy homeostasis in the immediate postnatal period of the fasted newborn rat, _in_: "The Physiological Development of the Fetus and Newborn", C. T. Jones and P. W. Nathanielsz, Academic Press, London, 1985.

34. F. Mayor and J. M. Cuezva, Hormonal and metabolic changes in the perinatal period, _Biol. Neonate_ 48:185 (1985).

35. J. M. Medina, J. M. Cuezva and F. Mayor, Non-gluconeogenic fate of lactate during the early neonatal period in the rat, _FEBS Lett._ 114:132 (1980).

36. J. M. Cuezva E. Fernández, C. Valcarce and J. M. Medina, The role of ATP/ADP ratio in the control of hepatic gluconeogenesis during the early neonatal period, _Biochim. Biophys. Acta_ 759:292 (1983).

37. C. Arizmendi, T. Caldes, M. Benito and J. M. Medina, Decreased glucose supply in the premature newborn rat and its relationship with the glucagon/insulin ratio, _IRCS med. Sci._ 12:592 (1984)

38. J. M. Cuezva, F. J. Moreno and J. M. Medina, Blood oxygen concentrations in premature newborn rats during the early neonatal period, _IRCS med. Sci._ 9:644 (1981).

39. J. M. Cuezva and J. M. Medina, Prematurity in the rat. III. Effect of oxygen supply, _Biol. Neonate_ 39:70 (1981).

40. C. E. Patterson, M. V. Koniki, W. M. Selig, C. M. Owens and R. A. Rohades, Integrated substrates utilization by perinatal lung, _Exper. Lung Res._ 10:71 (1986).

41. E. Fernández and J. M. Medina, Unpublished results.

42. M. A. Almeida J. P. Bolaños, E. Fernández and J. M. Medina, Unpublished results.

43. E. D. Saggerson, W. J. McAllister and H. S. Bath, Lipogenesis in rat brown adipocytes. Effect of insulin and noradrenaline, contributions from glucose and lactate as precursors and comparisons with white adipocytes, _Biochem. J._ 251:701 (1988).

44. T. Itoh and J. H. Quastel, Acetoacetate metabolism in infant and adult rat brain _in vitro_, _Biochem. J._ 116:641 (1970).

45. G. E. Shambaugh III, R. A. Koehler and N. Freinkel, Fetal fuels II:

contributions of selected carbon fuels to oxidative metabolism in rat conceptus, Am. J. Physiol. 233: E457 (1977).

46. J. Hellmann, R. C. Vannucci and E. E. Nardis, Blood-Brain barrier permeability to lactic acid in the newborn dog: Lactate as a cerebral metabolic fuel, Pediatr. Res. 16: 40 (1982).

47. C. A. Gleason, C. D. Rudolph, J. Bristow, J. Itskovitz and A. M. Rudolph, Lactate uptake by the fetal sheep liver, J. Develop. Physiol. 7: 177 (1985).

48. T. G. Ramsay, J. A. Sheahan and R. J. Martin, Comparison of lactate and glucose metabolism in the developing porcine placenta, Am. J. Physiol. 247: R755 (1984).

49. J. Fernades, R. Berger and G. P. A. Smit, Lactate as a cerebral metabolic fuel for glucose-6-phosphatase deficient children, Pediatr. Res 18: 335 (1984).

50. C. Arizmendi and J. M. Medina, Lactate as an oxidizable substrate for rat brain in vitro during the perinatal period, Biochem. J 214: 633 (1983).

51. E. Fernández and J. M. Medina, Lactate utilization by the neonatal rat brain in vitro. Competition with glucose and 3-hydroxybutyrate, Biochem. J. 234: 489 (1986).

52. E. Fernández J. P. Bolaños and J. M. Medina, Unpublished results.

53. E. Yavin and B. P. Zeigler, Regulation of phospholipid metabolism in differenciating cells from rat brain cerebral hemipheres in culture, J. Biol. Chem. 252: 260 (1977).

54. S. Tsakiris and G. Deliconstantinos, Influence of phosphatidyl serine on K^+/Na^+ stimulated ATPase and acetyl-colinesterase activities of dog brain synapthosomal plasma membrane, Biochem. J. 220: 301 (1984).

55. S. C. Lin and E. L. Way, Calcium activated ATPases in presynaptic nerve endings, J. Neurochem. 39: 1641 (1982).

UTILIZATION OF THE MAIN METABOLIC SUBSTRATES

BY FRESHLY ISOLATED CELLS FROM THE FETAL RAT BRAIN

Carlos Vicario, Carmen Arizmendi and José M. Medina

Departamento de Bioquímica y Biología Molecular
Facultad de Farmacia
Universidad de Salamanca, Spain

INTRODUCTION

Lactate utilization by the fetal rat brain has previously been studied in a preparation of brain slices (1), where it was found to be an important physiological substrate for oxidation and lipogenesis in comparison with glucose and ketone bodies (2). This *in vitro* system maintained brain architecture, although the presence of the blood brain barrier, and/or the thickness of the slice, could limit the accessibility of substrates and oxygen to brain cells. The aim of the present work was to study the utilization of the main metabolic substrates by freshly isolated fetal brain cells (2) under controlled environmental conditions.

RESULTS AND DISCUSSION

The oxidation to CO_2 and lipogenesis from glucose, lactate and β-Hydroxybutyrate were studied in freshly isolated fetal rat brain cells incubated in suspension. Brain cells were prepared from fetal rats, obtained immediately after cesarean section of pregnant rats at 21.5 days of gestation (21.7 d. for full gestation). When trypsin was used, in addition to mechanical disaggregation of the tissue, trypan blue exclusion gave a 95% of cell viability. The cells were incubated in suspension in an oxygen-saturated physiological medium (3) in the presence of cold and labelled substrates at the concentrations and specific radioactivities indicated in the legends to the figures. The amount of cell protein in the incubation flasks was 0.5-1 mg/0.5 ml, range where oxidation and lipogenesis from lactate showed linearity with cell protein (r=1.0 and r=0.98, respectively). During a 60 min incubation at $37°C$, 92% of lactate dehydrogenase activity remained in the cell pellet, indicating that cell viability was maintained throughout the experiment.

$^{14}CO_2$ produced from the oxidized substrates was trapped by hyamine and measured by liquid scintillation counting. Total lipids were extracted (4), washed until water soluble radiactivity was null, and the radioactivity determined. Results of the rate of oxidation to CO_2 of the different labelled substrates are shown in Figure 1. The total capacity of the cells for glucose oxidation to CO_2 was measured by using the uniformly labelled substrate. As shown in Figure 1a, glucose is converted to CO_2 at a rate

Endocrine and Biochemical Development of the Fetus and Neonate
Edited by J. M. Cuezva *et al.*
Plenum Press, New York, 1990

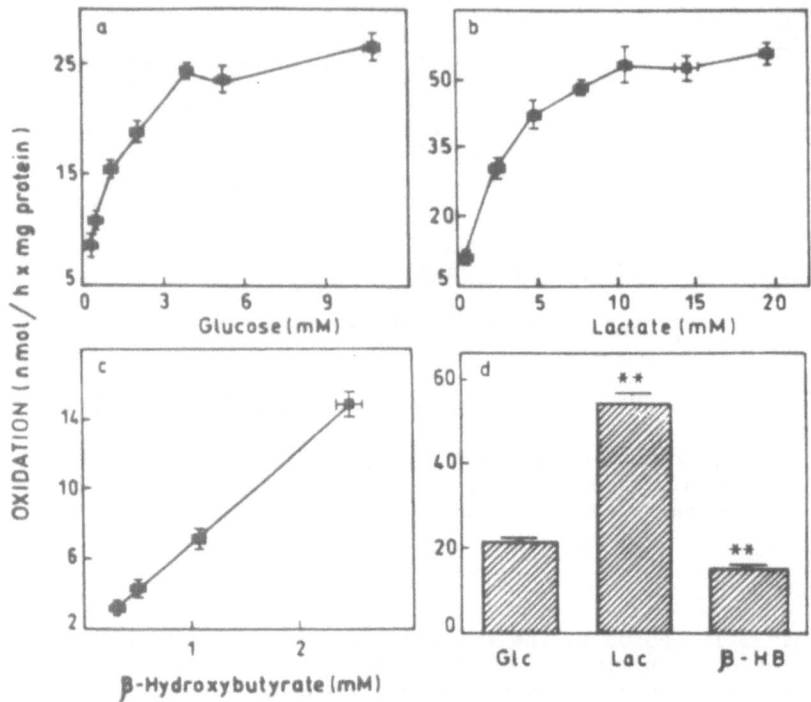

Fig. 1. Substrate oxidation rates to CO_2 by the fetal brain
cells. Dependence of (a) glucose, (b) lactate and
(c) DL- or D-β-hydroxybutyrate oxidation on
substrate concentration. (d) Comparison of oxidation
rates among the different substrates at optimal
concentrations (5.1 mM glucose, 11.8 mM lactate and
2.4 mM β-hydroxybutyrate). The labelled substrates
were 0.25-2 μCi of D-[U-^{14}C] glucose, L-[U-^{14}C]
lactate or D-3-hydroxy-[3-^{14}C] butyrate. Results are
means ± S.E.M. (n=5-10). ** significant (p<0.001).
Glc: glucose, Lac: lactate, β-HB: β-hydroxybutyrate.

depending on its supply to cells. This process achieved a Vmax of 24 nmol/h
x mg protein for a concentration close to 5 mM in the isolated fetal cells.

The adult rat brain synaptosomes also oxidized glucose in a concentration-dependent manner, attaining 34.8 nmol/h x mg protein at a glucose
concentration of 5 mM (results not shown). Although this is an adult
normoglycaemic concentration, it must be pointed out that the newborn rat
immediately after birth at 21.5 d. of gestation develops hypoglycaemia at
the 3.5 mM level (5). Hence, glucose oxidation in the early neonatal brain
would be in the part of the curve (Figure 1a) where a small change in
concentration leads to a large increase in oxidation, so that the energy
derived from this process would vary considerably.

Uniformly labelled glucose will produce CO_2 not only via the pyruvate
dehydrogenase complex reaction and in the tricarboxylic acid cycle, but
also through an active pentose phosphate pathway. In fact, the supply of
both NADPH, for lipogenesis (Figure 2), and riboses, for nucleic acids
synthesis, are metabolic pathways required at this stage of rapid brain
growth, differentiation and proliferation.

The rate of lipogenesis from glucose increased with its concentration (Figure 2a), and a plateau was reached at the same value observed for its oxidation (5 mM, Figure 1a). Between 5 and 10 mM, both oxidation and lipogenesis, remained constant; however, whether the saturation of these processes would have physiological significance for the fetal brain remains to be elucidated.

As shown in Figure 1b, lactate oxidation to CO_2 by the fetal brain cells increased with substrate concentration, and saturation at 53 nmol/h x mg protein is attained with physiological concentrations (10 mM) for the plasma of the newborn immediately after birth (5). Although the oxidative rate of both preparations cannot be compared in quantitative terms, the cells results qualitatively confirm those obtained with brain slices (1), suggesting indirectly an absence of blood brain barrier influence on the capacity for lactate uptake by the *in vivo* rat brain at this developmental age. Again, the comparison with adult rat brain synaptosomes incubated in the presence of 10 mM lactate led to a difference of 40 nmol/h x mg protein as compared with the fetal cells. An explanation for this could be the level of Pyruvate dehydrogenase complex activity (PDHc) in the fetal rat brain mitochondria that was about 44% of the adult value (6).

Lipogenesis from lactate (Figure 2b) was also dependent on a physiological range of substrate concentrations, and saturable. Like

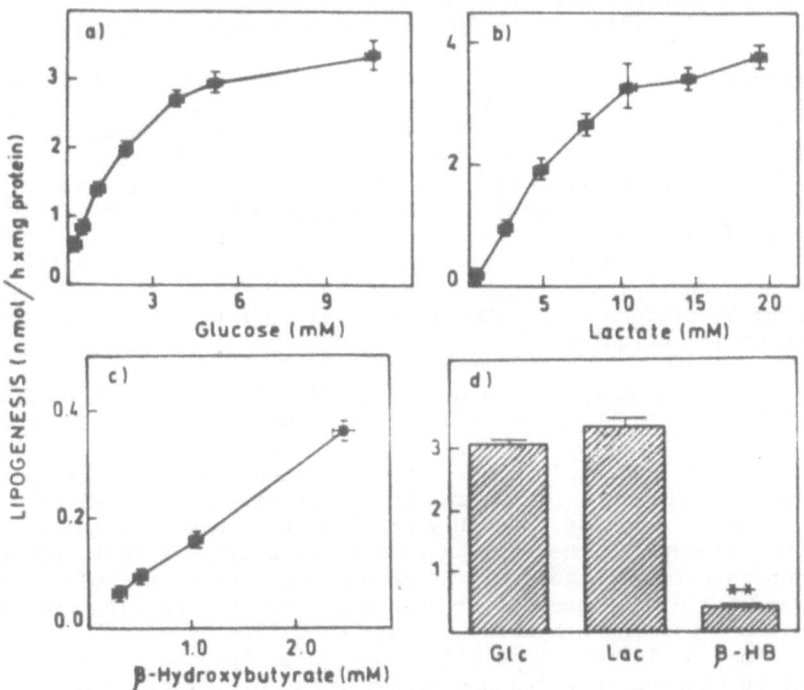

Fig. 2. Substrate incorporation rates into lipids by the fetal brain cells. Dependence of lipogenesis from (a) glucose, (b) lactate and (c) β-hydroxybutyrate on substrate concentration. (d) Comparison of lipogenesis rates among the different substrates at optimal concentrations (5.1 mM glucose, 11.8 mM lactate and 2.4 mM ß-hydroxybutirate). The labelled substrates were: 0.25-2 μCi of D-[U-14C] glucose, L-[U-14C] lactate or D-3-hydroxy-[3-14C] butyrate. Results are means ± S.E.M. (n=5- 10). ** significant (p<0.001). Abbreviations are as in Figure 1.

lactate oxidation, the rate of lipogenesis from this substrate ceases to increase close to 10 mM. This coincidence could only be explained if the amount of substrate required for the carrier and/or the limiting enzymes, common to both processes, to reach their maximal activities was around that concentration. Lactate transport to the isolated brain cells has not been studied, although it might be carried out by facilitated diffusion, and saturable at concentrations higher than 10 mM, as shown in other systems (7).

Lactate dehydrogenase (LDH) activity in the isolated fetal brain cells was 400 mU/mg protein (results not shown). The predominant LDH isoenzymes in the immature rat brain are those with M subunits (8), whose characteristics are a low affinity for the lactate oxidation but a high velocity in this reaction direction. Hence, this enzyme activity is unlikely to limit formation of pyruvate from lactate.

The total capacity for acetyl-CoA production through the Pyruvate dehydrogenase complex (PDHc) reaction has not yet been measured in the freshly isolated fetal brain cells. However, fully active dephosphorylated PDHc activity in isolated fetal rat brain mitochondria was about 50 mU/mg mitochondrial protein (6). Such an activity may account for the maximal rate of lactate oxidation (53 nmol/h x mg protein, Figure 1b) and maximal lipogenesis (3.5 nmol/h x mg protein, Figure 2b) by the isolated fetal rat brain cells. It might likely be that 10 mM pyruvate, from the oxidation of either glucose or lactate, would be the concentration required for the maximal activity of the Pyruvate dehydrogenase complex.

β-Hydroxybutyrate was shown to be a substrate for oxidation and lipogenesis in the brain cells at physiological concentrations (Figures 1 and 2c). This is in agreement with other results for the neonatal and fetal brain metabolism (2,9). Both pathways, oxidation and lipogenesis, were linear with the β-hydroxybutyrate concentration and were unsaturable even at 2.5 mM, which is a high concentration compared with the maximal value of 2.1 mM in blood of the 48 h-starved adult or the 1.5 mM of the suckling rat at 5 days of age (9). Whether saturation was achieved at concentrations in the range of pathological values, and/or whether ketone bodies utilization would depend on factors other than their availability to the fetal brain, remains to be investigated.

The comparison among the oxidative utilization of the three substrates is shown in Figure 1d. Glucose oxidation at a concentration of 5 mM was shown to be significantly lower than two-fold that saturating concentrations of lactate (p<0.001). On comparing the number of carbon atoms in CO_2 per mol of oxidized substrate, glucose gave a two-fold lower value than that of lactate, probably pointing to a preferential use of lactate over glucose by the early neonatal rat brain. In fact, plasma lactate concentration immediately after birth is about two-fold higher than glucose, and the difference has been shown to increase during the first postnatal hour (5). The β-hydroxybutyrate oxidative rate was significantly lower than that of lactate (p<0.001) and glucose (p<0.001), even when the concentration assayed in vitro was higher than physiological plasma levels immediately after birth.

Although care should be taken when extrapolating the in vitro results to the in vivo situation, it is tempting to speculate that during the transitional adaptation to the extrauterine life the brain would preferentially rely on lactate oxidation ratherthan on the other metabolic substrates for energy.

Maximal values of lipogenesis from the three substrates are shown in Figure 2d. As expected from the results obtained in brain slices (2), the rate of lipid synthesis from glucose will have to be two-fold lower than from lactate, in view of the number of carbon atoms from both substrates. However, under the present experimental conditions in isolated cells, lactate and glucose are converted into lipids at rates that are not statistically different (Figure 2d). An explanation for this could likely be the lack of glucose in the incubation medium to provide the NADPH required for lipogenesis from lactate. Hence, in the absence of added glucose, reducing equivalents would be supplied by endogenous brain carbohydrate stores, which are probably present only in a limited amount, and/or physically separated in the cells as compared with the brain slices (10).

β-Hydroxybutyrate is also a carbon donor for lipid synthesis in the isolated fetal brain cells (Figure 2c and d), although it is significantly less incorporated into lipids than the other two substrates ($p < 0.001$), and also, in comparison with the lipogenesis from the same substrate in the brain slices (2). The reason for this could be the limited supply of the NADPH available, as mentioned above for the results on lactate. Whether such an explanation is correct is currently being investigated in our laboratory.

ACKNOWLEDGEMENTS

The helpful discussions of Prof. J.B. Clark and Dr. G.D.A. Malloch are gratefully acknowledged. The technical assistance by J. Villoria is also acknowledged. C. Vicario is recipient of a fellowship from the P.F.P.I.- M.E.C., Spain. This work was partly supported by a grant from the D.G.I.C. Y.T., Spain.

REFERENCES

1. C. Arizmendi and J. M. Medina, Lactate as an oxidizable substrate for rat brain *in vitro* during the perinatal period, Biochem. J. 214:633 (1983).
2. E. Fernández and J. M. Medina, Lactate utilization by the neonatal rat brain *in vitro*. Competition with glucose and 3-hydroxybutyrate, Biochem. J. 234:489 (1986).
3. K. A. C. Elliot, in: "Handbook óf Neurochemistry", A. Lajtha, ed., Plenum Press, New York, vol 2, pp. 103 (1969).
4 J. Folch, M. Lees and G. H. Sloane Stanley, A simple method for the isolation and purification of total lipids from animal tissues, J. Biol. Chem. 226:497 (1957).
5. J. M. Cuezva, F. J. Moreno, J. M. Medina and F. Mayor, Prematurity in the rat I. Fuels and gluconeogenic enzymes, Biol. Neonate 37:88 (1980).
6. G. D. A. Malloch, L A. Munday, M. Olson and J. B. Clark, Comparative development of the pyruvate dehydrogenase complex and cytrate synthase in rat brain mitochondria, Biochem. J. 238:729 (1986).
7. P. Fafournoux, Ch. Demigne and Ch. Remesy, Carrier-mediated uptake of lactate in rat hapatocytes. Effects of pH and possible mechanisms for L-lactate transport, J. Biol. Chem. 260:292 (1985).
8. V. Bonavita, F. Ponte and G. Amore, Lactate dehydrogenase isoenzymes in the nervous tissue-IV. An ontogenic study on the rat brain, J. Neurochem. 11:39 (1964).
9. M. A. Page, H. A. Krebs and D. H. Williamson, Activities of enzymes of ketone body utilization by brain and other tissues of suckling rats, Biochem. J. 121:49 (1971).
10. V. W. Pentreath and M. A. Kai-Kai, Incorporation of (^{3}H) 2-deoxyglucose into glycogen in nervous tissues, Neuroscience 7:759 (1982)

THE USE OF PRIMARY CULTURE OF ASTROCYTES TO

STUDY GLIAL DEVELOPMENT. EFFECT OF ETHANOL

Consuelo Guerri, María Sancho-Tello,
Remedios Zaragoza and Jaime Renau-Piqueras

Instituto de Investigaciones Citológicas
Centro Asociado al CSIC
46010 Valencia, Spain

INTRODUCTION

The great complexity of the mammalian central nervous system (CNS) has made it difficult to elucidate the cellular mechanisms involved in its development and function. Primary culture of the various cells, which are derived from tissue taken directly from the organism, therefore presents an attractive approach for investigating the biochemical, morphological and functional aspects of these cells during differentiation (1). In addition, the direct effect of toxic agents can be evaluated, including the mechanism of action on specific cell types.

Astrocytes, constituting up to 50% of the cells in the brain cortex, play crucial roles in the nervous system including: a) regulation of neuronal function and ionic environment, b) form part of the blood brain barrier, c) act, during brain development, as a guide to assure the correct migration of neurons as well as the relationships between these cells and, d) also control extracellular brain fluid, oxidative metabolism and energy generation. Knowledge of characteristics and functions of astrocytes has increased during recent years due to the utilization of primary culture techniques (1).

We have used this approach to study glial development as well as to investigate the direct toxic effect of ethanol on astrocyte maturation. It is well known that ethanol alters the development of the CNS (2,3). Although how alcohol causes this effect is unknown, it has been shown to alter the proliferation and migration pattern of cortical neurons and has thus been related with ethanol-induced abnormalities in glial maturation (4).

In the present work we have analyzed the growth and differentiation of astrocytes *in vitro* by measuring synthesis of protein, of DNA and of RNA, the activity of several marker and functional enzymes and the development of the cytoskeleton. We have also studied the effect of ethanol on these parameters. The astrocytes were obtained from brain cortex of control and alcohol fed rat fetuses and the cells were cultured in the presence or absence of ethanol.

Endocrine and Biochemical Development of the Fetus and Neonate
Edited by J. M. Cuezva *et al.*
Plenum Press, New York, 1990

RESULTS

Growth

Primary cultures of rat cortex astrocytes were established as
described (5), using 21-day-old fetuses. The cultures were grown for 28
days from a seeding density of 1×10^6 cells per dish, using 35 mm plastic
tissue culture dishes. The cultures grew rapidly for 8-10 days, after which
the cell number increased slowly. This pattern corresponds to a typical
logarithmic growth and was reflected in the increase inof DNA content
(Figure 1). The amount of total protein in the cultures increased
continuously for 28 days although the increase was greater during the first
week (results not shown).

To analyze the kinetics of protein, DNA and RNA synthesis, cell
cultures were labelled with ^{3}H-leucine, ^{3}H-thymidine or ^{3}H-uridine,
respectively (6). The ^{3}H-leucine uptake, expressed as cpm/mg protein,
increased markedly during the first week and then remained constant until
day 15, followed by a decrease until the end of culture (data not shown).
A similar kinetic pattern was also observed for ^{3}H-uridine incorporation.
The pattern for DNA synthesis was different. ^{3}H-thymidine uptake increased
during the first days of culture reaching a peak at 7 days and then
decreased, leveling of after 15 days (Figure 1).

To evaluate the effect of alcohol on glial development, astrocytes
were cultured in the presence of 25 mM ethanol in the medium. There were

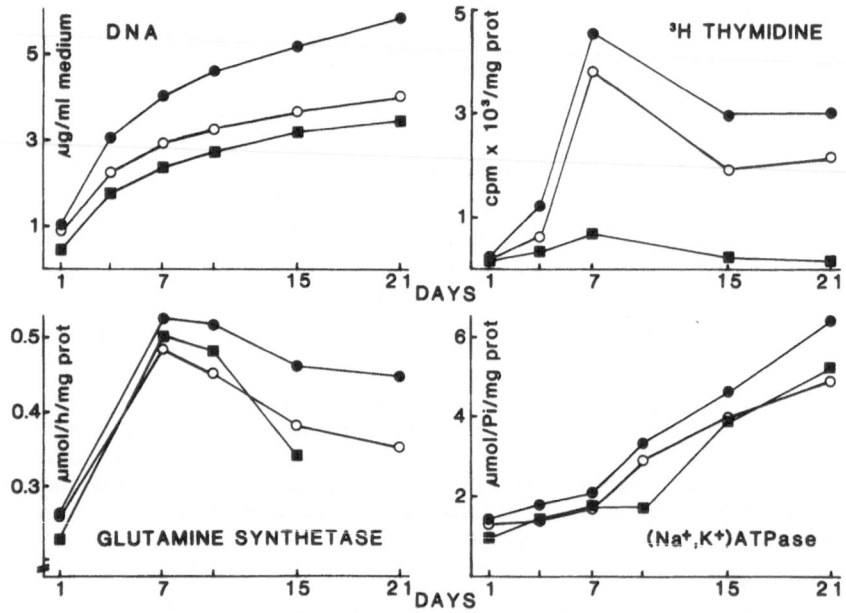

Fig. 1. Graphs showing the evolution of several parameters
analyzed in the study of astrocyte development in
primary culture including the effect of ethanol on
this evolution. Control cultures without (●) and
with 25 mM ethanol (o). Astrocytes obtained from
fetuses prenatally exposed to alcohol and cultured
in absence of ethanol (■). Each point represents
the average of 3-4 different cultures.

decreases in both protein and DNA content and also a significant reduction in the uptake of the radiolabelled precursors (Figure 1). Interestingly, when astrocytes obtained from fetuses prenatally exposed to alcohol were analyzed, there were reductions in cell growth as well as in DNA and protein content, compared with controls (Figure 1).

Enzyme Activities

The development of astrocytes in primary culture was also assessed by measuring the activity of several marker and functional enzymes. The specific activities of glutamine synthetase (GS) (Figure 1) and butyrylcholinesterase (BuChE) (data not shown), two enzymes largely confined to glia (7), had similar patterns during the culture period. Their activities increased sharply during the first week, followed by a slow decrease until the end of the culture.

As functional enzymes, the specific activities of (Na^+, K^+)ATPase (Figure 1), Ca^{++} ATPase and glutamate dehydrogenase (GDH) (data not shown) were measured. Both ATPases increased throughout the entire culture period but more notably between days 7 and 21. However GDH, an important enzyme involved in the interconversion between alfa-ketoglutarate and glutamate, increased significantly for 10 to 15 days but then declined (results not shown).

Alcohol at 25 mM in the culture medium caused a decrease in the activity of all the enzymes tested, except for GHD. This effect was also observed in astrocytes derived from alcoholic fetuses which were cultured in the absence of ethanol (Figure 1).

Glial Fibrillary Acidic Protein (GFAP) and Cell Morphology

Early in the culture most cells were polygonal in morphology (Figure 2A), whereas after 10 days the cultures were composed mainly of cells with a star-like form (Figure 2B). The first type corresponds to astroblasts and the second to mature astrocytes, the most frequent cell type in long-term cultures. These cells were arranged in confluent sheets with the cytoplasmic processes which grew from each individual cell forming a complex network.

Glial intermediate filaments (IF) are the most characteristic morphological feature of normal astrocytes both *in vivo* and *in vitro*. Although their function is unknown, they appear to serve important structural functions. GFAP, the major component in glial IF, has been used as a specific marker for glial cells, particularly in development of astrocytes *in vivo* as well as in primary cultures (8). We have therefore also analyzed the amount and distribution pattern of GFAP in astrocytes in primary culture, using immunocytochemical methods (immunofluorescence and immunogold) (5).

The proportion of GFAP-positive cells in the cultures ranged from 85 to 90%, indicating that almost all the cells were astrocytes. Images of 1-day-old astrocytes showed that fluorescence was restricted to small areas surrounding the nucleus (data not shown). In 4- and 7-day-old cells, the cytoskeleton showed a reticular pattern (Figure 2A) whereas in 14- and 21-day-old cultures, IF in most of the cells were arranged in parallel arrays (Figure 2B). The reticular pattern was found primarily in the flat cells whereas in mature astrocytes they were filamentous (Figure 2).

In electron microscopy studies, using postembedding techniques and cytoskeletal preparations (5), two IF distribution patterns were also shown. In the first the IF formed a reticular network (results not shown), whereas in the second they were disposed in parallel arrays (Figure 2C). When ethanol was present in the culture medium there were changes in the IF distribution patterns. Whereas in the 14-day-old control cultures, only 5% of the cells showed the immature reticular pattern, 50% of the cells cultured in the presence of 25 mM ethanol showed this pattern (results not shown). Interestingly, the reticular pattern was also dramatically increased (25%) in astrocytes obtained from fetuses prenatally exposed to ethanol even though cultured in the absence of this drug (results not shown).

To determine the amount of GFAP in astrocytes during their development, we have quantified, using scanning densitometry, the fluorescence in the cells, measured as relative absorbance (5). In control cultures this parameter increased regularly for 14 days followed by a decrease to day 21 (Figure 2D). Although the pattern was similar when the cells were incubated with alcohol, the relative absorbance decreased about 40-45% compared to the controls. This effect was also noted when astrocytes from alcohol exposed fetuses were analyzed (Figure 2D). Similar results

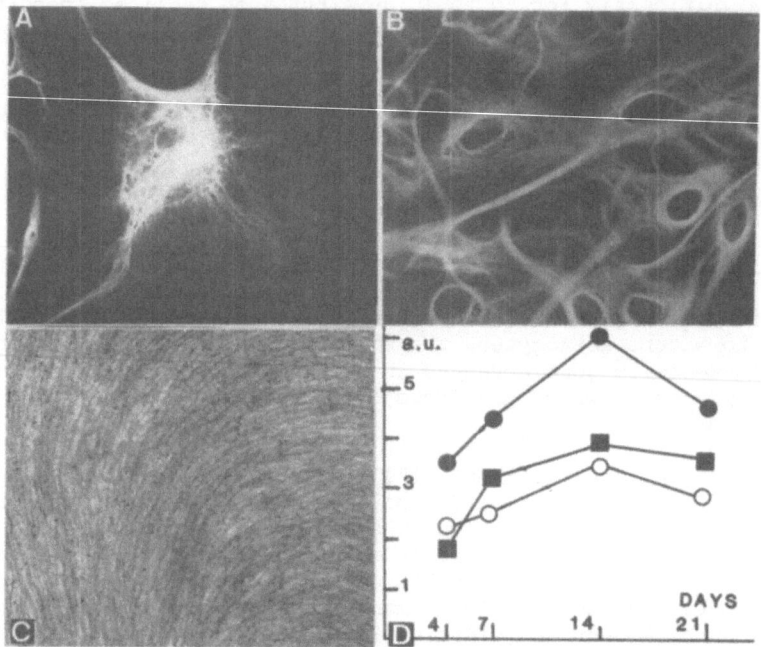

Fig. 2. (A,B) Brain cortex astrocytes in primary culture,
showing immunofluorescence due to staining with a
monoclonal antibody against GFAP tagged with a
fluorescent marker. (A) 4-day-old polygonal astroblast
showing a reticular distribution of IF. (B) Low
magnification micrographs of 14-day-old mature
astrocytes. (C) Ultrastructural localization of
anti-GFAP binding sites on 21-day-old astrocytes using
the immunogold technique and postembedding procedures.
The IF are disposed in parallel arrays. (D) Graphs
showing the relative absorbance of astrocytes prepared
for fluorescence microscopy and GFAP demonstration.
Relative absorbance (mean absorbance/mean area) is
expressed in a.u. Symbols same as Figure 1.

were obtained using the immunogold technique which permits to determine the anti-GFAP binding sites (Figure 2). The number of these sites in individual IF was reduced in the presence of ethanol. We have found, in addition, that ethanol interferes whith the reorganization of the anti-GFAP binding sites from clustered, in immature cells, to random, in mature astrocytes.

DISCUSSION

In the present work we have used primary cultures of astrocytes taken from brain cortex as a model to study the functional specialization of these cultures we have found striking increases in protein, DNA and RNA synthesis during the first week of culture indicating a rapid proliferation of cells during this period, followed by a reduction in the synthesis of these macromolecules, suggesting that most of the cells were in a maturation stage. These results agree with *in vivo* studies showing that protein and DNA synthesis are higher in brain of newborn animals, mainly due to astrocytes, and decline during the first month of growth. Further, our results also agree with data indicating that astroglial cells, unlike neurons, in adult brain cortex, retain the ability to proliferate. In fact, in our study some cells still were dividing at the end of the culture.

BuChE and particularly GS are specific markers for astrocytes and play important roles in the metabolism of these cells. Thus GS is not only an important enzyme in ammonia detoxification but it presumably also acts as a modulator of neuronal activity (9). Both enzymes show changes during astrocyte development *in vitro*, as shown in results, confirming data obtained *in vivo* (9).

Other important enzymes analyzed in this work were the ATPases which are involved in ontogenic development and ion homeostasis (7). The development of these enzymes *in vitro* parallel that of the cytoskeleton, with a maximum activity during the maturation period of astrocyte development.

Our determinations of GFAP during the culture evolution are similar to those obtained, during ontogeny, for the whole rat forebrain, mouse brain and primary astrocytic cultures, showing a 14-day maximum which could correspond to the burst of astroglial differentiation at the time of myelination (10). The evolution of GFAP content and distribution *in vitro* is parallel to the increase of gliofilaments and formation of cell processes, which occurs during astroglial maturation (10). These changes could be correlated with the development of certain functional enzymes as mentioned above. Our results are in agreement with the assumption that functional characteristics for astrocytes *in vitro* compare well with *in vivo* data and even appear to develop in parallel.

Concerning our observations on the effect of ethanol on astrocyte development, several conclusions can be emphasized: 1) ethanol at low concentration significantly reduces cell growth and division; 2) as demonstrated by the analyses of enzyme activities and GFAP content, ethanol causes a delay in the development of cells from astroblasts to mature fibrous astrocytes. This clearly indicates that this toxic agent also affects cell differentiation; 3) the toxic effects were also evident when astrocytes derived from fetuses prenatally exposed to ethanol were used, indicating that there had been damage to the astrocyte precursor cells *in vivo*.

Finally, since during development astroglial cells play a crucial role in CNS functions, regulating neuronal migration and cell interactions, the alcohol-induced alterations in astrocytes could be one of the primary mechanisms underlying the mental retardation and motor dysfunction which are characteristics of the fetal alcohol syndrome (2).

ACKNOWLEDGEMENTS

This research was partially supported by DGICYT (PB 87-0505) and by FISss (89/0002).

REFERENCES

1. E. Hansson, Primary astroglial cultures. A biochemical and functional evaluation, Neurochem. Res. 11:759 (1986).
2. A. P. Streissguth, S. Landesman-Dwyer, J. C. Martin and D. W. Smith, Teratogenic effects of alcohol in humans and laboratory animals, Science, 209:353 (1980).
3. C. Guerri, A. Esquifino, R. Sanchis and S. Grisolía, Growth, enzymes and hormonal changes in offspring of alcohol-fed rats, in: "The Mechanisms of Alcohol Damage in Utero", (Ciba Foundation), pp. 81, London. (1985).
4. M. W. Miller, Effects of alcohol on the generation and migration of cerebral cortical neurons, Science, 233:1308 (1986).
5. J. Renau-Piqueras, R. Zaragoza, P. De Paz, R. Baguena, L. Megias and C. Guerri, Effects of prolonged ethanol exposure on the glial fibrillary acidic protein-containing intermediate filaments of astrocytes in primary culture: A quantitative immunofluorescence and immunogold electron microscopic study, J. Histochem. Cytochem. 37:229 (1989)
6. J. Renau-Piqueras, M. Sancho-Tello, R. Zaragoza and C. Guerri, Ethanol on the development of astrocytes in primary culture, in: "Alcohol Toxicity and Free Radicals", R. Nordmann, C. Ribiere and H. Rouach, eds., pp. 269, Pergamon, Oxford. (1988).
7. L. Hertz, Astrocytes, in: "Handbook of Neurochemistry", A. Lajtha, ed., pp. 319, Plenum Press, New York. (1982).
8. L. Lazarides, Intermediate filaments: a chemically heterogeneous developmentally regulated class of proteins, Ann. Rev. Biochem. 51:219 (1982).
9. K. Hallermayer, C. Harmening and B. Hamprecht, Cellular localization and regulation of glutamine synthetase in primary cultures of brain cells from newborn mice, J. Neurochem. 37:43 (1981).
10. M. Sensenbrenner, G. Devillers, E. Bock and A. Porte, Biochemical and ultrastructural studies of cultures rat astrogial cells. Effect of brain extract and dibutyril cyclic AMP on glial fibrillary acidic protein and glial filaments, Differentiation, 17:51 (1980).

REGULATION OF PROTEIN SYNTHESIS DURING BRAIN

DEVELOPMENT. ROLE FOR INITIATION FACTOR 2

Matilde Salinas, Alberto Alcázar, María E. Martín
Maica Azuara and Juan L. Fando

Departamento de Investigación
Hospital Ramón y Cajal
28034 Madrid, Spain

INTRODUCTION

The decline in overall protein synthesis and concomitant increase in specific protein synthesis during brain development indicate the existence of mechanisms which control qualitative and quantitative modulations in protein biosynthesis as a function of physiological need. The decrease in the rate of protein synthesis in mammalian brain tissue during neural development is measurable at several levels of neural complexity including *in vivo*, cortical slices and cell-free systems. The aggregation of ribosomes forming polyribosomes also diminishes during brain development and an age-dependent decrease in the activity of protein initiation factors in rat brain has also been reported (1-3). These findings suggest that the control of protein synthesis during brain development may be exerted at the level of initiation.

There is no doubt that translational controls do contribute importantly to the functioning of eucaryotic cells. Such mechanisms operate on a variety of components of the translational apparatus, mainly at the initiation step. One of the earliest documented examples of regulation at the level of initiation occurs in mammaliam reticulocytes by phosphorylation of initiation factor 2 (eIF-2), crucial for the initiation of polypeptide chains. The kinase responsible for the phosphorylation of the factor is a heme-sensitive kinase (HCI) and there also exists in mammaliam cells another distinct kinase (DAI) that acts similarly. This kinase is activated by double-stranded RNA and is present in elevated amounts in interferon-treated cells. What role it plays in normal cells has not been established. Yet, the characterization ot two such kinases strongly hints that there may be others, activated by different agents but all exerting their regulatory effect by phophorylating eIF-2 (4).

Results from our group in the last 5 years have reinforced the participation of eIF-2 in the regulation of protein synthesis during brain development (3,5-7). The results we present in this article are concerned with the improvement of a cell-free system from brain tissue to study the molecular mechanism of the process, and the developmental changes in the activity of casein kinase II (CKII), a known eIF-2 (β) kinase in most of the systems studied.

Endocrine and Biochemical Development of the Fetus and Neonate
Edited by J. M. Cuezva *et al.*
Plenum Press, New York, 1990

RESULTS AND DISCUSSION

Protein synthesis in the mammaliam brain is impaired under a great variety of diverse conditions such as amino acid imbalance, hypoxia, cerebral ischemia, etc, and several cell-free systems from the brain have been described to study the mechanism which underlie these alterations. Low translational rates when compared with *in vivo* data, and very poor initiation of translation of endogenous mRNA, are the most serious pitfalls in these cell-free systems from brain tissue. To extend and confirm our previous data concerning the measurement of developmental changes in brain protein synthesis *in vitro*, we have optimized a system prepared from the postmitochondrial supernatant of rat brain, by introducing the following modifications: 1) Exclusion of chloride ions which inhibit the initiation step. 2) Tissue homogenization 1:2 (w/v). 3) Amino acid concentration of 25-50 μM in order to avoid inadequate charging of tRNAs due to the high Km values of some of the aminoacyl-tRNA synthetases. Under these conditions the method is also independent of endogenous amino acid concentrations. 4) High concentrations of ATP, GTP and energy regenerating system, since mammaliam cell-free systems lose activity very rapidly, if the nucleotides are not maintained in adequate levels. 5) A temperature of 30°C was used instead of 37°C to extend the period of time during which the system remains active.

Protein syntesis rate obtained with the new conditions is higher than that previously described (3) and similar to that obtained with cultured cells. The system initiates translation of either endogenous or exogenous mRNA, 40-50% of the activity is due to reinitiation of protein synthesis, as shown by inhibition with edeine and aurintricarboxylic acid, and the

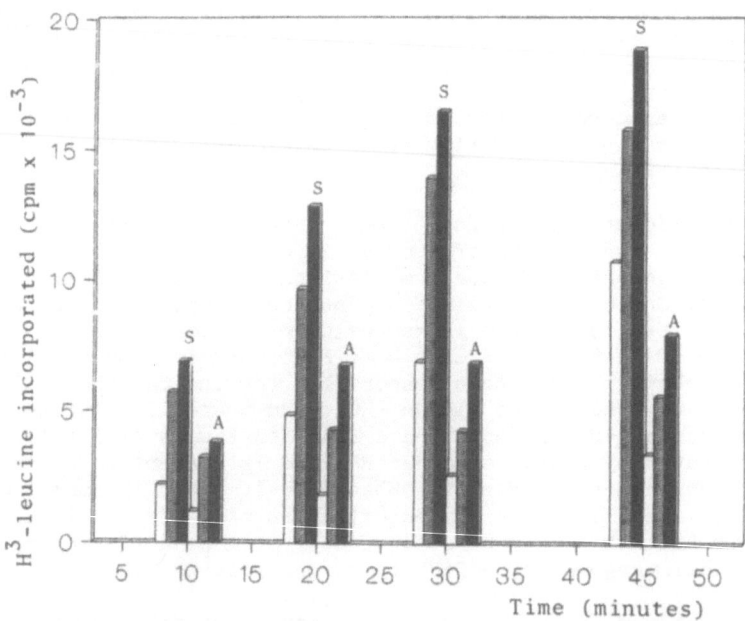

Fig. 1. Protein synthesis as a function of time and amino acid concentration. Protein synthesis was determined in postmitochondrial supernatant from the brain of suckling (S) and adult (A) animals, at four points of time at different amino acid concentrations: □ 25μM, ▨ 50μM and ■ 75μM. Each point represents the mean of 6-10 values (S.E.M. approximately 10% of the mean, not shown).

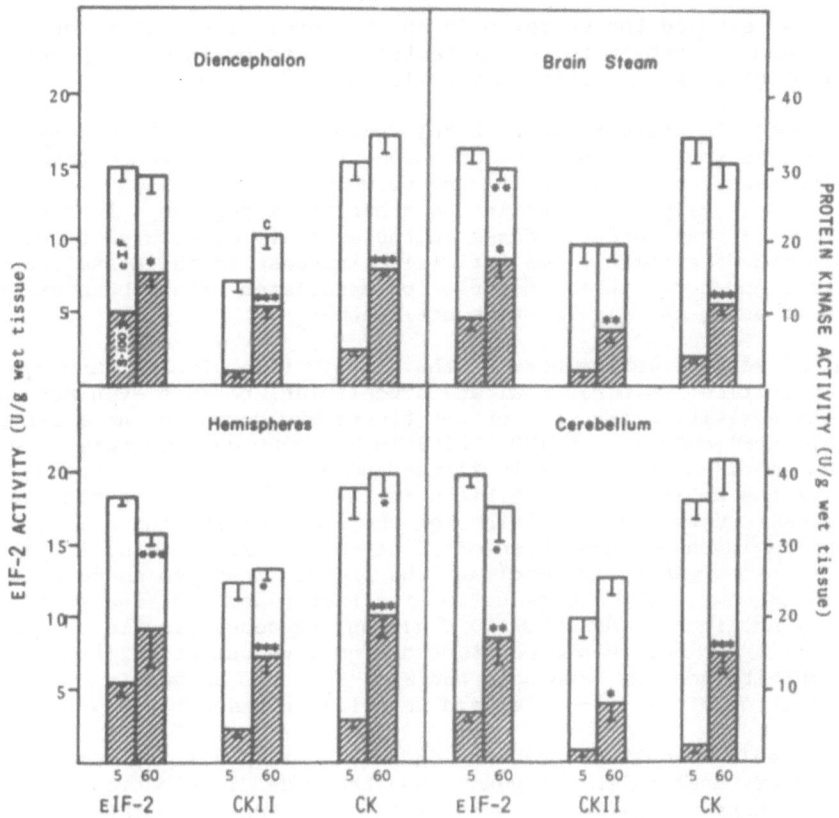

Fig. 2. Total activity (soluble plus associated with ribosomes)
of eIF-2 factor, total casein kinase and casein kinase II
during development. Empty bars represent the activity
(expresed in U/g of tissue) in cIF fraction, meanwhile
cross bars represent the activity present in the soluble
fraction. Statistical difference between 5 and 60 day-old
animals for each of the fractions and activities: *p<0.05;
**p<0.01; ***p<0.001. Superscript letters show differences
between 5 and 60 day-old animals for total activity (S-
100 plus cIF fractions): C.p<0.05.

addition of globin mRNA results in the synthesis of a protein which
comigrates with commercial rabbit globin. Globin synthesis is completely
inhibited in the presence of initiation inhibitors (results not shown).

As can be seen in Figure 1 the method clearly detects the different
rate in protein synthesis obtained with the postmitochondrial fractions
from the brain of suckling (4-10 day- old) and adult (60 day-old) animals.
It may be a useful tool for studying the translational mechanisms that
regulate the decrease of protein synthesis not only in the developing
brain, but also in the aging process or under conditions that disrupt
protein synthesis in this tissue.

We have demonstrated that eIF-2 from the brain is phosphorylated on
its β subunit by casein kinase II (CK II) (8). Due to the lack of
information about the ontogeny-related changes of CK II activity in brain
tissue and the special attention given in recent years to the possible
participation of phosphorylation of the β subunit of the factor on its

function, we studied the subcellular and regional distribution of casein kinase II and its substrate, eIF-2 factor, during brain development. The measurement of both activities was realized as described (3,6).

The specific activity of both the factor and the protein kinase is much higher in the protein fraction associated with ribosomes (crude initiation fraction, cIF) than in the soluble fraction (S-100) and slightly higher in the hemispheres than in the other three regions, diencephalon, cerebellum and brain stem. Changes in the activity of both proteins are in parallel with development, the activities increase in the postmicrosomal supernatant and decrease in the fraction associated with ribosomes from suckling to adult animals (results not shown).

The close parallelism between the distribution of eIF-2 and casein kinase activities, both regional and subcellular, becomes even more evident when total activities per gram of wet tissue are compared. As shown in Figure 2, total activity (S-100 + cIF) in 5 and 60 day-old rats is almost the same for eIF-2, total casein kinase and casein kinase II (total activity minus activity not inhibited in the presence of heparin) in hemispheres, cerebellum and brain stem and only casein kinase II shows a higher value in the diencephalon of 60 day-old rats. The distribution of the activity between the cytosol and the protein fraction associated with ribosomes is completely analogous for the factor and the kinase. As the activities are increasing in S-100 fraction and decreasing in cIF fraction, total activity is maintained constant during development. Similar changes in the distribution has been reported also for this enzyme in the differentiation of 3T3-L1 cells, and for other kinases.

The almost ubiquitous presence of casein kinase II in mammalian and avian species, drosophila and yeast, and the wide distribution of its substrates among the different subcellular fractions, support the hypothesis that this kinase participates in total cell metabolism integration, although no physiological modulators have yet been found for CK II (9). Our data showing a different distribution of the enzyme in the soluble and the ribosomal fraction, depending on the biosynthetic activity of the tissue, may help to explain the mode of action of the kinase, since in recent years great importance has been given to the distribution of regulatory proteins in soluble and membrane-bound, nuclear or ribosomes-bound fractions as a possible mechanism of action for many of them.

In 1981 it was reported that phosphorylation of β subunit of eIF-2 factor increased its capacity to form ternary complex with GTP and met-tRNAi. The data were consistent with the assumption that phosphorylation of β subunit to a certain level was a prerequisite for its activity. It has also been claimed more recently that β subunit is necessary to allow optimal interaction between eIF-2 and the guanine nucleotide exchange factor, GEF. It has been also been demonstrated that GEF phosphorylation by CK II increased its activity (10).

Our findings reveal this type of close relationship in the ontogeny-related changes for the two proteins that reinforced the involvement of CK II in the regulation of protein synthesis during development by phosphorylation of eIF-2(β) factor. The posible regulation of protein chain initiation by two distinct phosphorylation/dephosphorylation mechanisms with opposite effects: negative, by the phosphorylation of eIF-2(α) by HCI, DAI or similar; and positive through the phosphorylation of eIF-2(β) and GEF by casein kinase II, is a exciting model for translational control during development.

ACKNOWLEDGMENTS

This work was supported by grants B23/85 from the CAYCIT, Ministerio de Educación y Ciencia and 88/1055 from the FISSs, Ministerio de Sanidad y Consumo (Spain). A. Alcázar acknowledges a fellowship from the FISSs (86/490).

REFERENCES

1   T.C.Johnson, Regulation of protein synthesis during postnatal maturation of the brain, J.Neurochem 27:17 (1976).

2. D.S.Dunlop, W.Van Elden and A.Lajtha, A method for measuring brain protein synthesis in young and adult rats, J.Neurochem. 29:939 (1977).

3. C.Cales, J.L.Fando, C.Azuara and M.Salinas, Developmental studies of the first step of the initiation of brain protein synthesis, role for initiation factor 2, Mech.Ageing Dev. 33:147 (1986).

4. V.Pain, Initiation of protein synthesis in mammalian cells, Biochem.J. 235:625 (1986)

5. C.Cales, M.Salinas and J.L.Fando, Functional heterogeneity of GEF-free initiation factor 2 purified from suckling and adult rat brain, FEBS Lett. 190:307 (1985)

6. C.Cales, J.L.Fando, T.Fernandez, A.Alcázar and M.Salinas, Initiation factor 2 isolated from rat brain contains kinase activities responsible for its phosphorylation, Neurosci.Lett, 61:333 (1985).

7. C.Cales, J.L.Fando, A.Alcázar and M.Salinas, Differential subcellular distribution of guanine nucleotide exchange factor in suckling and adult rat brain, Neurosci.Lett. 87:271 (1988).

8. A.Alcázar, E.Mendez, J.L.Fando and M.Salinas, Specific phosphorylation of the β subunits of eIF-2 factor from brain by three different protein kinases, Biochem.Biophys.Res.Commun. 153:313 (1988).

9. G.M.Hathaway and J.A.Traugh, Casein kinase II, Methods Enzymol. 99:317 (1983).

10. J.N.Dholakia and A.J.Wahba, Phosphorylation of the guanine nucleotide exchange factor from rabbit reticulocytes regulates its activity in polypeptide chain initiation, Proc.Natl.Acad.Sci.USA. 85:51 (1988).

LIPID METABOLIC INTERACTIONS IN THE MOTHER

DURING PREGNANCY AND THEIR FETAL REPERCUSSIONS

Emilio Herrera, Miguel A. Lasunción, Diego Gómez-Coronado, Antonia Martín and Bartolomé Bonet

Departamento de Bioquímica e Investigación
Universidad de Alcalá de Henares and
Hospital Ramón y Cajal
28034 Madrid, Spain

LIPID METABOLISM IN THE MOTHER

Aside from the products of conception, maternal body weight increase during gestation corresponds to increased mass in certain maternal structures and a specific accumulation of fat, as shown both in human pregnancy (1) and in the rat (2-4). This fat accumulation occurs during the first part of gestation (2,4) and is related to both maternal hyperphagia, since it is not found in food restricted (5,6) or food deprived rats (2,7), and enhanced lipogenesis and unmodified or even augmented extrahepatic lipoprotein lipase (8,9). The tendency to accumulate fat ceases during late gestation (1,2,4) because the maternal lipid metabolism changes to a catabolic condition as shown by the increased lipolysis (10, 11) and the reduced circulating triglycerides uptake by adipose tissue (12). The latter is a consequence of reduced adipose tissue lipoprotein lipase activity (4,13-15). These changes, together with the hepatic overproduction of triglycerides (16-18) and enhanced absorption of dietary lipids (19) boosts the circulating triglyceride concentration in the mother during late gestation (4,8,13,20-26).

Table 1. Plasma lipidic components in 20-day pregnant and nonpregnant rats

	VIRGIN	20-DAY PREGNANT
FFA (μM)	282±24	438±49**
Glycerol (μM)	81±16	152±21*
Triglycerides (mg/dl)	77±6	390±61***
VLDL-triglycerides (mg/dl)	38±7	233±25***

Determinations were carried out as previously reported (4). Mean ± SEM; n = 5-8 rats/group. Statistical comparison of pregnant and virgin rats: *p<0.05; **p<0.02; ***p<0.001.

Endocrine and Biochemical Development of the Fetus and Neonate
Edited by J. M. Cuezva *et al.*
Plenum Press, New York, 1990

213

The combined changes occurring in maternal adipose tissue, liver and intestinal activity during late gestation are responsible for the increments in circulating lipid metabolites concentration. As shown in Table 1, in the 20-day pregnant rat plasma level of free fatty acids (FFA), glycerol and triglycerides are greatly enhanced as compared to age and sex matched virgin controls. As also shown in Table 1, maternal hypertriglyceridemia mainly corresponds to an increase in circulating VLDL-triglycerides, the values of which parallel the increase found in plasma triglycerides in the same animal. However rising levels of chylomicrons in circulation may also contribute to maternal hypertriglyceridemia.

Enhanced plasma VLDL-triglyceride concentrations during late pregnancy in both the rat and human agree with several previous reports including our own (4,26). However, it is not well established whether this change implies an alteration in the intrinsic composition of the VLDL particles or a change in their number. We previously found an enhanced triglycerides/cholesterol ratio and a modified elution profile from heparin-Sepharose column chromatography of VLDLs from 20-day pregnant rats as compared to virgin controls (4,12). Because these differences may also imply different metabolic behaviour of these VLDL particles, the subject is evaluated below.

As shown in Figure 1, the percent composition of VLDL particles remain unchanged in 20-day pregnant rats despite the increase in lipoprotein concentration comented above. As shown in Figure 2, VLDL diameter and particle mass do not differ between 20-day pregnant rats and virgin controls, whereas ther seem to be differences in the apoprotein content. As also shown in Figure 2, it appears that, apo E/apo C-II and apo E/apo C-III ratios are lower in VLDL from 20-day pregnant rats whereas apo C-II/apo C-III ratio is higher than in those from virgin controls.

Since differences in the apolipoprotein content in lipoprotein particles could affect their efficiency as substrates for specific enzymes (27) and their consequent catabolism, the differences we found in the VLDL apoprotein composition between pregnant and virgin rats forced us to examine whether there were also differences in their behaviour as substrates for the two immediate key enzymes of their catabolism, lipoprotein lipase and hepatic lipase (28). As shown in Figure 3, it appeared that when the same amount of VLDL-triglycerides from 20-day pregnant and virgin rats were offered either to the purified lipoprotein lipase or to

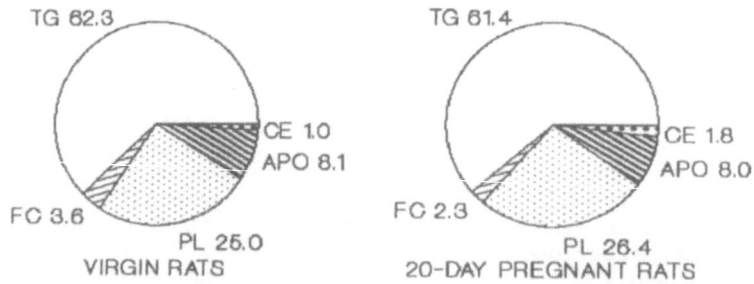

Fig. 1. Percent composition of rat VLDL. Values correspond to a pool of plasma coming from 4 rats/group. TG= Triglycerides; FC= Free cholesterol; CE= Cholesteryl esters; PL= Phospholipids; Apo= Apoproteins. VLDL were isolated by sequential ultracentrifugation for the corresponding determinations as previously shown (4).

214

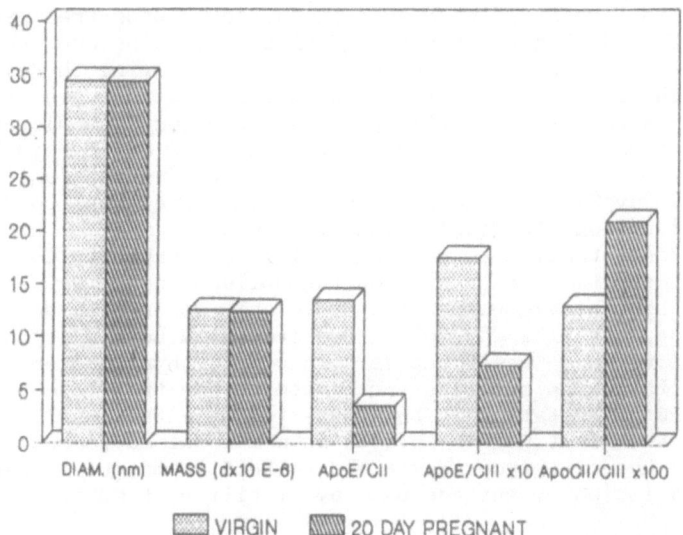

Fig. 2. Diameter, mass and apoprotein ratios of VLDL from virgin
and 20-day pregnant rats. VLDL particle diameter and mass
were estimated (121) by assuming that they are spherical.
For apoprotein determinations VLDL were delipidized with
acetone:ethanol (1:1 by vol) and proteins resuspended in
0.01 M Tris pH 8.6 containing 8 M urea and 0.01 mM dithiothreitol. Sixty to eighty μg of VLDL protein were
applied to 7.5% acrylamide gels containing ampholytes
pH 4-6.5, which were run at 2000 V for 150 min. After
staining with Coomassie brilliant blue R-250, the gels
were used for densitometric scanning at 560 nm in a
Beckman DU-8 spectrophotometer. Relative areas of each
band were used to calculate the apoprotein mass ratios.

the hepatic lipase, their respective hydrolitic efficiency was very similar
with substrate coming from either rat groups. Actually,
although no statistical comparison between the groups is given, it appeared
that in the case of lipoprotein lipase the amount of VLDL-triglycerides
hydrolized is even slightly higher when the particles came from pregnant
rather than virgin rats (Figure 3), clearly indicating that the catalytic
efficiency of this enzyme is by no means reduced in the former. We may
therefore conclude that abundance of VLDL particles in maternal circulation
during late pregnancy is not a consequence of their modified apoprotein
composition, which could have altered their characteristics as adequate
substrates for the enzymes responsible for their catabolism. They are,
however, an adequate floating lipidic source for use wherever required (see
below).

Consequences on Late Gestation of Maternal Lipidic Accumulation During
the First Half of Pregnancy

We believe that changes in lipid metabolism during late gestation are
not only a direct consequence of the metabolic adaptations occurring during
that stage but may also be extensively influenced by the maternal capacity
to accumulate lipid stores during the first half of pregnancy. There are no
previous studies to directly support this hypothesis, but we had carried
out two different experiments that are consistent with it. The first one
corresponds to unpublished data from our recently reported study with
thyroidectomized pregnant rats (29).

As shown in Figure 4, increments in both conceptus-free maternal body and liver weights during the first 12 days of gestation are greatly impaired in pregnant thyroidectomized rats not receiving thyroid hormone treatment. Such reduction in the mass of maternal structures in hypothyroid pregnant rats is maintained until the end of the gestational period even when rats are treated with daily substitution doses of thyroid hormones (1.8 μg of thyroxine/100 g body wt.) from day 12 of gestation to term (Figure 4). At day 21 of gestation these animals are known to show significant reductions in fetal weight (29) and in maternal circulating triglyceride concentration as well as an impaired capacity to mobilize endogenous fat depots with starvation (unpublished results). As also shown in Figure 4, when animals were hypothyroid only during the second half of gestation (from day 12 to day 20) no reductions in both maternal liver and conceptus-free weights were found as compared to thyroidectomized rats receiving the thyroxine substitution treatment for the whole period (from day 0 to day 20 of gestation). These data indicate that sufficient maternal stores during the first half of gestation are required to sustain maternal structures and hypertriglyceridemia and to fulfill metabolic needs during late pregnancy including the adequate availability of substrates for the fetus.

Figure 5 summarizes a second experiment that was addressed in establishing the role of increased fat accumulation during the first part

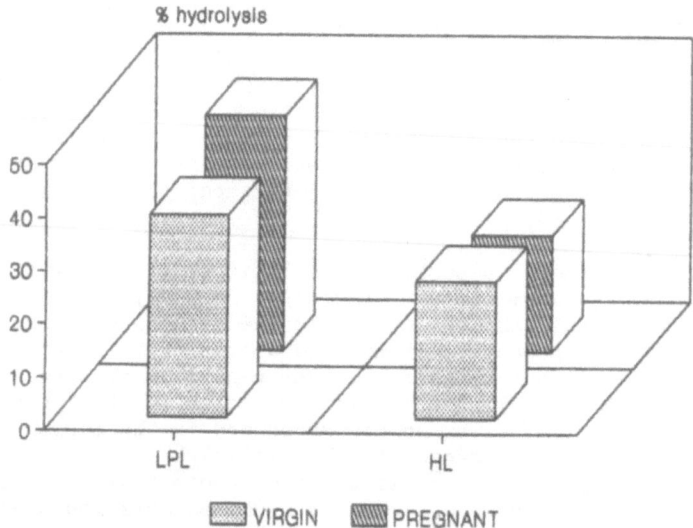

Fig. 3. Hydrolysis of tri(^{3}H)-olein labelled VLDL from virgin and 20-day pregnant rats by purified lipoprotein lipase (LPL) and hepatic lipase (HL). Purified VLDL from a pool of plasma from 5 virgin or from 20-day pregnant rats were labelled *in vitro* with tri(^{3}H)-olein (122) and amounts corresponding to 82 or 90 nmoles/ml of triglyceride from each preparation were respectively incubated at 37°C for 2 h in the presence of either lipoprotein lipase or hepatic lipase purified from human post-heparin plasma by heparin-Sepharose affinity chromatography. Hydrolysis was estimated as the appearance of ^{3}H-free fatty acids in the media, and expressed as % of the initial tri (^{3}H)-olein.

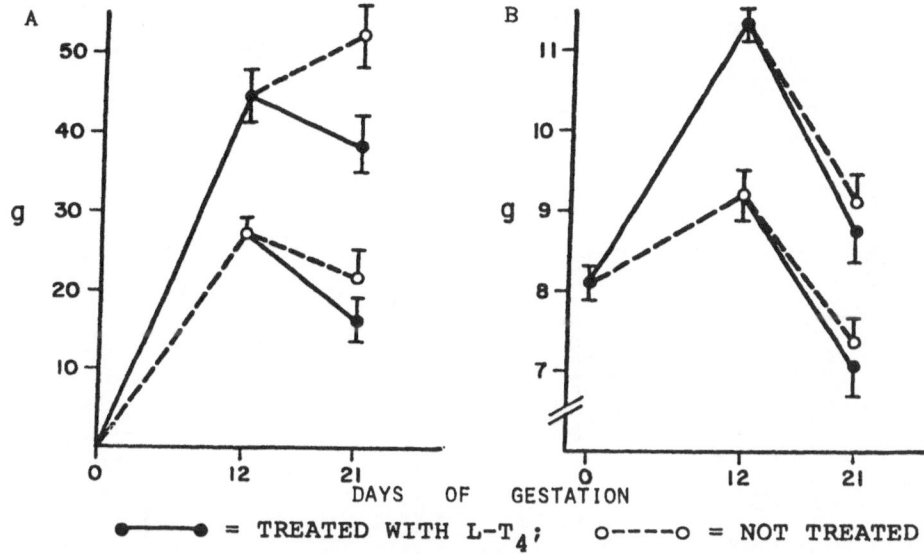

Fig. 4. Conceptus-free maternal body weight (A) and liver weight (B) in untreated thyroidectomized rats (o---o) or rats treated with 1.8 μg thyroxine/100 g body wt for different gestational periods (●——●). Rats weighing 180-200 g were surgically thyroidectomized at day 0 of gestation (the day that sperm appeared in vaginal smears) and intraperitoneal treatments with either thyroxine or saline were given, starting the day after surgery during the days indicated in the Figure.

of pregnancy in the metabolic adaptations occurring during late gestation. Rats were made diabetic by the intravenous administration of streptozotocin (45 mg i.v./Kg body wt.) prior mating, at which time they were divided into three groups: 1) *Controls*, which received a daily subcutaneous replacement insulin therapy (1.5 IU/100 g body weight); 2) *D12-20* that received such treatment from day 0 until the 12th day of gestation, being kept diabetic between days 12 and 20 ; or 3) *D* that did not receive any treatment throughout pregnancy. All animals were studied at day 20th of gestation. As shown in Figure 5, when compared to *Controls*, both *D* and *D12-20* rats were highly hyperglycemic, although, as expected, blood glucose levels were higher in the former. Lumbar fat pad weight was, however, greatly reduced in animals from group *D* and stable in those from group *D12-20*, indicating that diabetes during the first half of gestation, but not during the second, impairs maternal capacity to maintain augmented lipidic stores.

As also shown in Figure 5 plasma triglyceride levels were greatly augmented in both *D* and *D12-20* animals. The values in the latter were even higher than in the former. Since no lipid stores were available in the *D* animals the efficient lipolytic activity required to sustain the endogenous triglycerides overproduction was impossible, and these findings indicate that hypertriglyceridemia in these animals is mainly caused by enhanced reductions in the use of circulating triglycerides by extrahepatic tissues. On the contrary, in the *D12-20* animals, where the lumbar fat pad weight remained at the same level as in *Controls*, their enhanced hypertriglyceridemia must be the result of a greatly augmented lipolytic activity which could be sustained by the normal fat stores they had accumulated during the first half of gestation, when they were not diabetics.

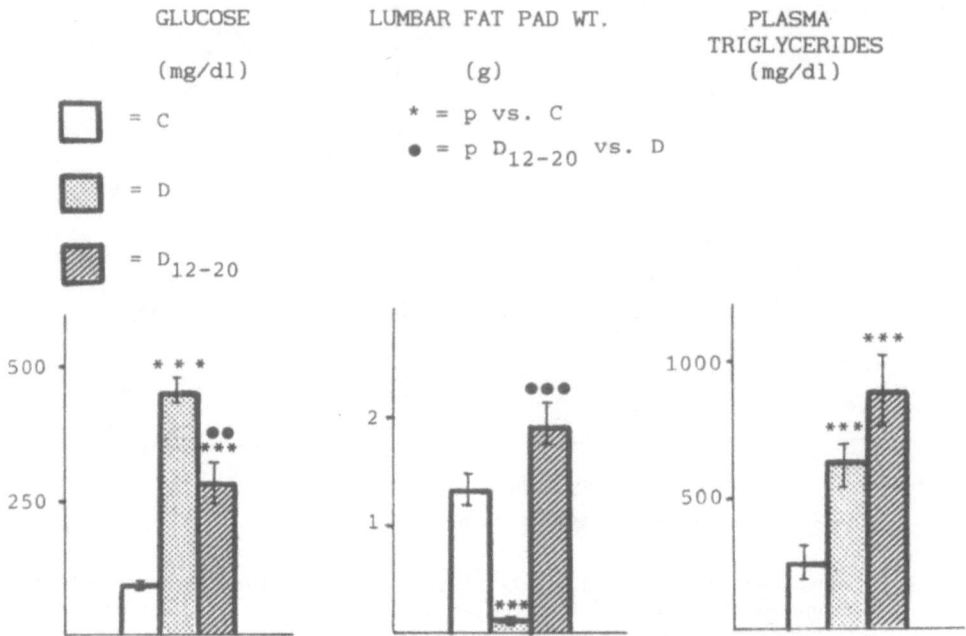

Fig. 5. Plasma glucose and triglyceride concentration and lumbar fat pad weight in 20 day-pregnant streptozotocin treated rats. Animals were intravenously treated with 45 mg/kg body weight of streptozotocin prior mating. One group was treated daily with subcutaneous replacent insulin therapy from the time of receiving the streptozotocin until sacrifice (*Controls*). Another group received the replacement insulin therapy only up to the 12th day of gestation, and therefore were maintained diabetic from that time until sacrifice (*D12-20*), and a third group that did not receive any therapy (*D*). Determinations of glucose and triglycerides were carried out as previously (4).

In summary, present findings indicate that maternal fat store accumulation during the first half of gestation has a pivotal importance on the development of maternal hypertriglyceridemia during late gestation. The question now is to discover what the role of these adaptations in the mother and fetus is.

CONSEQUENCES OF MATERNAL HYPERLIPIDEMIA

Increases in maternal circulating glycerol levels (30) caused by active lipolytic activity allow the use of this metabolite as a preferential gluconeogenetic substrate (31), and so contributes to the maintainance of sufficient glucose production for fetal and maternal tissues. Increments in circulating levels of triglycerides facilitate their use by the mammary glands prior to parturition, and this process is driven by the increased lipoprotein lipase activity found in this tissue at very late gestation (14,15). No other parts of the increase in circulating lipidic components in the fed mother during late gestation seem to directly benefit her metabolic needs. This increase, however, may benefit the fetus since this gestational period coincides with the rate of maximal fetal accretion, when the substrate, metabolic fuel, and essential component requirements of the fetus are greatly enhanced.

The lipid component may also constitute a circulating fuel store for both mother and fetus, which is easily accessible under conditions of food deprivation, and may explain the well-known finding of enhanced ketogenesis in the mother under fasting conditions (24,25,32,33). This hypothesis is sustained by the increased arrival of FFA in the liver as a result of greatly enhanced adipose tissue lipolysis (10, 11) and, by the increase in liver lipoprotein lipase activity recently published (4) which would facilitate maternal liver use of circulating triglycerides as ketogenic substrates.

The enhanced arrival of ketone bodies in fasted maternal tissues allows the ketone bodies to be used as metabolic fuels, and may spare other more limited and essential substrates, like amino acids and glucose, for transport to the fetus. The fetus also receives maternal ketone bodies through the placenta and their use plays an important role in the fetal metabolic economy under conditions of maternal food deprivation.

TRANSFER OF LIPIDIC PRODUCTS TO THE FETUS

Having described the metabolic adaptations in the mother which cause her hyperlipidemic condition during late gestation, we shall review current bibliography treating the mechanism controling the placental transfer of FFA, glycerol and ketone bodies, the three maternal lipid metabolism products that most easily reach the fetus. Understanding the placental transfer process and respective metabolic fates of these products in the fetus should improve our grasp of the fetal consequences of rising lipid levels in maternal circulation.

Since the level of FFA, glycerol and ketone bodies in maternal circulation are a consequence of adipose tissue lipolytic activity, which is intensely enhanced in the starved late pregnant rat (10,30), the maternal/fetal ratio of these metabolites under this particular condition must be examined.

Table 2 summarizes the comparison of plasma levels in 48 h starved virgin rats and 19-day pregnant rats and their fetuses. It can be seen that whereas fetal FFA and glycerol levels are much lower than in the mothers, ketone bodies have a similar value. In an initial approach, these maternal/fetal concentration differences reflect the efficiency or magnitude of the placental transfer process.

Nutrient placental transfer may be accomplished by means of different mechanisms, such as facilitated diffusion, active transport and simple diffusion (35-38). Simple diffusion seems to be the common and unique mechanism for the lipidic-derived moieties, although some qualifications must be made. Simple diffusion is carried out from a region with a high concentration to one with a low concentration, and the rate of transfer is directly proportional to the concentration gradient and decreases with molecular size and hydrosolubility (39). Other factors can also specifically affect the efficiency of placental nutrient transfer (39-43): uterine and umbilical blood flows; the intrinsic placental metabolism (utilization versus production); and placental structure. The contribution of some factors which involve simple diffusion, like blood flow, is the same with any nutrient, but the contribution of other factors varies with each nutrient and must be considered specifically.

Free Fatty Acids

In addition to the essential fatty acids which support growth (44) and brain development (45) the fetus also needs the non-essential lipids from

Table 2. Circulating levels of FFA, glycerol and
ketone bodies in 48 h starved virgin, and
19-day pregnant rats and their fetuses

	FFA (μM)	Glycerol (μM)	Ketone bodies (μM)
Virgin rats	443±21	153±10	982±67
Pregnant rats	739±59***	264±16***	2569±210***
Fetuses	45±12+++	36±13+++	2470±157
Maternal/ fetal ratio	16.6±0.8	7.0±0.5	1.02±0.08

Determinations were carried out as previously shown
(4). Mean ± SEM of 6-8 rats/group. Statistical comparison of pregnant versus virgin rats: ***, p<0.001
and of fetuses vs. their mothers, +++, p<0.001. Not
significant, p>0.05.

the mother. They are stored in fetal body fat, and become an important
substrate during early post-natal life (46). Body fat at term can represent
a substantial percentage of body weight (10% in the guinea pig and 16% in
man) (47), and the fetus cannot synthesize enough fatty acids to satisfy
these requirements. Using sheep (48) and rats (49), the initial studies on
placental lipid transfer suggested that the amount of fatty acids
transferred from the mother to the fetus was minimal, but subsequent
investigations in species like the rabbit (50), the guinea-pig (51,52), the
rat (54) and different primates (53), suggest that the transfer of fatty
acids across the placenta can surpass the normal fetal lipid storage
requirements (55).

Like other fats, fatty acids are insoluble in water and must be
transported in the blood either as albumin-bound FFA, or else in their
esterified form (triglycerides, phospholipids and esterified cholesterol)
which associates with other lipids and proteins to form lipoproteins.
Maternal FFA, esterified fatty acids that have been hydrolyzed at the
placental level, and unmodified lipoproteins make up the potential sources
for the fatty acids that reach the fetus.

Placental transfer of fatty acids varies considerably between
different species. In general, fatty acid transfer is limited when the
placental barrier is constitued by multiple maternal and fetal cell layers
(sheep, pig and cat), whereas the net flux can be high when the placental
barrier is only constituted by fetal layers (man, rabbit, rat and guinea-
pig). The relative fatty acid concentrations entering the fetus from the
placenta in these species reflect the circulating maternal free fatty acid
concentrations with the common exception of arachidonic acid (56). Since a
higher proportion of arachidonic acid has been found in fetal plasma than
in maternal plasma in both ruminant (57) and non-ruminant species (58-61),
it has been inferred that arachidonic acid synthesis by the placenta must
contribute heavily to the fetal supply of this fatty acid (62-65).

Current evidence suggests that fatty acid transfer across the placenta
is not selective and since essential and non-essential fatty acids use the
same transfer mechanism, most investigators in maternal-fetal transfer
experiments use ^{14}C-palmitic acid to represent all the fatty acids. The
quantity of fatty acid transferred varies considerably between species

(66), but in all instances the transplacental non-esterified fatty acid gradient grossly regulates the system.

In vitro studies using perfused placenta and/or cultured trophoblast cells must be employed to study the mechanism of FFA transfer in the human placenta, and correlate FFA levels in maternal and umbilical blood. A more direct approach may be used with experimental animals to study the placental transfer of FFA. We infused $(1-{}^{14}C)$-palmitate tracer through the left uterine artery of 20-day pregnant rats for 20 min before comparing the amount of label in the placentas and fetuses from the left uterine horn with that found in the ones from the right horn (67) as we have done before with different substrates (68-71). While the left uterine horn received the tracer directly, it reached the right horn after dilution in the mother's circulation, and so the amount of substrate transferred to the fetus can be calculated as a function of the values for the maternal FFA concentration, the difference in radioactive levels in fetuses between the left and right uterine horns, and the left uterine blood flow. Results have been recently reported by us (72), and they may be summarized as follows: the estimated FFA transfer was significantly above zero, indicating that it is significant. Absolute value of FFA transfer appears to be higher than levels previously found for glycerol even though lower than those found for glucose or alanine (72).

Around 50% of the ${}^{14}C$-lipids retained in the placentas after tracer infusion corresponded to esterified fatty acids, so a certain proportion of the FFA's reaching the placenta must have been esterified. Although the participation of fatty acid esterification in the FFA transfer process has not been ascertained as yet, an active placental capacity to form esterified fatty acids from maternal FFA has been described in man and other species (73-76). The presence of an active enzymatic glyceride hydrolytic system (phospholipase and triacylglycerol lipase), which would ensure rapid triglyceride and phospholipid turnover, indicates that the esterification/hydrolysis cycle in the placental cells is one type of placental FFA transport, as Szabo et al. (77) and Hummel et al. (78) already proposed.

Free fatty acid contribution by maternal circulating triglycerides in the rat (79,80), the rabbit (81), the guinea-pig (82,83) and man (84) have been demonstrated. However, the passage of intact triglyceride across the placenta has not been detected yet.

Lipoprotein lipase activity has been detected in the placentas of all the species studied (4,15,80,81,85,86). Placental triglyceride hydrolysis with a direct transfer of released non-esterified fatty acids to the fetus has been considered as a source for this activity, but direct studies with in situ perfused guinea-pig placenta have shown that this accounts for a very small percentage of all the fatty acid transferred to the fetus (83). Therefore, maternal triglyceride-rich lipoproteins that have been broken down in the placenta by lipoprotein lipase are of minor quantitative importance as a fetal source of fatty acids under normal conditions. However, under conditions of exaggerated maternal hypertriglyceridemia this system of fatty acid supply from esterified maternal fats in the presence of sustained placental lipoprotein lipase activity, become much more important as has been proposed in streptozotocin diabetic rats (80,87).

Glycerol

The active lipolytic activity of maternal adipose tissue causes persistent elevation of plasma glycerol levels during late gestation (30, 31). Although there are interspecies differences, the values for plasma glycerol concentration are generally higher than in the fetus (Table 2).

The maternal/fetal glycerol gradient is greater in those species with epitheliochorial placenta (ruminants) (88,89) than in those with a hemochorial placenta (90-92).

Few experimental studies of placental glycerol transfer have been made with any species. Although the low weight and uncharged molecule of glycerol facilitate placental transfer, the quantity of glycerol actually transferred is notably less than that of other metabolites with similar molecular characteristics such as glucose or L-alanine (71,93). In contrast with the carrier-mediated process followed by these two metabolites, placental glycerol transfer is carried out by simple diffusion (88,94).

The umbilical glycerol balance in sheep indicates that fetal uptake is very low, accounting for no more than 1.5% of all the total oxygen consumed by the fetus (89). Despite the very favorable gradient (90), trans-placental glycerol transfer in man has been impossible to detect. When comparing different substrates, and by using an *in situ* placental infusion technique in the rat, we have found that the level of glycerol transfer is far below that the levels of transferred glucose and alanine (71). We have also found that the fetal-placental unit rapidly processes glycerol into lactate and lipids (91), and believe this rapid utilization may actively contribute to maintaining the consistently high glycerol gradient between maternal and fetal blood (33,90-93).

Accelerated turnover of maternal glycerol seems to be influenced by the high liver glycerolkinase activity which facilitates fast phosphorylation with subsequent conversion into glucose (30,31). Although this mechanism indirectly benefits the fetus by providing glucose from maternal adipose tissue breakdown product, it may limit the availability of sufficient extra glycerol molecules for transfer to the fetus. This hypothesis is supported by our earlier findings after hepatectomy and nephrectomy in pregnant rats where glucose synthesis was negligible and the maternal transfer of glycerol to the fetus was augmented (95).

Consequently, and besides the intrinsic mechanism, placental glycerol transfer seems to be limited by the effective and rapid maternal kidney cortex and liver utilization of this substrate in gluconeogenesis. Although the fetal-placental unit actively uses glycerol, helping to maintain a favorable transfer gradient, the quantitative and physiological role for glycerol in the fetus, except as a preferential substrate for fetal liver glyceride glycerol synthesis (91), seems to be limited under normal conditions. However, under conditions of exaggeratedly elevated maternal glycerol levels, the placental transfer of glycerol could become much more important as a substrate supply for the fetus, but this possibility has not been researched sufficiently.

Ketone Bodies

The plasma levels of ketone bodies during fasting in late pregnancy are greatly increased (24,25,32,96-100), and the same elevation occurs in pregnant diabetics even when they are not fasting (101,102). These conditions coincide with enhanced FFA arrival in the liver consequent to hightened adipose tissue lipolytic activity. Although some maternal tissues (for example: skeletal muscle) use them as alternative substrates under conditions of limited glucose availability secondary to hypoglycemia and/or reduced insulin levels or sensitivity, ketone bodies may easily cross the placental barrier for fetal use as energetic fuels and lipogenic substrates (103-107).

Placental permeability of any compound is affected not only by molecular size and lipid solubility but also by electrical charge. At the physiological pH, 7.4, most of the two main ketone bodies, β-hydroxy-butyrate and acetoacetate, would take dissociated or ionized form, which would retard their diffusion across the placenta to the fetus. In spite of this, man (91,99,108-110), rat (24,32,111) and sheep (98,112), all species studied, rising maternal ketone bodies are followed by rising fetal plasma levels, indicating efficient placental transfer since fetal liver ketogenesis is practically negligible (113).

Although ketone body transfer through the placenta is performed by simple diffusion, it has a high unspecific component (88), and its efficiency varies with species. Whereas the maternal-fetal gradient for ketone bodies is above 10 in the sheep (98,112), in man it is about 2 (91), and in the rat it is close to 1 (24,33,111) (Table 2), indicating that the amount of ketone bodies transported to the fetus is much higher in non-ruminant than in ruminant species. These interspecies differences also affect the particular fetal ketone body oxidative metabolism since in the case of sheep, the ketone body oxidative metabolism account for no more than 3% of the total fetal oxygen consumption (98,114), whereas in fetal rat brain and liver it has been shown that β-hydroxybutyrate can adequately replace the glucose deficit during fasting hypoglycemia (104). This suggests that ketone bodies are much more important for the oxidative metabolism in non-ruminant fetuses under fasting conditions than they are for ruminants.

The enzymes 3-hydroxybutyrate dehydrogenase (E.C. 1.1.1.30), 3-oxoacid-CoA transferase (E.C. 2.8.3.5) and acetyl-CoA acetyltransferase (E.C. 2.3.1.9), which are the key ones to ketone body utilization, have been found in brain and other tissues of both human and rat fetuses (105,115-117). *In vitro* β-hydoxybutyrate oxidation in the human and rat brain depends on substrate concentration rather than maternal nutritional state (103, 104, 107). Fetal kidney, heart, liver and placenta have been demonstrated to oxidize ketone bodies (105, 118), and some have used ketone bodies as substrates for fatty acid and cholesterol synthesis, as has been shown in *in vivo* experiments in the rat brain, liver, placenta and lung after ^{14}C-β-hydroxybutyrate administration to pregnant animals (119). Starvation during the last days of gestation (120), or high fat feeding (117), create maternal hyperketonemia which rises ketone-body metabolism enzyme activity in fetal tissues and protects the fetal metabolic economy by the preferential use of ketone bodies as both oxidative fuels and carbon donors in the anabolic processes.

The maternal ketone bodies that cross the placenta are used by the fetus in non-ruminant species as preferential substrates for both oxidation and lipogenesis, thereby allowing the other substrates (glucose, lactate and amino acids) to be consumed in different pathways. Since both placental transfer and utilization of ketone bodies are concentration dependent, the quantitative contribution of these lipidic products to the fetal metabolism is significant only under conditions (starvation, high fat diet, diabetes, ...) that produce maternal hyperketonemia.

SUMMARY AND FINAL REMARKS

Maternal body fat accumulation during the first part of gestation is mainly sustained by hyperphagia and enhanced lipogenesis. This condition is of pivotal importance in the maintenance of the metabolic adaptations that take place during late gestation. Provoking hypothyroidism or diabetes in pregnant rats during the first 12 days of gestation only, blocks fat accumulation, hinders maternal hypertriglyceridemia and catabolic responses during the second gestational phase and impairs normal fetal growth.

Increased adipose tissue lipolysis, reduced uptake of circulating triglycerides secondary to decreased adipose tissue lipoprotein lipase activity, hepatic overproduction of triglycerides and enhanced absorption of dietary lipids result in the maternal circulating triglycerides rise found during late gestation.

Besides fulfilling her own metabolic needs, sustained maternal hyperlipidemia during late pregnancy is very important for fetal development. Placentally transferred free fatty acids are directly needed by the fetus to sustain the continuous growth of its structures and accumulate circulating lipids and the fat depot. The fetus also benefits from the two other products of maternal lipid metabolism that we have considered: glycerol and ketone bodies.

Placental transfer of maternal glycerol is quantitatively small, but it is used by the fetal liver for the fatty acid esterification and actively contributes to the fetal triglyceride synthesis. The fetus mainly benefits from maternal hyperglycerolemia in a secondary manner. Glycerol is a preferential substrate for maternal gluconeogenesis, and since the fetal oxidative metabolism is fuelled principally by maternal glucose crossing the placenta, the use of this maternal adipose tissue lipolysis product for glucose synthesis actively contributes to fetal glucose supply.

Maternal ketone bodies easily cross the placenta and are efficiently used as either carbon fuels for the oxidative metabolism or as lipogenic substrates by the fetus, specially in non-ruminant species. Since all these processes are concentration-dependent, they do not become relevant unless conditions of maternal hyperketonemia exist. Under healthy physiological conditions they constitute an important support for the fetal metabolism when the availability of other substrates is more limited as is the case during periods of maternal starvation. In this situation brain development seems to be specially preserved as a result of the capacity of the fetus to use these metabolites.

Maternal hypertriglyceridemia during late gestation is also important as preparation for lactation. The level of circulating triglyceride-rich lipoproteins is not as high as would be expected, or may even decline, because of their augmented removal by the mammary gland for milk synthesis.

We may therefore conclude that although placental transfer of lipidic products has normally been undervalued, sustained maternal hyperlipidemia during late pregnancy is of pivotal importance for the offspring not only during their intrauterine life but also during suckling. This hyper-lipidemia is, however the result of several dynamic and intrincate metabolic adaptations, and any deviation in them may directly modify the maternal lipoprotein profile and under pathological conditions the alteration may be permanent.

ACKNOWLEDGMENTS

Present work was carried out in part with a grant from the U.S.-Spain Joint Committee for Scientific and Technological Cooperation (CCA-8510/061). We thank Carol F. Warren from the I.C.E. of Alcalá de Henares University for her editorial help.

REFERENCES

1.   F.E.Hytten and I. Leitch, "The physiology of human pregnancy", (2nd ed.) Blackwell Scientific, Oxford (1971).

2. G. H. Beaton, J. Beare, M. H. Ryv and E. W. McHewry, Protein metabolism in the pregnant rat, _J. Nutr._ 54: 291 (1954).

3. P. López-Luna, T. Muñoz and E. Herrera, Body fat in pregnant rats at mid and late gestation, _Life Sci._ 39: 1389 (1986).

4. E. Herrera, M. A. Lasunción, D. Gómez-Coronado, A. Aranda, P. López-Luna and I. Maier, Role of lipoprotein lipase activity on lipoprotein metabolism and the fate of circulating triglycerides in pregnancy, _Am. J. Obstet. Gynecol._ 156: 1575 (1988).

5. S. A. Lederman and P. Rosso, Effects of food restriction on maternal weight and body composition in pregnant and non-pregnant rats, _Growth_ 44: 77 (1980).

6. B. J. Moore and J. A. Brassel, One cycle of reproduction consisting of pregnancy, lactation, and recovery: effects on carcass composition in ad libitum-fed and food-restricted rats, _J. Nutr._ 114: 1548 (1984).

7. J. M. Fain and R. O. Scow, Fatty acid synthesis _in vivo_ in maternal and fetal tissues in the rat, _Am. J. Physiol._ 210: 19 (1966).

8. R. H. Knopp, M. A. Boroush and J. B. O'Sullivan, Lipid metabolism in pregnancy. II. Postheparin lipolytic activity and hypertriglyceridemia in the pregnant rat, _Metabolism_ 24: 481 (1975).

9. P. K. Kinnunen, H. A. Unnerus, T. Ranta, C. Ehnholm, E. A. Nikkila and M. Seppala, Activities of post-heparin plasma lipoprotein lipase and hepatic lipase during pregnancy and lactation, _Eur. J. Clin. Invest._ 10: 469 (1980).

10. R. H. Knopp, E. Herrera and N. Freinkel, Carbohydrate metabolism in pregnancy. VIII. Metabolism of adipose tissue isolated from fed and fasted pregnant rats during late gestation, _J. Clin. Invest._ 49: 1438 (1970).

11. J. M. Chaves and E. Herrera, _In vitro_ glycerol metabolism in adipose tissue from fasted pregnant rats, _Biochem. Biophys. Res. Commun_ 85: 1299 (1978).

12. E. Herrera, D. Gómez-Coronado and M. A. Lasunción, Lipid metabolism in pregnancy, _Biol. Neonate_ 51: 70 (1987).

13. S. Otway and D. S. Robinson, Significance of changes in tissue clearing-factor lipase activity in relation to the lipemia of pregnancy, _Biochem. J._ 106: 677 (1968).

14. M. Hamosh, T. R. Clary, S. S. Chernick and P. O. Scow, Lipoprotein lipase activity in adipose tissue and mammary tissue and plasma triglyceride in pregnant and lactating rats, _Biochim. Biophys. Acta_ 210: 473 (1970).

15. I. Ramirez, M. Llobera and E. Herrera, Circulating triacylglycerols, lipoproteins and tissue lipoprotein lipase activities in rat mothers and offspring during the perinatal period: effect of postmaturity, _Metabolism_ 32: 333 (1983).

16. J. L. Humphrey, M. T. Childs, A. Montes and R. H. Knopp, Lipid metabolism in pregnancy. VII. Kinetics of chylomicron triglyceride removal in the fed pregnant rat, _Am. J. Physiol._ 239: E81 (1980).

17. R. K. Kalkhoff, S. K. Bhatia and M. L. Matute, Influence of pregnancy and sex steroids on hepatic triglyceride biosynthesis, _Diabetes_ 21: suppl 1 pp. 365 (1972).

18. I. Wasfi, I. Weinstein and M. Heimberg, Hepatic metabolism of (1-14C) oleate in pregnancy, _Biochim. Biophys. Acta_ 619: 471 (1980).

19. J. Argiles and E. Herrera, Appearance of circulating and tissular 14C-lipids after oral 14C-tripalmitate in the late pregnant rat, _Metabolism_ 38: 104 (1989).

20. L. Stemberg, P. Dagenais-Perusse and M. Dreyfuss, Serum proteins in parturient mother and newborn: an electrophoretic study, _Can. Med. Ass. J._ 74: 49 (1956).

21. R. H. Knopp, A. Montes and M. R. Warth, Carbohydrate and lipid metabolism in normal pregnancy. Food and nutrition board: Laboratory indices of nutritional status in pregnancy, pp. 35, (National Academy of Sciences, Washington) (1978).

22. M. Russ, H. A. Eder and D. P. Barr, Protein-lipid relationships in human plasma. III. In pregnancy and the newborn, J. Clin. Invest. 33:1662 (1954).

23. A. T. Konttinen, T. Pyorala and E. Carpen, Serum lipid pattern in normal pregnancy and preeclampsia, J. Obstet. Gynaec. Br. Commonw. 71:453 (1964).

24. R. O. Scow, S. S. Chernick and M. S. Brinley, Hyperlipemia and ketosis in the pregnant rat, Am. J. Physiol. 206:796 (1964).

25. E. Herrera, R. H. Knopp and N. Freinkel, Carbohydrate metabolism in pregnancy. VI. Plasma fuels insulin liver composition, gluconeo-genesis and nitrogen metabolism during late gestation in the fed and fasted rat, J. Clin. Invest. 48:2260 (1969).

26. J. Argiles and E. Herrera, Lipids and lipoproteins in maternal and fetus plasma in the rat, Biol. Neonate 39:37 (1981).

27. R. L. Jackson, S. Tajima, T. Yamamura, S. Yokoyama and A. Yamamoto, Comparison of apolipoprotein C-II-deficient triacylglycerol-rich lipoproteins and trioleylglycerol/phosphatidylcholine-stabilized particles as substrates for lipoprotein lipase, Biochim. Biophys. Acta 875:211 (1986).

28. E. Herrera, M. A. Lasunción, D. Gómez-Coronado and E. Orozco, Papel de la lipoproteina lipasa en el metabolismo de las lipoproteinas, con especial referencia a las VLDL, Drugs of Today 24 (Spl. 1): 65 (1988).

29. B. Bonet and E. Herrera, Different response to maternal hypothyroidism during the first and second half of gestation in the rat, Endocrinology 122:450 (1988).

30. J. M. Chaves and E. Herrera, In vivo glycerol metabolism in the pregnant rat, Biol. Neonate 37:172 (1980).

31. A. Zorzano, M. A. Lasunción and E. Herrera, Role of availability of substrates on hepatic and renal gluconeogenesis in the fasted late pregnant rat, Metabolism 35:297 (1986).

32. R. O. Scow, S. S. Chernick and B. B. Smith, Ketosis in the rat fetus, Proc. Soc. Exptl. Biol. Med. 98:833 (1958).

33. J. Girard, P. Ferré, M. Gilbert, A. Kervran, R. Assan and E. Marliss, Fetal metabolic response to maternal fasting in the rat, Am. J. Physiol. 232: E456 (1977).

34. X. Testar, M. Llobera and E. Herrera, Increase with starvation in the pregnant rat of the liver lipoprotein lipase activity, Biochem. Soc. Transac. 13:134 (1985).

35. J. Dancis and H. Schneider, Physiology of the placenta, in: "Human Growth", Falkner and Tanner (eds.), Plenum Publishing, New York, vol. 1, pp. 355 (1978).

36. H. N. Munro, S. J. Pilistine and M. E. Fant, The placenta in nutrition, A. Rev. Nutr. 3:97 (1983).

37. M. Palacín, M. A. Lasunción and E. Herrera, Transporte de metabolitos a través de la placenta, Rev. Esp. Pediatr. 40(3):163 (1984).

38. D. L. Yudilevich and J. H. Sweiry, Transport of amino acids in the placenta, Biochim. Biophys. Acta 822:169 (1985).

39. E. P. Hill and L. D. Longo, Dynamics of maternal-fetal nutrient transfer, Federation Proc. 39:239 (1980).

40. R. Baur, Morphometric data and questions concernig placental transfer, Placenta (suppl 2):35 (1981).

41. H. Fox, The correlation between placental structure and transfer function, in: "Placental Transfer", G. V. P. Chamberlain and A. W. Wilkinson (eds.), Pitman Medical Ed. London, pp. 15 (1979).

42. F. H. Morris jr., Placental factors conditioning fetal nutrition and growth, Am. J. Clin. Nutrition 34:760 (1981).

43. G. Meschia, Substrate availability and fetal growth, in: "Abnormal fetal growth: biological bases and consequences", F. Naftolin (ed): Dahlem Knoferenzen, Berlin, pp. 221 (1978).

44. G.O.Burr and M.M. Burr, On the nature and role of the fatty acids essential in nutrition, J.Biol.Chem. 82:345 (1929).
45. A.G.Hassam and M.A. Crawford, The differential incorporation of labelled linoleic, g-linolenic, dihomo-g-linolenic and arachidonic acids into the developing brain, J.Neurochem. 27:967 (1976).
46. D.Hull, Total fat metabolism, in: "Fetal Physiology and Medicine", Beard, and G. Nathanielz (ed),. Saunders, London, pp. 105 (1976).
47. E.M.Widdowson, Chemical composition of newly born mammals, Nature Lond 166:626 (1950).
48. E.James, G. Meschia and F.C. Battaglia, A-V differences of free fatty acids and glycerol in the ovine umbilical circulation, Proc.Soc. Exper.Biol.Med. 138:823 (1971).
49. Z.Koren and E. Shafrir, Placental transfer of free fatty acids in the pregnant rat, Proc.Soc.Exper.Biol.Med. 116:411 (1964).
50. M.C.Elphick, D.G. Hudson and D. Hull, Transfer of fatty acids across the rabbit placenta, J.Physiol. 252:29 (1975).
51. M.S.Hershfield and A.M. Nemeth, Placental transport of free palmitic and linoleic acids in the guinea pig, J.Lipid.Res. 9:460 (1968).
52. T.Bohmer and R.J. Havel, Genesis of fatty liver and hyperlipemia in the fetal guinea pig, J.Lipid.Res. 16:454 (1975).
53. O.W.Portman, R.E. Behrman and P. Soltys, Transfer of free fatty acid across the primate placenta, Amer.J.Physiol. 216:143 (1969).
54. L.Hummel, W. Schirrmeister, T. Zimmerman and H. Wagner, Studies on the lipid metabolism using 14C-1-palmitate in fetal rats, Biol.Neonate 24:298 (1974).
55. C.T.Jones, Lipid metabolism and mobilization in the guinea pig during pregnancy, Biochem.J. 156:357 (1976).
56. D.Hull and J.P. Stammers, Placental transfer of fatty acids, Biochem. Soc.Trans. 13:821 (1985).
57. R.C.Noble, J.H. Shand and D.T. Calvert, The role of the placenta in the supply of essential fatty acids to the fetal sheep: studies of lipid compositions at term, Placenta 3:287 (1982).
58. O.W.Portman, R.E. Behrman and P. Soltys, Transfer of free fatty acids across the primate placenta, Am.J.Physiol. 216:143 (1969).
59. S.Satomi and I. Matsuda, Microsomal desaturation of linoleic into g-linolenic acid in livers of fetal, suckling and pregnant rats, Biol.Neonate 22:1 (1973).
60. C.Phanteliadis and U. Troll, Fatty acid pattern of the serum lipids (normal values in mothers, placentae and infants), Zeitschrift fur Ernaehrungwissenschaft 15:305 (1976).
61. M.C.Elphic and D. Hull, The transfer of free fatty acids across the rabbit placenta, J.Physiol. 264:751 (1977).
62. J.H.Shand and R.C. Noble, Ag and A6-desaturase activities of the ovine placenta and their role in the supply of fatty acids to the foetus, Biol.Neonate 36:298 (1979).
63. J.H.Shand and R.C. Noble, The metabolism of 18:0 and 18:2 (n-6) by the ovine placenta at 120 and 180 days of gestation, Lipids 16:68 (1981).
64. J.H.Shand and R.C. Noble, Incorporation of linoleic and arachidonic acids into ovine placental phospholipids in vitro, Biol.Neonate 48: 299 (1985).
65. D.C.Kuhn and M. Crawford, Placental essential fatty acid transport and prostaglandin synthesis, Prog.Lipid.Res. 25(1-4):345 (1986).
66. D.Hull and M. Elphick, Transfer of fatty acids, in: "Placental Transfer", Chamberlain and Wilkinson (eds.), Pitman Medical, London, pp. 159 (1979).
67. M.A.Lasunción, X. Testar, M. Palacín, R. Chieri and E. Herrera, Method for the study of metabolite transfer from rat mother to fetus, Biol. Neonate 44:85 (1983).
68. M.Palacín, M.A. Lasunción, R. Martín del Río and E. Herrera, Placental formation of lactate from transferred L-alanine and its impairment

by aminooxyacetate in the late-pregnant rat, <u>Biochim.Biophys.Acta</u> 841:90 (1985).

69. E.Herrera, M. Palacín, A. Martín and M.A. Lasunción, Relationship between maternal and fetal fuels and placental glucose transfer in rats with maternal diabetes of varying severity, <u>Diabetes</u> 34(2):42 (1985).

70. M.Palacín, M.A. Lasunción, A. Martín and E. Herrera, Decreased uterine blood flow in the diabetic pregnant rat does not modify the augmented glucose transfer to the fetus, <u>Biol.Neonate</u> 48:197 (1985).

71. M.A.Lasunción, J. Lorenzo, M. Palacín and E. Herrera, Maternal factors modulating nutrient transfer to fetus, <u>Biol.Neonate</u> 51:86 (1987).

72. E.Herrera, M.A. Lasunción and M. Asunción, Placental transport of free fatty acids, glycerol and ketone bodies, <u>in</u>: "Neonatal and Fetal Medicine: Physiology and Pathophysiology", R. Polin and W.Fox, eds., Grune & Stratton, Inc., New York, (in press).

73. M.C.Elphick and D. Hull, Incorporation *in vivo* of 1-14C-palmitic acid into plancental and fetal liver lipids of the rabbit, <u>Biol.Neonate</u> 32:24 (1977).

74. Ch.R.Thomas, Placental transfer of non-esterified fatty acids in normal and diabetic pregnancy, <u>Biol.Neonate</u> 51:94 (1987).

75. R.A.Coleman, Placental metabolism and transport of lipid, <u>Fed.Proc.</u> 45(10):2519 (1986).

76. R.A.Coleman and E.B. Haynes, Synthesis and release of fatty acids by human trophoblast cells in culture, <u>J.Lipid.Res.</u> 28:1335 (1987).

77. A.J.Szabo, R. Lellis and R.D. Grimaldi, Triglyceride synthesis by the human placenta. I. Incorporation of labelled palmitate into plancental triglycerides, <u>Am.J.Obstet.Gynec.</u> 115:257 (1973).

78. L.Hummel, W. Schirmeister, T. Zimmermann and H. Wagner, Quantitative studies on the metabolism of placental triglycerides and phospholipids in the rat, <u>ActaBiol.Med.Germ.</u> 32:311 (1974).

79. L.Hummel, A. Schwartze, W. Schirrmeister and H. Wagner, Maternal plasma triglycerides as a source of fetal fatty acids, <u>ActaBiol. Med.Germ.</u> 35:1635 (1976).

80. E.Shafrir and V. Barash, Placental function in maternal-fetal fat transport in diabetes, <u>Biol.Neonate</u> 51:102 (1987).

81. M.C.Elphick and D. Hull, Rabbit placental clearing-factor lipase and transfer to the fetus of fatty acids derived from triglycerides injected into the mother, <u>J.Physiol.</u> 273:475 (1977).

82. C.R.Thomas and C. Lowy, The clearance and placental transfer of free fatty acids and triglycerides in the pregnant guinea pig, <u>J.Develop. Physiol.</u> 4:163 (1982).

83. C.R.Thomas and C. Lowy, The interrelationships between circulating maternal esterified and non-esterified fatty acids in pregnant guinea pigs, and their relative contributions to the fetal circulation, <u>J.Develop.Physiol.</u> 9:203 (1987).

84. M.C.Elphick, G.M. Filshie and D. Hull, The passage of fat emulsion across the human placenta, <u>Br.J.Obstet.Gynaecol.</u> 85:610 (1978).

85. C.R.Thomas, C. Lowy, R.J. St Hillaire and J.D. Brunzell, Studies on the plancental hydrolysis and transfer of lipids to the fetal guinea pig, <u>in</u>: "Fetal nutrition, metabolism and immunology: role of the placenta", R.K. Miller and H.A. Thied (ed), Plenum Press, New York, pp. 135 (1984).

86. J.E.Rotherwell and M.C. Elphick, Lipoprotein lipase activity in human and guinea pig placenta, <u>J.Develop.Physiol.</u> 4:153 (1982).

87. R.H.Knopp, M.R. Warth, D. Charles, M. Childs, J.B. Li, H. Mabuchi and M.I. Van Allen, Lipoprotein metaboism in pregnancy, fat transport to the fetus, and the effects of diabetes, <u>Biol.Neonate</u> 50:297 (1986).

88. A.E.Seeds, L.S. Leung, J.J. Stys, K.E. Clark and P.T. Russell, Comparison of human and sheep chorion laeve permeability to glucose, beta-hydroxybutyrate and glycerol, <u>Am.J.Obstet.Gynecol.</u> 138:604 (1980).

89. E. James, G. Meschia and F.C. Battaglia, A-V differences of free fatty acids and glycerol in the ovine umbilical circulation, Proc.Soc. Expt.Biol.Med. 138:823 (1971).
90. V. Sabata, H. Wolf and S. Lausmann, The role of free acids, glycerol, ketone bodies and glucose in the energy metabolism of the mother and fetus during delivery, Biol.Neonate 13:7 (1968).
91. M. Palacín, M.A. Lasunción and E. Herrera, Lactate production and absence of gluconeogenesis from placental transferred substrates in fetuses from fed and 48-h starved rats, Pediat.Res. 22:6 (1987).
92. M. Gilbert, Origin and metabolic fate of plasma glycerol in the rat and rabbit fetus, Pediat.Res. 11:95 (1977).
93. F.C. Battaglia and C. Meschia, Principal substrates of fetal metabolism, Physiol.Rev. 58: 499 (1978.).
94. J.M. Bissonette, Studies in vivo of glucose transfer across the guinea-pig placenta, Placenta (suppl 2):155 (1981).
95. T. Mampel, F. Villarroya and E. Herrera, Hepatectomy-nephrectomy effects in the pregnant rat and fetus, Biochem.Biophys.Res.Commun 131:1219 (1985).
96. E.M. Mackay and R.H. Barnes, Fasting ketosis in the pregnant rat as influenced by adrenalectomy, Proc.Soc.Exptl.Biol.Med. 34:682 (1936).
97. D.H. Williamson, Regulation of the utilization of glucose and ketone bodies by brain in the perinatal period, in: "Early diabetes in early life", R.A. Camerini-Davalos and H.S. Cole (ed), New York, Academic Press, pp. 195 (1975).
98. F.H. Morris, R.D.H. Boyd, E.L. Makowski, G. Meschia and F.C. Battaglia, Umbilical V-A differences of acetoacetate and beta hydroxybutyrate in fed and starved ewes, Proc.Soc.Exp.Biol.Med. 145:879 (1974).
99. Y.J. Kim and P. Felig, Maternal and amniotic fluid substrate levels during caloric deprivation in human pregnancy, Metabolism 21:507 (1971).
100. P. Felig and V. Lynch, Starvation in human pregnancy: hypoglycemia, hypoinsulinemia, and hyperketonemia, Science 170:990 (1970).
101. B. Persson and N.O. Lunell, Metabolic control of diabetic pregnancy. Variations in plasma concentrations of glucose, free fatty acids, glycerol, ketone bodies, insulin and human chorionic somato-mammotropin during the last trimester, Am.J.Obstet.Gynecol. 122:737 (1975).
102. M.I. Drury, A.T. Green and J.M. Stronge, Pregnancy complicated by clinical diabetes mellitus, Obstet. Gynecol. 49:519 (1977).
103. G.E. Shambaugh III, S.C. Mrozak and N. Freinkel, Fetal fuels, I: utilization of ketones by isolated tissues at various stages of maturation and maternal nutrition during late gestation, Metabolism 26:623 (1977).
104. G.E. Shambaugh III, A. Koehler and N. Freinkel, Fetal fuels, II: contributions of selected carbon fuels to oxidative metabolism in rat conceptus, Am.J.Physiol. 233:E457 (1977).
105. G.E. Shambaugh III, R.R. Koehler and H. Yokoo, Fetal fuels, III: ketone utilization by fetal hepatocytes, Am.J.Physiol. 235(3):E330 (1978).
106. M.A. Mage and D.H. Williamson, Enzymes of ketone-body utilisation in human brain, Lancet II:66 (1971).
107. P.A.J. Adam, N.L. Raiha and M. Kekomaki, Oxidation of glucose and D-hydroxybutyrate by the early human fetal brain, Acta Paediat. Scand. 64:17 (1975).
108. P. Paterson, J. Sheath, P. Taft and C. Wood, Maternal and fetal ketone concentrations in plasma and urine, Lancet 1:862 (1967).
109. A.L. Smith and J. Scanlon, amniotic fluid D(-)-B-hydroxybutyrate and the dysmature newborn infant, Am.J.Obstet.Gynecol. 115:569 (1973).
110. P. Felig and V. Lynch, Starvation in human pregnancy: hypoglycemia, hypoinsulinemia, and hyperketonemia, Science 170:990 (1970).
111. Ll. Arola, A. Palou, X. Remesar and M. Alemany, Effects of 24 hour

starvation on plasma composition in 19 and 21 day pregnant rats and their foetuses, Horm.Metab.Res. 14:364 (1982).

112. M.Miodovnik, J.P. Lavin, D.J. Harrington, L.S. Leung, A.E. Seeds and K.E. Clark, Effect of maternal ketoacidemia on the pregnant ewe and the fetus, Am.J.Obstet.Gynecol. 144:585 (1982).

113. L.P.K.Lee and I.B. Fritz, Hepatic ketogenesis during developmet, Can. J.Biochem. 49:599 (1971).

114. R.D.Boyd, F.H. Morriss jr., G. Meschia, E.L. Makowski and F.C. Battaglia, Growth of glucose and oxygen uptakes by fetuses of fed and starved ewes, Am.J.Physiol. 225:897 (1973).

115. M.A.Page and D.H. Williamson, Enzymes of ketone body utilization in human brain, Lancet II:66 (1971).

116. M.S.Patel, C.A. Johnson, R. Rajan and O.E. Owen, The metabolism of ketone bodies in developing human brain: Development of ketone-body utilizing enzymes and ketone bodies as precursors for lipid synthesis, J.Neurochem. 25:905 (1975).

117. Ch.Dierks-Ventling, Prenatal induction of ketone-body enzymes in the rat, Biol.Neonate 19:426 (1971).

118. D.H.Williamson, Ketone body metabolism and the fetus, in: "Fetal Growth Retardation", F.A. Van Assche, W.B. Robertson and M.C. Renaer (eds.), Churchill Livingstone, Edinburg, pp. 29 (1981).

119. D.W.Seccombe, P.G.R. Harding and F. Possmayer, Fetal utilization of maternally derived ketone bodies for lipogenesis in the rat, Biochim.Biophys.Acta 438:402 (1977).

120. M.M.Thaler, Effects of starvation on normal development of beta-hydroxybutyrate dehydrogenase activity in fetal and newborn rat brain, Nature New Biology 236:140 (1972).

121. T.G.Redgrave and L.A. Carlson, Changes in plasma very low density and low density lipoprotein content, composition and size after a fatty meal in normal and hypertriglyceridemic men, J.Lipid Res. 20:217 (1979).

122. D.E.Brenneman and A.A. Spector, Utilization of ascites plasma very low density lipoprotein triglycerides by Ehrlich cells, J.Lipid Res. 15:309 (1974).

SURFACTANT LIPIDS AND PROTEINS IN

THE PERINATAL AND ADULT LUNG

Henk P. Haagsman, Joseph J. Batenburg, Cecile Clercx*,
Math J.H. Geelen and Lambert M.G. van Golde

Laboratory of Veterinary Biochemistry
and *Small Animal Clinic
University of Utrecht
Utrecht, The Netherlands

INTRODUCTION

The alveolar surfaces of the lungs are coated with a highly surface-active material. This material, pulmonary surfactant, consists for approx. 90% of lipids and contains, in addition, several specific proteins (1). Pulmonary surfactant is manufactured in the type II pneumocytes (2), one of the epithelial cell types in the alveolar walls. The type II pneumocytes secrete the surfactant into the fluid layer that covers the alveolar epithelium. The surfactant lipids can subsequently spread as a monolayer at the air-liquid interface and, by decreasing surface tension, protect the alveoli against collapse at end-expiration. By reducing the contractile force in the curved air-liquid interface, pulmonary surfactant also precludes alveolar edema (3). Although the major physiologic function of pulmonary surfactant is undoubtedly to confer mechanical stability to the alveoli, there are important suggestions that the surfactant system also plays a role in several pulmonary defense mechanisms (for review see ref.4).

In preparation for extrauterine life, the fetal lung must produce adequate amounts of pulmonary surfactant for appropriate postnatal functioning. It is well known that there is a spurt in surfactant production in the prenatal lung during the terminal period of gestation (4-7). The accumulation of abundant amounts of surfactant in the airways coincides with the ability of the neonate to establish regular air-breathing. A deficit of surfactant severely jeopardizes the alveolar gas-exchange processes. This becomes manifest in the respiratory distress syndrome of the neonate, an important clinical condition that is essentially caused by a shortage of surfactant due to immaturity of the lungs (8).

In the present review we will first briefly summarize current knowledge concerning the lipid components of the pulmonary surfactant system and the changes that occur in the lipid composition of surfactant during perinatal development of the lung. In that section, we will also consider possible alterations that may occur in the phospholipid composition of the surfactant system during ageing. Subsequently a brief discourse follows of the fascinating progress that has been made in the past few years with respect to our understanding of the composition,

Endocrine and Biochemical Development of the Fetus and Neonate
Edited by J. M. Cuezva *et al.*
Plenum Press, New York, 1990

properties and physiologic functions of the specific surfactant proteins. The review will then focus on the mechanisms that are involved in the biosynthesis of dipalmitoylphosphatidylcholine (DPPC), the major surface-active compound of pulmonary surfactant (9). In the final sections we will outline current thoughts concerning possible regulatory steps in the formation of this prominent surfactant lipid. For comprehensive information concerning a variety of other aspects of the pulmonary surfactant system, several recent reviews could be consulted (4-7, 10-13).

THE LIPID COMPONENTS OF THE PULMONARY SURFACTANT SYSTEM

There are two major surfactant pools in the lung. The surfactant fraction that can be harvested by bronchoalveolar lavage represents the extracellular pool. This extracellular pool comprises a variety of morphologically different complexes, such as the actual monolayer at the air-water interface, tubular myelin, and surfactant components destined to be cleared from the alveolar spaces (see Figure 1 for a schematic overview). The intracellular surfactant pool resides in the lamellar bodies, the characteristic inclusion organelles in the alveolar type II cells. It is generally assumed that these organelles represent intra-cellular reservoirs of surfactant before this material is secreted into the alveolar space (14). This notion is corroborated by the fact that the phospholipid composition of lamellar bodies is very similar to that of extracellular surfactant. This is illustrated in Figure 2 for lamellar

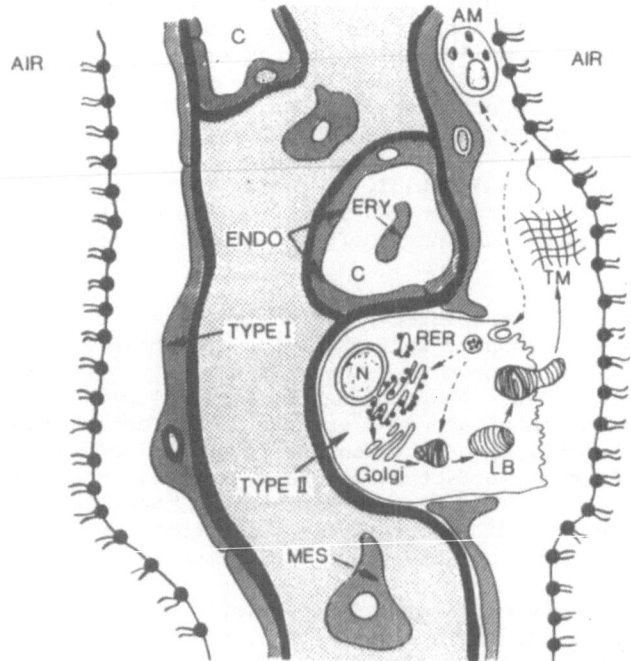

Fig. 1. Schematic cross-section of an alveolar wall. TYPE I: type I pneumocyte, TYPE II type II pneumocyte, ENDO: endothelial cell, MES: mesenchymal cell, ERY: erythrocyte, C: capillary, N: nucleus, RER: rough endoplasmic reticulum, LB: lamellar body, TM tubular myelin, AM: alveolar macrophage.

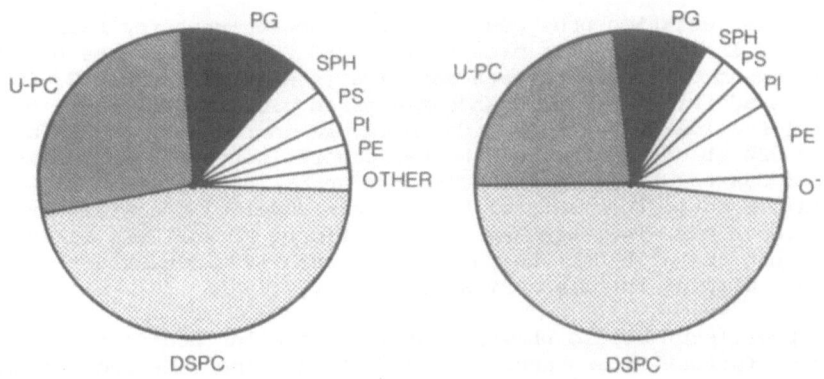

Fig. 2. Phospholipid composition of extracellular surfactant (15)
(left) and lamellar bodies (16) (right) of adult human
lung. DSPC: disaturated phosphatidylcholine, U-PC: unsaturated phosphatidylcholine, PG: phosphatidyl-glycerol,
SPH: sphingomyelin, PS: phosphatidylserine, PI: phosphatidylinositol, PE: phosphatidylethanolamine.

bodies (15) and extracellular surfactant (16) purified from adult human
lung.

There appears to be little variation among mammalian species as far as
surfactant lipid composition is concerned. Of the lipids in surfactant,
80-90% is phospholipid and cholesterol is the most abundant neutral lipid
(1). As illustrated in Figure 2 phosphatidylcholine (PC) is the major
phospholipid class in surfactant and about 60% of the PC molecules contain
two saturated acyl residues. This disaturated PC (DSPC) is largely DPPC.
This particular species thus comprises almost half of the surfactant
phospholipids.

There is no questioning that DPPC is the principal surface-active
component of pulmonary surfactant that is responsible for decreasing
surface tension to values below 10mN/m at low lung volumes (3). It should
be emphasized, however, that DPPC is not a surfactant-specific lipid as is
often quoted. Abundant levels of this particular PC species are also
present in other cell types of the lung (17) as well as in several other
tissues such as brain (18). Monoenoic PC molecules represent the bulk of
the unsaturated PC species in surfactant. Although the function of DPPC in
surfactant is well documented, there is less certainty about the role of
the other surfactant lipids. It is usually presumed that the unsaturated PC
molecules promote rapid adsorption and spreading of surfactant at the air-
liquid interface (1).

Surprising is the atypically high proportion of phosphatidylglycerol
(PG) in surfactant. In most adult mammalian species PG accounts for up to
10% of the surfactant lipids. There are several lines of evidence (19,20)
which suggest that phosphatidylinositol (PI) can replace PG without
affecting the most important physiological and physico-chemical properties
of surfactant. Interestingly, pulmonary surfactant contains much lower
proportions of phosphatidylethanolamine, phosphatidylserine and sphingomyelin than whole lung tissue.

Surfactant Lipid Composition during Perinatal Development

It has been reported for a number of species that increasing
quantities of lung tissue and surfactant phospholipids are produced by the

fetal lung as gestation progresses. In the rabbit lung, for example, there is a 10-fold increase in the amount of PC and DSPC that can be harvested by lavage per gram of tissue between day 27 of gestation and day 31 (term) (21,22). An additional 10-fold increase takes place in the first day of neonatal life. The proportion of PC in lavage increases from 29% at fetal day 27 to 82% at day 2 after birth, whereas the percentage of sphingomyelin drops from 38% to 2% within the same time span. This implies that the so-called L/S ratio (Lecithin or PC to Sphingomyelin) increases from 0.8 to 40 (21). There are also increases in lung tissue PC and DSPC during this perinatal period, although the changes are much less dramatic than for these phospholipids in pulmonary surfactant (22).

The profile of acidic phospholipids in lamellar bodies and in extra-cellular surfactant alters around term. At that time, the proportion of PG increases whereas that of PI diminishes. The exact timing of this change from PI to PG synthesis varies among different mammalian species. In human amniotic fluid the percentage of PI starts to decline at 35 weeks of gestation with a concomitant increase in the percentage of PG (23). The appearance of PG in human amniotic fluid shortly before term is, in addition to the aforementioned increased L/S ratio in amniotic fluid, a useful criterium to indicate fetal lung maturity (23). In rat surfactant, like in that of man, PG also shows a developmental rise before birth (24). However, in species such as rabbit (25) and sheep (26) the switch-over from PI to PG formation occurs shortly after birth. The fact that term neonates of these species do not show respiratory distress is another indication that the presence of PG per se is not a prerequisite for optimal functioning of surfactant or at least that its purported function can be replaced by PI. These two acidic phospholipids share CDPdiacylglycerol as a common precursor. There is general agreement that decreased availability of myo-inositol around term is a major factor that channels CDPdiacylglycerol into PG instead of into PI (27-29).

Surfactant Lipid Composition during Ageing

As described in the preceding section the changes in amount and composition of surfactant lipids during perinatal development of the lung have been extensively documented. However, very little information is available about possible alterations in the lipids of the pulmonary surfactant system during ageing. Studies with whole lung tissue of various animals (30) showed that the levels of DSPC hardly changed upon ageing. It should be emphasized, however, that these studies did not include separate analysis of the surfactant system. This is important, as the pulmonary surfactant system contains only 30% of the total amount of DSPC present in whole lung (17). Recently we compared the phospholipid composition of extracellular surfactants isolated from bronchoalveolar lavage of young beagle dogs (3-7 months) with that of surfactants obtained from old beagles (>12 years). Figure 3 shows that the differences are not very pronounced. However, there is nevertheless a significant (P<0.05) increase in the percentage of PC, a clear trend (P=0.06) towards a higher proportion of PG, and significantly (P<0.05) lower percentages of phosphatidylserine and sphingomyelin in surfactant of old dogs.

Despite age related changes in lung morphology, lung mechanics, and lung volumes in dogs, these animals, unlike man, do not show differences in blood gases during ageing (31). This suggests that compensating mechanisms may develop in the ageing dog which prevent impairment of gas exchange efficiency. One of these ventilatory compensating mechanisms could be an "improved" quality of surfactant. In this respect it is of interest to mention that surface tension measurements with surfactant from both groups of dogs demonstrated that the surfactant of the old dogs reached a

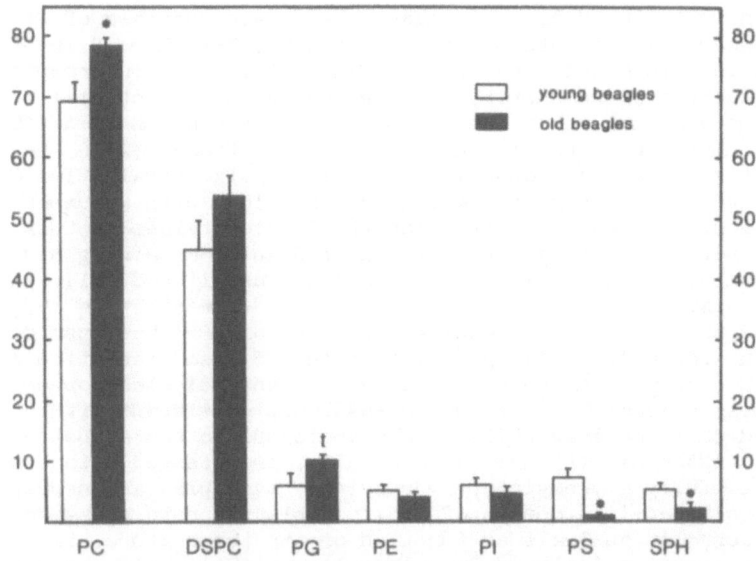

Fig. 3. Phospholipid composition of surfactant isolated from
bronchoalveolar lavage of young and old beagles. For
abbreviations see Figure 2.

significantly (P< 0.02) lower minimal surface tension than surfactant of
the young dogs (not shown).

SURFACTANT SPECIFIC PROTEINS

    As mentioned above, pulmonary surfactant consists of a variety of
macromolecular complexes comprised of lipids and specific proteins. The
primary structures of three different families of surfactant-associated
proteins from several species have been determined. According to recently
proposed nomenclature these proteins will be referred to as SP-A, SP-B, and
SP-C (13). The structure, properties, and possible functions of these
proteins will be discussed.

Surfactant-associated Protein A (SP-A)

    This glycoprotein was first described by King and coworkers in 1973
(32). It is the most abundant surfactant protein. The molecular weight of
SP-A is 28,000-36,000 under denaturing and reducing conditions. The
molecular weight heterogeneity is due to variable degrees of glycosylation
(33- 36). SP-A is acidic and has isoelectric points ranging from pH 4.4 to
pH 5.6 (37). The presence of sialic acid contributes to this charge
heterogeneity. SP-A is a dimer under nonreducing, denaturing conditions.
SP-A has been localized to several forms of surfactant: intracellularly in
lamellar bodies of type II cells (38,39), and, extracellularly, in tubular
myelin (39- 41). SP-A is synthesized by type II cells (34,35), and possibly
also by Clara cells (38,41). The coding sequence for human SP-A has been
localized to chromosome 10 (42,43).

    Structure. The primary structure has been determined for human (44,
45), dog (46), rat (47), and rabbit (48) SP-A and is highly conserved in
these species. A short NH2-terminal region of 7 (human) to 10 (dog) amino
acids contains a cysteine which forms an interchain disulfide bridge (46).

This region is followed by a collagen-like domain composed of 24 (23 for the human sequence) repeats of the sequence Gly-X-Y where Y is often hydroxyproline. This sequence of repeating triplets is interrupted in the 13th repeat. SP-A is susceptible to degradation with bacterial collagenase (46,49). Two distinct cDNAs have been reported for human SP-A (45). The nucleotide differences result in 6 amino acid differences in the protein. All variant residues are found in the collagen-like domain. The COOH-terminal region has a remarkable sequence homology with a number of Ca^{2+}-dependent lectins (50). In particular the four cysteines in this domain are conserved in these lectins. The pattern of disulfide pairing has recently been elucidated: residues 135-226 and 204-218 are linked (51). The same pattern of disulfide bonding has been described for other Ca^{2+}-dependent lectins (52,53). The macromolecular structure of SP-A has been determined by electron microscopy (54), gel filtration (55), sedimentation analysis (56), and non-denaturing polyacrylamide gradient gel electrophoresis (51). The molecular mass of SP-A is 650,000-845,000, consistent with a structure of 6 linked collagen-like helices, and analogous to the structure of complement factor C1q (Figure 4). CHO cells, transfected with a single SP-A gene, successfully glycosylate, hydroxylate, assemble, and secrete active SP-A. The collagen-like domain of the recombinant protein, as determined by circular dichroism analysis as function of the temperature, is less stable compared to proteins isolated from lung lavage (51). This observation would be in line with the presence of more than one gene product in the assembled native proteins.

   Properties. SP-A binds to phospholipids (57), causes phospholipid aggregation in the presence of Ca^{2+} (33,58,59) and promotes Ca^{2+}-dependent adsorption of phospholipids to an air-fluid interface in the presence of surfactant proteins SP-B and SP-C (60). Aggregation of SP-A requires the self-association of the oligomeric structure of 6 linked collagen-like triple helices. SP-A may have an important role in generating and

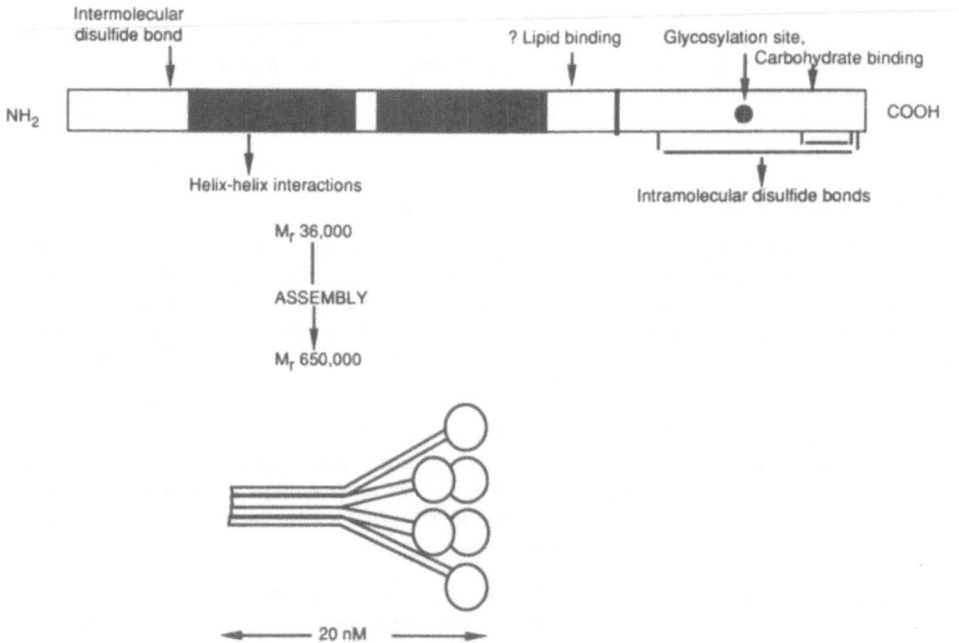

Fig. 4. Model of SP-A monomer and assembly into the native
   oligomeric state.

236

stabilizing the structure of tubular myelin, a subfraction of surfactant that probably forms the surface film (10) (Figure 1). The carbohydrate-binding properties, predicted by sequence analysis, have been demonstrated experimentally (61). Binding to immobilized monosaccharides is also Ca^{2+}-dependent. SP-A is indeed a Ca^{2+}-binding protein and binds 2 mol Ca^{2+}/mol monomer as determined by equilibrium dialysis (H.P. Haagsman, unpublished results). Surfactant is efficiently recycled between intracellular and extracellular pools (Figure 1) (62-64). Results from recent studies suggest that SP-A may regulate surfactant homeostasis. SP-A enhances lipid uptake by type II cells (65) and inhibits stimulated lipid secretion from these cells (66-68). Type II cells bind radiolabeled SP-A with high affinity (69). SP-A also binds with a high affinity to alveolar macrophages, in line with recently found biological effects of SP-A on these cells (70).

Developmental aspects. SP-A becomes detectable in human amniotic fluid after 30-32 weeks gestation and increases toward term (71-72). In the lungs of rat fetuses SP-A is first detected at day 18 (73). A marked increase in the level of translatable mRNA for SP-A was found in rabbit lung tissue of 28-day fetal rabbits (48,74). Cortisol and cAMP increase both the accumulation of SP-A and its mRNA in fetal rabbit lung explants. SP-A was not detected in fetal human lung explants from second trimester abortuses but increased during culture, exceeding adult levels after 5 days (75). In cultures treated with dexamethasone the increase of both SP-A and its mRNA was accelerated. However, an inhibiting effect of dexamethasone has also been reported (76). The presence of a potential glucocorticoid receptor binding site in close proximity to the SP-A gene suggests that these hormones may regulate this gene (44).

The Small Hydrophobic Surfactant Proteins

Surfactant preparations, prepared through organic solvent extraction, have been used successfully to prevent and treat respiratory distress syndrome in newborns (11,77). The efficacy of these preparations could be attributed to the presence of hydrophobic proteins that copurify with the lipids (60,78-81). These proteins promote the rate of formation of the phospholipid surface film *in vitro* (60,79-81). SP-A has a cooperative effect on phospholipid adsorption when the components are reassembled with a stoichiometry comparable to natural surfactant (60). There are at least two groups of hydrophobic surfactant proteins: SP-B and SP-C (13).

SP-B. Mature SP-B is a cysteine-rich, basic hydrophobic peptide with a molecular weight of approximately 18,000 (60,82-84). The protein has a molecular weight of 7,000 in the reduced form (60,82-85). SP-B is processed from a glycosylated precursor of approximately 42,000 daltons (60,82-83). Type II cells and Clara cells contain SP-B mRNA (86). The coding sequence of human SP-B is located on chromosome 2 (87). No SP-B was demonstrated in human fetal lung of 16-21 weeks of gestation (88). Induction of synthesis of SP-B occurs rapidly in explant culture and is enhanced by addition of dexamethasone (88).

SP-C. The primary structure of human SP-C has been determined from the nucleotide sequence of cDNA (89,90). SP-C is a very hydrophobic peptide characterized by a stretch of 6 contiguous valine residues. It is produced as a large primary translation product of approximately 21,000 daltons (89,90). The mature peptide as isolated from lavage exhibits N-terminal trimming (91,92). Comparison of the primary structure of human SP-C with the primary structures of canine (89), bovine (92), and porcine (92) SP-C reveals a variable hydrophilic N-terminal part and a conserved hydrophobic region. The secondary structure of canine SP-C was determined by circular dichroism. In the presence of DPPC and PG SP-C has 54% α, 26% β, and 20% unordered structure (H.P. Haagsman, unpublished results). The structure of

SP-C is reminiscent to that of signal peptides. The SP-C gene locus was assigned to chromosome 8 (90). Like SP-B, SP-C was expressed in human fetal lung explants (88). Increased protein content was associated with increased mRNA for both SP-B and SP-C (88).

BIOSYNTHESIS OF THE MAJOR SURFACTANT LIPID

Most biochemical studies on the biosynthesis of surfactant lipids have been focussed on the formation of DPPC, the principal surface-active component of surfactant. The older studies have been largely carried out with preparations derived from whole lung tissue such as lung slices, homogenates of lung tissue, and microsomes isolated from whole lung homogenates. The results of these studies should be interpreted with some caution, as the surfactant synthesis is limited to the type II cells which comprise only 10-15% of the total cell population of adult lung. Fortunately techniques have been developed to isolate type II pneumocytes both from adult and from fetal lung (for references to these methods see for example ref. 7). In the subsequent paragraphs we will emphasize studies on the synthesis of DPPC that have been performed with pure preparations of isolated type II cells.

Substrates Required for the Synthesis of DPPC in Adult and Fetal Lung

As will be discussed below there is now abundant evidence that type II cells can synthesize DPPC both by direct synthesis *de novo* via the CDPcholine route and by remodeling of unsaturated PC species that have been assembled via the CDPcholine pathway. The following building stones are required for the formation of DPPC via either of these routes: glycerol-3-phosphate, choline, and fatty acids.

It is generally assumed that the type II pneumocyte of the adult lung acquires most of its glycerol-3-phosphate by reduction of dihydroxyacetone phosphate (Figure 5, reaction 1). The latter intermediate is derived by glycolytic breakdown of glucose that is taken up by the type II cells from the circulation. In the type II cells of the fetal lung intracellular glycogen stores appear to be the chief source of glycerol-3-phosphate that is required for surfactant lipid synthesis (93,94) (for review see refs. 5-7,22). As phosphofructokinase is the only regulatory enzyme between glycogen derived glucose-6-phosphate and dihydroxyacetone phosphate, it is attractive to speculate that this enzyme may play an important regulatory role in regulating the onset of surfactant synthesis in the prenatal lung. This speculation is endorsed by recent findings of Heesbeen et al. (95), who demonstrated that phosphofructokinase was the only regulatory enzyme of glycolysis of which the activity in the fetal type II cell increased at the time of glycogen breakdown and the onset of surfactant lipid synthesis.

Choline can be efficiently taken up by the type II cells from the circulation. Interestingly, choline kinase, which converts choline into choline phosphate (Figure 5, reaction 5), is highly enriched in primary cultures of type II cells when compared to whole lung (96).

The blood also supplies the type II cells with fatty acids that are needed for surfactant lipid synthesis, but these cells can also synthesize fatty acids *de novo* with lactate as an important source of acetyl groups (97). Although the relative importance of endogenously synthesized fatty acids for surfactant lipid formation remains unknown, there is evidence that lipogenesis is of particular importance to supply the palmitate that is required for the formation of surfactant lipids in the prenatal lung (98). In slices of fetal rat lung the synthesis of fatty acids, estimated from the incorporation of $^{3}H_2O$, peaked at days 21 and 22 of gestation (97)

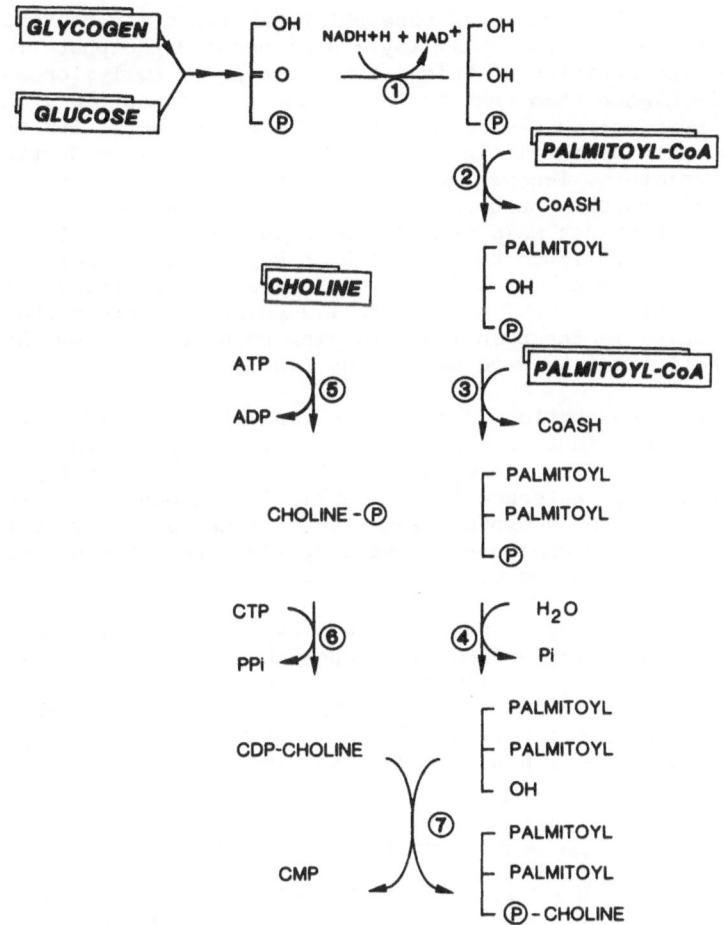

Fig. 5. Biosynthesis of dipalmitoylphosphatidylcholine
via synthesis *de novo*.

(term is day 22). A similar developmental profile with a peak in activity
at day 21 was reported for fatty acid synthase (99). There is disagreement
in the literature on the profile of acetyl-CoA carboxylase in perinatal rat
lung. Maniscalco et al. (97) found that the activity of this enzyme also
peaked at day 21, whereas Pope and Rooney (99) reported that the enzyme
showed a gradual decline in its activity during the perinatal development.
Both groups used an assay in which $^{14}CO_2$ was incorporated into acid-stable
product presumed to be malonyl-CoA. It is well known, however, that in
crude extracts other carboxylation reactions may proceed as well. Bijleveld
et al. (100) developed a new assay to measure acetyl-CoA carboxylase. In
this assay $[1-^{14}C]$acetyl-CoA is converted by the enzyme into $[^{14}C]$malonyl-
CoA that is subsequently removed by conversion into labelled fatty acid by
adding an excess of purified fatty acid synthase to the assay medium. With
this assay, we reinvestigated the profile of acetyl-CoA carboxylase in
developing rat lung. The results shown in Figure 6 show that not only fatty
acid synthase but also acetyl-CoA carboxylase displays a peak in activity
in fetal rat lung at one day before term.

Synthesis *de novo* of DPPC

The first step in the synthesis *de novo* of DPPC involves the action of glycerol-3-phosphate acyltransferase which catalyzes the conversion of glycerol-3-phosphate into 1-palmitoyl-sn-glycerol-3-phosphate (Figure 5, reaction 2). The activity of this enzyme in type II cell microsomes is substantially higher than that in whole-lung microsomes (101). Like in other cell types, glycerol-3-phosphate acyltransferase limits the rate of phosphatidic acid synthesis in type II cells, as the introduction of the second acyl-moiety by 1-acylglycerol-3-phosphate acyl-transferase (Figure 5, reaction 3) proceeds much faster than the first acylation step. In principle, phosphatidic acid can also be produced by acylation of dihydroxy-acetone phosphate followed by reduction of acyldihydroxyacetone phosphate to 1-acyl-glycerol-3-phosphate. However, there is now strong evidence from studies with other tissues that glycerol 3-phosphate rather than dihydroxy-acetone phosphate is the principal starting point for the synthesis of diacylglycerolipids (reviewed by Kennedy (102)).

Dipalmitoylphosphatidic acid can be converted into 1,2-dipalmitoyl-sn-glycerol by phosphatidate phosphatase (reaction 4) which is followed by the final step involving the formation of DPPC from 1,2-dipalmitoyl-sn-glycerol by choline-phosphotransferase (reaction 7). The second substrate for this reaction, CDP-choline is synthesized from choline by the sequential action of choline kinase and choline-phosphate cytidylyltransferase, respectively (Figure 5, reactions 5 and 6).

In an earlier study we reported that type II cell microsomes have the capability of synthesizing dipalmitoylphosphatidic acid from glycerol 3-phosphate and palmitoyl-CoA (101). As illustrated in Figure 7 (open bars) more than 70% of the phosphatidic acid that is synthesized upon incubation of type II cell microsomes with [^{14}C]glycerol 3-phosphate and palmitoyl-CoA as single exogenous acyl donor represents disaturated species. As phospha-

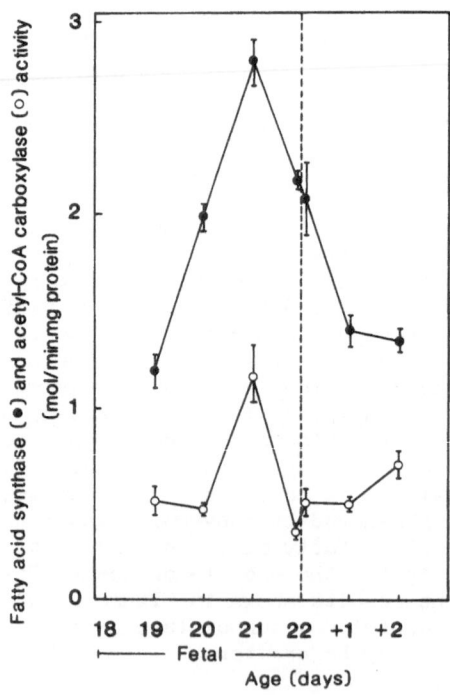

Fig. 6. Activity of fatty acid synthase and acetyl-CoA carboxylase in homogenates of perinatal lung as a function of development.

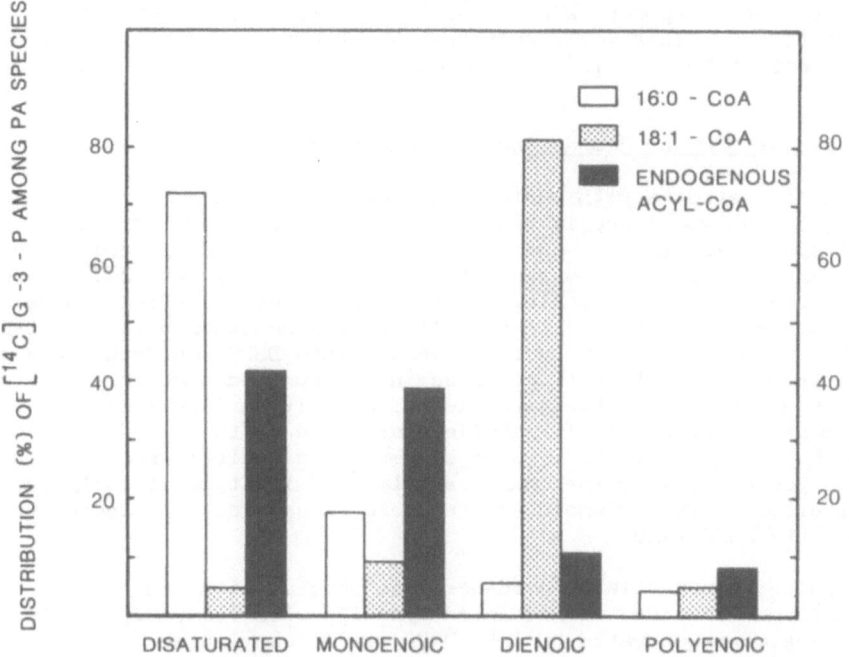

Fig. 7. Distribution of label from [^{14}C]glycerol 3-phosphate among
species of phosphatidic acid (PA) after incubation with
type II cell microsomes in the presence of 16:0-CoA
(palmitoyl-CoA), 18:1-CoA (oleoyl-CoA) or a mixture of
acyl-CoA of the same composition as that of endogenous
acyl-CoA in type II cell microsomes. For further details
see reference 105.

tidate phosphatase (103) and cholinephosphotransferase (104) of type II
cells show little if any specificity towards various molecular species of
their substrates, type II cells thus do have the potential to generate
DPPC directly via the route delineated in Figure 5. However, we also
demonstrated that the proportion of disaturated phosphatidic acid that was
formed from radioactively labelled glycerol 3-phosphate, strongly depended
on the composition of the exogenously added acyl-CoAs. If oleoyl-CoA was
supplied as single exogenous acyl donor, the formation of disaturated
phosphatidic acid was reduced to less than 5% and the majority of the
produced phosphatidic acid comprised monoenoic and dienoic species (Figure
7, dotted bars). Most interesting is, of course, the question which species
of phosphatidic acid will be produced if the type II cell microsomes are
incubated with labelled glycerol 3-phosphate and a mixture of acyl-CoAs of
the same composition as that of endogenous acyl-CoAs. We found that
microsomes of type II cells contain 49% palmitoyl-CoA, 2% myristoyl-CoA,
21% stearoyl-CoA, 5% palmitoleoyl-CoA, 16% oleoyl-CoA, 5% linoleoyl-CoA,
and 2% arachidonoyl-CoA (105). Incubation of type II cell microsomes with
[^{14}C]glycerol 3-phosphate and a mixture of acylCoAs of this composition
yielded phosphatidic acid of which approx. 40% of the species were
disaturated (Figure 7, solid bars). As parallel experiments had shown that
palmitoyl-CoA incorporated 4 times faster into disaturated phosphatidic
acid than the other saturated acyl-CoA (stearoyl-CoA), it could be
estimated that 26% of the synthesized phosphatidic acid consisted of
dipalmitoyl-phosphatidic acid. This would imply, assuming again that
phosphatidate phosphatase and cholinephosphotransferase do not show
substrate specificity with respect to the fatty acid composition of their

substrates, that approx. 26% of the PC species synthesized *de novo* represents DPPC. This would mean that about 45% of the surfactant PC could be synthesized by type II cell microsomes via direct synthesis *de novo* (105).

Synthesis of DPPC by Remodeling of Unsaturated PC

Earlier studies with whole lung (for review see refs. 4 and 7) and with isolated type II cells had already provided strong evidence that direct synthesis *de novo* cannot account for the entire production of surfactant DPPC and that a part of this DPPC must be provided by remodeling of unsaturated PC. The findings of the type II cell studies can be summarized as follows: (i) if type II cells were exposed to labelled palmitate, the radioactivity incorporated into DSPC was found predominantly at the 2-position but that recovered in disaturated diacylglycerol was randomly distributed between the 1- and 2-position (104,106). The enrichment of radioactively labelled palmitate at the 2-position of DSPC cannot be explained if DSPC would be generated entirely from disaturated diacylglycerol. (ii) pulse-chase experiments with type II cells provided direct evidence that dienoic and trienoic PC species were indeed transformed into DSPC (106).

Most of the available evidence from studies with whole-lung preparations (107,108) and isolated type II cells (104,106,109,110) favors remodeling of unsaturated PC into DPPC via a deacylation-reacylation mechanism (Figure 8). Studies with whole lung (for review see refs. 4 and 7) indicated that the deacylation step can be catalyzed either by a microsomal Ca^{2+}-requiring phospholipase A2 or by a cytosolic Ca^{2+}-independent phospholipase A2, which remove the unsaturated acyl-group (in the example of Figure 8 a linoleate) from the 2- position of an unsaturated PC molecule. Unfortunately, phospholipase A2 activities of type II cells have not yet been adequately characterized. The resulting lysophosphatidylcholine, which will be largely 1-palmitoyl-lysophosphatidylcholine, can then be reacylated with palmitoyl-CoA to form DPPC (Fig 8, reaction 2). The activity of the enzyme catalyzing this reaction, lysophos-phatidylcholine acyltransferase, is highly enriched in type II pneumocytes when compared to whole lung and the type II cell enzyme displays pronounced specificity towards palmitoyl-CoA as substrate (109,110). An enrichment of this enzyme in type II cells and its preference for palmitoyl-CoA as substrate has also been reported for fetal lung (111).

An interesting alternative for removal of the unsaturated acyl-moiety from the 2-position was proposed from studies with subcellular fractions from whole lung (108,112,113). As shown in Figure 8 a lysophosphatidylcholine acyltransferase operating in backward direction can catalyze the release of the acyl-group from the 2-position of an unsaturated PC yielding 1-palmitoyl-lysophosphatidylcholine and an unsaturated acyl-CoA. This reaction can be driven by cytosolic CoA-SH and by cytosolic lysophospholipids acting as an acceptor for the released unsaturated acyl-CoA (108, 113). It remains to be explored whether such CoA-mediated transacylation reactions also occur in type II cells.

Potentially Important Regulatory Steps in Surfactant PC Synthesis

As this topic has been reviewed extensively (4,5,7,12) a brief discourse will suffice. There is ample evidence that choline-phosphate cytidylyltrans-ferase (Figure 5, reaction 6) plays an important role in the regulation of surfactant PC synthesis: (i) Pulse-label and pulse-chase experiments with both adult and fetal type II cells using radioactively labelled choline as a precursor showed that the formation of CDPcholine is limited in these cells (114-116). This conclusion was strongly endorsed by

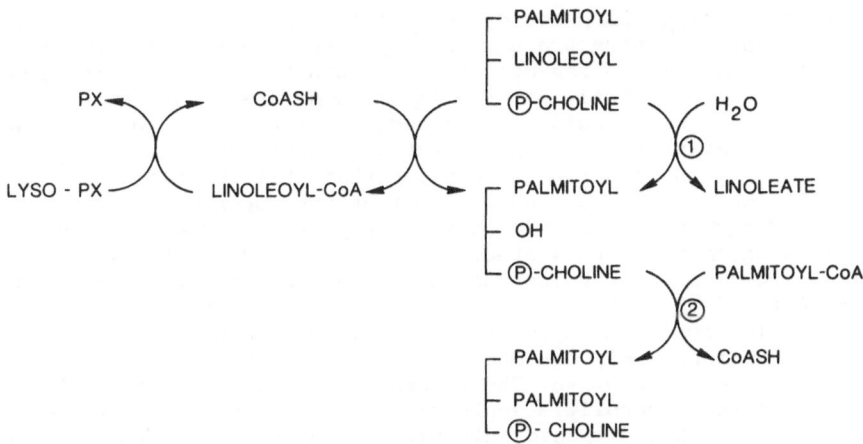

Fig. 8. Possible mechanisms operating in type II cells in the
remodeling of unsaturated phosphatidylcholines into
dipalmitoylphosphatidylcholine. Lyso-PX: lysophospho-
lipid.

measurement of the pool sizes of choline and its intermediates in fetal and
adult type II pneumocytes (115-116). Kinetic analysis of the data on the
pool sizes showed that both choline kinase (Figure 5, reaction 6) and
choline-phosphate cytidylyltransferase catalyze non-equilibrium reactions
in the type II cell, whereas the reaction catalyzed by cholinephosphotransferase (Figure 5, reaction 7) seems to operate near equilibrium. As choline
is very rapidly phosphorylated in type II cells and as the pool size of
choline phosphate is much larger than that of choline, it seems plausible
to conclude that the rate of choline incorporation into PC is largely
controlled by choline-phosphate cytidylyltransferase (ii) several
investigators have shown a close correlation between the developmental
increase of choline-phosphate cytidylyltransferase in the fetal lung and
the enhanced rate of PC synthesis as term approaches, whereas such
developmental increases are generally not observed for choline kinase (for
review see refs. 5-7) (iii) there is abundant evidence (5-7) that the
conversion of choline phosphate into CDPcholine is an important target for
a variety of hormones that promote surfactant synthesis in the developing
lung.

Experiments with a variety of different cell types suggest that
choline-phosphate cytidylyltransferase is a so-called ambiquitous enzyme
(reviewed by Vance and Pelech (117)), which occurs in an inactive cytosolic
form and an active membrane-associated form. There is convincing evidence
from studies with a variety of systems that stimulation of PC synthesis is
accompanied by translocation of the cytidylyltransferase from the cytosol
to microsomes. It has been suggested that this translocation may be
controlled by reversible phosphorylation of the enzyme and/or mediated by
fatty acids (117). Although there is also evidence from whole-lung studies
(118) that favors such translocation mechanism for the regulation of
choline-phosphate cytidylyltransferase, the results of experiments with
type II cells designed to address this question are less conclusive. These
studies (119-121) indicated that enhanced synthesis of PC synthesis in type
II cells exposed to either fatty acids (119-121) or dexamethasone (120),
was indeed accompanied by augmented activity of microsome-associated
choline-phosphate cytidylyltransferase. However, there was little evidence
for a parallel decrease in the cytosolic activity of this enzyme.

Although choline-phosphate cytidylyltransferase undoubtedly plays a crucial role in regulating the synthesis of surfactant PC in the type II cells, it is important to remember that it is not very useful to speak of the regulation of an overall pathway in terms of one single rate-limiting enzyme. The final goal should be to assess quantitatively the contribution to overall regulation that is exerted at each individual step of the pathway (102). It is important to realize that this contribution may vary between different cell types and may further vary upon changing external conditions. Haagsman et al. (96) recently reported that at low choline availability the rate of PC synthesis was determined by the supply of CDPcholine. However, at choline concentrations in the upper physiological range the rate of PC synthesis appeared to depend on the availability of diacylglycerols. Important sites for control in this branch of the pathway (Figure 5, reactions 2-4), could be glycerol-3-phosphate acyltransferase and phosphatidate phosphatase. The former enzyme catalyzes the first committed step in glycerolipid synthesis and the latter may control the rate at which phosphatidic acid molecules are channelled into the route leading to PC.

ACKNOWLEDGEMENTS

Research in the authors laboratories was supported by the Netherlands Foundation for Chemical Research (SON) with financial aid from the Netherlands Organization for Scientific Research (NWO) and by a C. and C. Huygens stipend (to H.P.H.) from NWO. We thank Mrs. Dorothé Beer for preparation of the manuscript.

REFERENCES

1. R.J.King, Pulmonary surfactant, J.Appl.Physiol. 53:1 (1982).
2. G.Chevalier and A.J.Collet, In vivo incorporation of choline-3H and galactose-3H in alveolar type II pneumocytes in relation to surfactant synthesis. A quantitative radioautographic study in mouse by electronmicroscopy, Anat.Rec. 174:289 (1972).
3. J.A.Clements, Function of the alveolar lining, Am.Rev.Respir.Dis. 115: 67 (1977).
4. L.M.G. van Golde, J.J.Batenburg and B.Robertson, The pulmonary surfactant system: biochemical aspects and functional significance, Physiol.Rev. 68:374 (1988).
5. S.A.Rooney, The surfactant system and lung phospholipid biochemistry, Am.Rev.Respir.Dis. 131:439 (1985).
6. F.Possmayer, Biochemistry of pulmonary surfactant during fetal development and in the perinatal period, in: "Pulmonary Surfactant," B.Robertson, L.M.G.van Golde and J.J. Batenburg, eds., pp. 295, Elsevier, Amsterdam, (1984).
7. M.Post and L.M.G van Golde, Metabolic and developmental aspects of the pulmonary surfactant system, Biochim.Biophys.Acta 947:249 (1988).
8. P.M.Farrell and M.E.Avery, Hyaline membrane disease, Am.Rev.Respir. Dis. 111:657 (1975).
9. R.J.King and J.A.Clements, Surface active materials from dog lung II. Composition and physiological correlations, Am.J.Physiol. 223:715 (1972).
10. J.R.Wright and J.A. Clements, Metabolism and turnover of lung surfactant, Am.Rev.Respir.Dis. 135:426 (1987).
11. A.Jobe and M.Ikegami, Surfactant for the treatment of respiratory distress syndrome, Am.Rev.Respir.Dis. 136:1256 (1987).
12. J.L.Harwood, Lung surfactant, Prog.Lipid.Res. 26:211 (1987).
13. F.Possmayer, A proposed nomenclature for pulmonary surfactant-associated proteins, Am.Rev.Respir.Dis. 138:990 (1988).

14. J.Goerke, Lung surfactant, Biochim.Biophys.Acta 344:241 (1974).

15. M.Post, J.J.Batenburg, E.A.J.M.Schuurmans, C.D.Laros and L.M.G. van Golde, Lamellar bodies isolated from adult human lung tissue, Exp. Lung.Res. 3:17 (1982).

16. M.Hallman, R.Spragg, J.H.Harrell, K.M.Moser and L.Gluck, Evidence of lung surfactant abnormality in respiratory failure: Study of bronchoalveolar lavage phospholipids, surface activity, phospholipase activity, and plasma myoinositol, J.Clin.Invest. 70:673 (1982).

17. S.L.Young, S.A.Kremers, J.S.Apple, J.D.Crapo and G.W.Brumley, Rat lung surfactant kinetics: biochemical and morphometric correlation, J.Appl.Physiol. 51:248 (1981).

18. A.Montfoort, L.M.G. van Golde and L.L.M. van Deenen, Molecular species of lecithins from various animal tissues, Biochim.Biophys.Acta 231:335 (1971).

19. O.S.Beppu, J.A.Clements and J.Goerke, Phosphatidylglycerol-deficient lung surfactant has normal properties, J.Appl.Physiol. 55:496 (1983).

20. M.Hallman, G.Enhorning and F.Possmayer, Composition and surface activity of normal and phosphatidylglycerol-deficient lung surfactant, Pediatr.Res. 19:286 (1985).

21. S.A.Rooney, T.S.Wai-Lee, L.Gobran and E.K.Motoyama, Phospholipid content, composition and biosynthesis during fetal lung development in the rabbit, Biochim.Biophys.Acta 431:447 (1976).

22. S.A.Rooney, Biochemical development of the lung, in: "The Biological Basis of Reproductive and Developmental Medicine," J.B. Warshaw, ed., pp. 239, Elsevier, New York (1983).

23. M.Hallman, M.Kulovich, E.Kirkpatrick, R.G.Sugarman and L.Gluck, Phosphatidylinositol and phosphatidylglycerol in amniotic fluid: indices of lung maturity, Am.J.Obstet.Gynecol. 125:613 (1976).

24. J.Egberts and W.A.Noort, Gestational age-dependent changes in plasma inositol levels and surfactant composition in the fetal rat, Pediatr.Res. 20:24 (1986).

25. M.Hallman and L.Gluck, Formation of acidic phospholipids in rabbit lung during perinatal development, Pediatr.Res. 14:1250 (1980).

26. B.J.Benson, J.A.Kitterman, J.A.Clements, E.J.Mescher and W.H.Tooley, Changes in phospholipid composition of lung surfactant during development in the fetal lamb, Biochim.Biophys.Acta 753:83 (1983).

27. M.Hallman and B.L.Epstein, Role of myo-inositol in the synthesis of phosphatidylglycerol and phosphatidylinositol in the lung, Biochem. Biophys.Res.Commun. 92:1151 (1980).

28. J.J.Batenburg, W.Klazinga and L.M.G. van Golde, Regulation of phosphatidylglycerol and phosphatidylinositol synthesis in alveolar type II cells isolated from adult rat lung, FEBS Lett. 147:171 (1982).

29. J.E.Bleasdale, N.E.Tyler, F.N.Busch and J.G.Quirk, The influence of myo-inositol on phosphatidylglycerol synthesis by rat type II pneumocytes, Biochem.J. 212:811 (1983).

30. M.J.Engle and P.M.Farrell, Lung phospholipids in aging primates, in: "Behavior and Pathology of Aging in Rhesus Monkeys," pp. 275, Alan R. Liss, New York (1985).

31. J.L.Mauderly and F.F.Hahn, The effects of age on lung function and structure of adult animals, Adv.Vet.Sc.Comp.Med. 26:35 (1982).

32. R.J.King, D.J.Klass, E.G.Gikas and J.A.Clements, Isolation of apoproteins from canine surface active material, Am.J.Physiol. 224: 788 (1973).

33. S.Hawgood, B.J.Benson and R.L.Hamilton, Effects of a surfactant-associated protein and calcium ions on the structure and surface activity of lung surfactant lipids, Biochemistry 24:184 (1985).

34. T.E.Weaver, W.M.Hull, G.Ross and J.A.Whitsett, Intracellular and oligomeric forms of surfactant-associated apolipoprotein(s) A in the rat, Biochim.Biophys.Acta 827:260 (1985).

35. J. A. Whitsett, T. Weaver, W. Hull, G. Ross and C. Dion, Synthesis of surfactant-associated glycoprotein A by rat type II epithelial cells. Primary translation products and post-translational modification, <u>Biochim. Biophys. Acta</u> 828:162 (1985).

36. J. Floros, D. S. Phelps, S. Kourembanas and H. W. Taeusch, Primary translation products, biosynthesis and tissue specificity of the major surfactant protein in rat, <u>J. Biol. Chem.</u> 261:828 (1986).

37. D. S. Phelps and H. W. Taeusch, A comparison of the major surfactant-associated proteins in different species, <u>Comp. Biochem. Physiol.</u> 82B:441 (1985).

38. S. R. Walker, M. C. Williams and B. Benson, Immunocytochemical localization of the major surfactant apoproteins in type II cells, Clara cells, and alveolar macrophages of rat lung, <u>J. Histochem. Cytochem.</u> 34:1137 (1986).

39. J. J. Coalson, V. T. Winter, H. M. Martin and R. J. King, Colloidal gold immunoultrastructural localization of rat surfactant, <u>Am. Rev. Respir. Dis.</u> 133:230 (1986).

40. M. C. Williams and B. Benson, Immunocytochemical localization and identification of the major surfactant protein in adult rat lung, <u>J. Histochem. Cytochem.</u> 29:291 (1981).

41. J. U. Balis, J. F. Paterson, J. E. Paciga, E. M. Haller and S. A. Shelley, Distribution and subcellular localization of surfactant-associated glycoproteins in human lung, <u>Lab. Invest.</u> 52:657 (1985).

42. J. H. Fisher, F. T. Kao, C. Jones, R. T. White, B. J. Benson and R. J. Mason, The coding sequence for 32,000 dalton pulmonary surfactant-associated protein is located on chromosome 10 and identifies two separate restriction fragment-length polymorphisms, <u>Am. J. Hum. Genet.</u> 40:503 (1987).

43. G. Bruns, H. Stroh, G. M. Veldman, S. A. Latt and J. Floros, The 35 kd pulmonary surfactant-associated protein is encoded on chromosome 10, <u>Hum. Genet.</u> 76:58 (1987).

44. R. T. White, D. Damm, J. Miller, K. Spratt, J. Schilling, S. Hawgood, B. Benson and B. Cordell, Isolation and characterization of the human pulmonary surfactant apoprotein gene, <u>Nature</u> 317:361 (1985).

45. J. Floros, R. Steinbrink, K. Jacobs, D. Phelps, R. Kriz, R. Recny, L. Sultzman, S. Jones, H. W. Taeusch, H. A. Frank and E. F. Fritsch, Isolation and characterization of cDNA clones for the 35-kDa pulmonary surfactant-associated protein, <u>J. Biol. Chem.</u> 261:9029 (1986).

46. B. Benson, S. Hawgood, J. Schilling, J. Clements, D. Damm, B. Cordell and R. T. White, Structure of canine pulmonary surfactant apoprotein: cDNA and complete amino acid sequence, <u>Proc. Natl. Acad. Sci. USA</u> 82:6379 (1985).

47. K. Sano, J. Fisher, R. J. Mason, Y. Kuroki, J. Schilling, B. Benson and D. Voelker, Isolation and sequence of a cDNA clone for the rat pulmonary surfactant-associated protein (PSP-A), <u>Biochem. Biophys. Res. Commun.</u> 144:367 (1987).

48. V. Boggaram, K. Qing and C. R. Mendelson, The major apoprotein of rabbit pulmonary surfactant. Elucidation of primary sequence and cyclic AMP and developmental regulation, <u>J. Biol. Chem.</u> 263:2939 (1988).

49. G. F. Ross, R. H. Notter, J. Meuth and J. A. Whitsett, Phospholipid binding and biophysical activity of pulmonary surfactant-associated protein (SAP)-35 and its non-collagenous COOH-terminal domains, <u>J. Biol. Chem.</u> 261:14283 (1986).

50. K. Drickamer, Two distinct classes of carbohydrate-recognition domains in animal lectins, <u>J. Biol. Chem.</u> 263:9557 (1988).

51. H. P. Haagsman, R. T. White, J. Schilling, K. Lau, B. J. Benson, J. Golden, S. Hawgood and J. A. Clements, Studies of the structure of the lung surfactant protein, SP-A. Native size, intramolecular disulfide bonding and circular dichroic analysis of canine, human and recombinant SP-A, <u>Am. J. Physiol.</u> (in press).

52. K. Muramoto and H. Kamiya, The amino-acid sequence of a lectin of the acorn barnacle Megabalanus rosa, <u>Biochim.Biophys.Acta</u> 874:285 (1986).

53. Y. Giga, A. Ikai and K. Takahashi, The complete amino acid sequence of echinoidin, a lectin from the coelomic fluid of the sea urchin Anthocidaris crassipina. Homologies with mammalian and insect lectins, <u>J.Biol.Chem.</u> 262:6197 (1987).

54. T. Voss, H. Eistetter, K. P. Schäfer and J. Engel, Macromolecular organization of natural and recombinant lung surfactant protein SP 28-36. Structural homology with the complement factor C1q, <u>J.Mol. Biol.</u> 201:219 (1988).

55. H. P. Haagsman, S. Hawgood and B. J. Benson, The major surfactant protein, SP-A, is a 650 kDa Ca^{2+}-binding protein, <u>Am.Rev.Respir.Dis.</u> 137:277 (1988).

56. R. J. King, Comparative properties of recombinants formed with SP-A and SP-C, <u>Prog.Respir.Dis.</u> (in press).

57. R. J. King and M. C. Macbeth, Physicochemical properties of dipalmitoyl-phosphatidylcholine after interaction with an apolipoprotein of pulmonary surfactant, <u>Biochim.Biophys.Acta</u> 557:86 (1979).

58. R. J. King and M. C. MacBeth, Interaction of the lipid and protein components of pulmonary surfactant. Role of phosphatidylglycerol and calcium, <u>Biochim.Biophys.Acta</u> 647:159 (1981).

59. H. Efrati, S. Hawgood, M. C. Williams, K. Hong and B. J. Benson, Divalent cation and hydrogen ion effects on the structure and surface activity of pulmonary surfactant, <u>Biochemistry</u> 26:7986 (1987).

60. S. Hawgood, B. J. Benson, J. Schilling, D. Damm, J. A. Clements and R. T. White, Nucleotide and amino acid sequence of pulmonary surfactant protein SP 18 and evidence for cooperation between SP 18 and SP 28-36 in surfactant lipid adsorption, <u>Proc.Natl.Acad.Sci.USA</u> 84:66 (1987).

61. H. P. Haagsman, S. Hawgood, T. Sargeant, D. Buckley, R. T. White, K. Drickamer and B. J. Benson, The major lung surfactant protein, SP 28-36, is a calcium-dependent, carbohydrate-binding protein, <u>J.Biol.Chem.</u> 262: 13877 (1987).

62. M. Hallman, B. L. Epstein and L. Gluck, Analysis of labeling and clearance of lung surfactant phospholipids in rabbit.Evidence of bidirectional flux between lamellar bodies and alveolar lavage, <u>J.Clin.Invest.</u> 68: 742 (1981).

63. H. Jacobs, A. Jobe, M. Ikegami and D. Conaway, The significance of regulation of surfactant phosphatidylcholine, <u>J.Biol.Chem.</u> 258:4159 (1983).

64. M. W. Magoon, J. R. Wright, A. Baritussio, M. C. Williams, J. Goerke, B. J. Benson, R. L. Hamilton and J. A. Clements, Subfractionation of lung surfactant. Implications for metabolism and surface activity, <u>Biochim.Biophys.Acta</u> 750:18 (1983).

65. J. R. Wright, R. E. Wager, S. Hawgood, L. Dobbs and J. A. Clements, Surfactant apoprotein Mr=26,000-36,000 enhances uptake of liposomes by type II cells, <u>J.Biol.Chem.</u> 262:2888 (1987).

66. L. G. Dobbs, J. R. Wright, S. Hawgood, R. Gonzalez, K. Venstrom and J. Nellenbogen, Pulmonary surfactant and its components inhibit secretion of phosphatidylcholine from cultured rat alveolar type II cells, <u>Proc.Natl.Acad.Sci.USA</u> 84:1010 (1987).

67. W. R. Rice, G. F. Ross, F. M. Singleton, S. Dingle and J. A. Whitsett, Surfactant-associated protein inhibits phospholipid secretion from type II cells. <u>J. Appl.Physiol.</u> 63:692 (1987).

68. Y. Kuroki, R. J. Mason and D. R. Voelker, Pulmonary surfactant apoprotein A structure and modulation of surfactant secretion by rat alveolar type II cells, <u>J.Biol.Chem.</u> 263:3388 (1988).

69. Y. Kuroki, R. J. Mason and D. R. Voelker, Alveolar type II cells express a high-affinity receptor for pulmonary surfactant protein A, <u>Proc. Natl.Acad.Sci.USA</u> 85:5566 (1988).

70. L.M.G. van Golde, J.F. van Iwaarden, J.J.Batenburg and J.Verhoef, Metabolic aspects of pulmonary surfactant: synthesis by alveolar type II cells and possible interactions with alveolar macrophages, in: "Surfactant and the Respiratory Tract," L.Ekelund, ed., pp. 3, Elsevier, Amsterdam (1989).

71. S.A.Shelley, J.U.Balis, J.E.Paciga, R.A.Knuppel, E.H.Ruffolo and P. J.Bouis, Surfactant "apoproteins" in human amniotic fluid: an enzyme-linked immunosorbent assay for the prenatal assessment of lung maturity, Am.J.Obstet.Gynecol. 144:224 (1982).

72. Y.Kuroki, H.Takahashi, Y.Fukuda, M.Mikawa, A.Inagawa, S.Fujimoto and T.Akino, Two-site "simultaneous" immunoassay with monoclonal antibodies for the determination of surfactant apoproteins in human amniotic fluid, Pediatr.Res. 19:1017 (1985).

73. S.L.Katyal and G.Singh, An enzyme-linked immunoassay of surfactant apoproteins. Its application to the study of fetal lung development in the rat, Pediatr.Res. 17:439 (1983).

74. C.R.Mendelson, C.Chen, V.Boggaram, C.Zacharias and J.M.Snyder, Regulation of the synthesis of the major surfactant apoprotein in fetal rabbit lung tissue, J.Biol.Chem. 261:9938 (1986).

75. P.L.Ballard, S.Hawgood, H.Liley, G.Wellenstein, L.W.Gonzales, B. Benson, B.Cordell and R.T.White, Regulation of pulmonary surfactant apoprotein SP28-36 gene in fetal human lung, Proc.Natl.Acad.Sci.USA 83:9527 (1986).

76. J.A.Whitsett, T.Pilot, J.C.Clark and T.E.Weaver, Induction of surfactant protein in fetal lung. Effects of cAMP and dexamethasone on SAP-35 RNA and synthesis, J.Biol.Chem. 262:5256 (1987).

77. T.Fujiwara, H.Maeta, S.Chida, T.Morita, Y.Watabe and T.Abe, Artificial surfactant therapy in hyaline membrane disease, Lancet, 1:55 (1980).

78. P.J.R.Phizackerly, M.H.Town and G.E.Newman, Hydrophobic proteins of lamellated osmiophilic bodies isolated from pig lung, Biochem.J. 183:731 (1979).

79. Y.Tanaka, T.Takei and Y.Kanazawa, Lung surfactants. II. Effects of fatty acids, triacylglycerols and protein on the activity of lung surfactant, Chem.Pharm.Bull. 31:4100 (1983).

80. J.A.Whitsett, B.L.Ohning, G.Ross, J.Meuth, T.Weaver, B.A.Holm, D.L. Shapiro and R.H.Notter, Hydrophobic surfactant-associated protein in whole lung surfactant and its importance for biophysical activity in lung surfactant extracts used for replacement therapy, Pediatr.Res. 20:460 (1986).

81. S.H.Yu and F.Possmayer, Reconstitution of surfactant activity by using the 6 kDa apoprotein associated with pulmonary surfactant, Biochem. J. 236:85 (1986).

82. S.W.Glasser, T.R.Korfhagen, T.Weaver, T.Pilot-Matias, J.L.Fox and J.A. Whitsett, cDNA and deduced amino acid sequence of human pulmonary surfactant-associated proteolipid SPL(Phe), Proc.Natl.Acad.Sci.USA 84:4007 (1987).

83. K.A.Jacobs, D.S.Phelps, R.Steinbrink, J.Fisch, R.Kriz, L.Mitsock, J.P. Dougherty, H.W.Taeusch and J.Floros, Isolation of a cDNA clone encoding a high molecular weight precursor to a 6-kDa pulmonary surfactant-associated protein, J.Biol.Chem. 262:9808 (1987).

84. S.D.Revak, T.A.Merritt, E.Degryse, L.Stefani, M.Courtney, M.Hallman and C.G.Cochrane, Use of human surfactant low molecular weight apoproteins in the reconstitution of surfactant biological activity, J.Clin.Invest. 81:826 (1988).

85. T.Curstedt, J.Johansson, J.Barros-Söderling, B.Robertson, G.Nilsson, M.Westberg and H.Jörnvall, Low-molecular mass surfactant protein type 1. The primary structure of a hydrophobic 8-kDa polypeptide with eight half-cystine residues, Eur.J.Biochem. 172:521 (1988).

86. D.S.Phelps and J.Floros, Localization of surfactant protein synthesis in human lung by in situ hybridization, Am.Rev.Respir.Dis. 137:939 (1988).

87. P.A.Emrie, C.Jones, T.Hofmann and J.H.Fisher, The coding sequence for the human 18,000-dalton hydrophobic pulmonary surfactant protein is located on chromosome 2 and identifies a restriction fragment length polymorphism, Somatic Cell Mol.Gen. 14:105 (1988).

88. J.A.Whitsett, T.E.Weaver, J.C.Clark, N.Sawtell, S.W.Glasser, T.R. Korfhagen, and W.M.Hull, Glucocorticoid enhances surfactant proteolipid Phe and pVal synthesis and RNA in fetal lung, J.Biol. Chem. 262:15618 (1987).

89. R.G.Warr, S.Hawgood, D.I.Buckley, T.M.Crisp, J.Schilling, B.J.Benson, P.L.Ballard, J.A.Clements and R.T.White, Low molecular weight human pulmonary surfactant protein (SP5): Isolation, characterization, and cDNA and amino acid sequences, Proc.Natl.Acad.Sci.USA 84:7915 (1987).

90. S.W.Glasser, T.R.Korfhagen, T.E.Weaver, J.C.Clark, T.Pilot-Matias, J.Meuth, J.L.Fox and J.A.Whitsett, cDNA, deduced polypeptide structure and chromosomal assignment of human pulmonary surfactant proteolipid, SPC(pVal), J.Biol.Chem. 263:9 (1988).

91. R.W.Olafson, U.Rink, S.Kielland, S.-H.Yu, J.Chung, P.G.R.Harding and F.Possmayer, Protein sequence analysis studies on the low molecular weight hydrophobic proteins associated with bovine pulmonary surfactant, Biochem.Biophys.Res.Commun. 148:1406 (1987).

92. J.Johansson, H.Jörnvall, A.Eklund, N.Christensen, B.Robertson and T.Curstedt, Hydrophobic 3.7 kDa surfactant polypeptide: structural characterization of the human and bovine forms, FEBS Lett. 232:61 (1988).

93. M.C.Williams and R.J.Mason, Development of the type II cell in the fetal rat lung, Am.Rev.Respir.Dis. 115(suppl):37 (1977).

94. J.R.Bourbon, M.Rieutort, M.J.Engle and P.M.Farrell, Utilization of glycogen for phospholipid synthesis in fetal rat lung, Biochim. Biophys.Acta 712:382 (1982).

95. E.C.Heesbeen, G.Rijksen, J.J.Batenburg, L.M.G. van Golde and G.E.J. Staal, Phosphofructokinase in alveolar type II cells isolated from fetal and adult rat lung, Biochim.Biophys.Acta 1002:388 (1989)

96. H.P.Haagsman, E.A.J.M.Schuurmans, J.J.Batenburg and L.M.G. van Golde, Synthesis of phosphatidylcholine in ozone-exposed alveolar type II cells isolated from adult rat lung. Is glycerolphosphate acyltransferase a rate-limiting enzyme?, Exp.Lung Res. 14:1 (1988).

97. W.M.Maniscalco, J.N.Finkelstein and A.B.Parkhurst, De novo fatty acid synthesis by freshly isolated type II epithelial cells, Biochim. Biophys.Acta 751:462 (1983).

98. C.E.Patterson, K.S.Davis and R.A.Rhoades, Regulation of fetal lung disaturated phosphatidylcholine synthesis by de novo palmitate supply, Biochim.Biophys.Acta 958:60 (1988).

99. T.S.Pope and S.A.Rooney, Effects of glucocorticoid and thyroid hormones on regulatory enzymes of fatty acid synthesis and glycogen metabolism in developing fetal rat lung, Biochim.Biophys.Acta 918: 141 (1987).

100. C.Bijleveld and M.J.H.Geelen, Measurement of acetyl-CoA carboxylase activity in isolated hepatocytes, Biochim.Biophys.Acta 918:274 (1987).

101. J.J.Batenburg, J.N. den Breejen, R.W.Yost, H.P.Haagsman and L.M.G. van Golde, Glycerol 3-phosphate acylation in microsomes of type II cells isolated from adult rat lung, Biochim.Biophys.Acta 878:301 (1986).

102. E.P.Kennedy, The biosynthesis of phospholipids, in: "Lipids and Biomembranes. Past, Present and Future," J.A.F.Op den Kamp, B. Roelofsen and K.W.A.Wirtz,eds., pp. 171, Elsevier, Amsterdam (1986).

103. C.A.Crecelius and W.J.Longmore, Phosphatidic acid phosphatase activity in subcellular fractions derived from adult rat type II pneumocytes in primary culture, Biochim.Biophys.Acta 750:447 (1983).

104. M.Post, E.A.J.M.Schuurmans, J.J.Batenburg and L.M.G. van Golde, Mechanisms involved in the synthesis of disaturated phosphatidyl-

choline by alveolar type II cells isolated from adult rat lung, Biochim.Biophys.Acta 750:68 (1983).

105. J.N. den Breejen, J.J.Batenburg and L.M.G. van Golde, The species of acyl-CoA in subcellular fractions of type II cells isolated from adult rat lung and their incorporation into phosphatidic acid, Biochim.Biophys.Acta 1002:277 (1989).

106. R.J.Mason and J.Nellenbogen, Synthesis of saturated phosphatidyl-choline and phosphatidylglycerol by freshly isolated rat alveolar type II cells. Biochim.Biophys.Acta 794:392 (1984).

107. J.M.Vereyken, A.Montfoort and L.M.G. van Golde, Some studies on the biosynthesis of the molecular species of phosphatidylcholine from rat lung and phosphatidylcholine and phosphatidyletanolamine from rat liver, Biochim.Biophys.Acta 260:70 (1979).

108. J.G.Nijssen and H van den Bosch, Cytosol-stimulated remodeling of phosphatidylcholine in rat lung microsomes, Biochim.Biophys.Acta 875:450 (1986).

109. J.J.Batenburg, W.J.Longmore, W.Klazinga and L.M.G. van Golde, Lysolecithin acyltransferase in adult rat alveolar type II epithelial cells, Biochim.Biophys.Acta 573:136 (1979).

110. C.A.Crelius and W.J.Longmore, Acyltransferase activities in adult rat type II pneumocyte-derived subcellular fractions, Biochim.Biophys. Acta 795:238 (1984).

111. A.C.J. de Vries, J.J.Batenburg and L.M.G. van Golde, Lysiphosphatidyl-choline acyltransferase and lysophosphatidylcholine: lysophosphat-idylcholine acyltransferase in alveolar type II cells from fetal rat lung, Biochim.Biophys.Acta 833:93 (1985).

112. S.Stymne and A.K.Stobart, Involvement of acyl-exchange between acyl-CoA and phosphatidylcholine in the remodelling of phosphatidyl-choline in microsomal preparations of rat lung, Biochim.Biophys.Acta 837:239 (1985).

113. J.G.Nijssen and H. van den Bosch, Coenzyme A-mediated transacylation of sn-2 fatty acids from phosphatidylcholine in rat lung microsomes, Biochim.Biophys.Acta 875:458 (1986).

114. M.Post,J.J.Batenburg, E.A.J.M.Schuurmans and L.M.G. van Golde, The rate-limiting step in the biosynthesis of phosphatidylcholine by alveolar type II cells from adult rat lung, Biochim.Biophys.Acta 712:390 (1982).

115. M.Post, J.J.Batenburg, B.T.Smith and L.M.G. van Golde, Pool sizes of precursors of phosphatidylcholine formation in adult rat lung type II cells, Biochim.Biophys.Acta 795:552 (1984).

116. M.Post, J.J.Batenburg, L.M.G. van Golde and B.T.Smith, The rate-limiting reaction in phophatidylcholine synthesis by alveolar type II cells isolated from fetal rat lung, Biochim.Biophys.Acta 795:558 (1984).

117. D.E.Vance and S.L.Pelech, Enzyme translocation in the regulation of phosphatidylcholine biosynthesis, Trends Biochem.Sci. 9:17 (1984).

118. P.A.Weinhold, M.E. Rounsifer, S.E.Williams, P.G.Brubaker and D.A. Feldman, CTP: phosphorylcholine cytidylyltransferase in rat lung. The effect of free fatty acids on the translocation of activity between microsomes and cytosol, J.Biol.Chem. 259:10315 (1984)

119. E.E.Aeberhardt, C.T.Barrett, S.A.Kaplan and M.L.Scott, Stimulation of phosphatidylcholine synthesis by fatty acids in fetal rabbit type II pneumocytes, Biochim.Biophys.Acta 875:6 (1986).

120. M.Post, Maternal administration of dexamethasone stimulates choline-phosphate cytidylyltransferase in fetal type II cells, Biochem.J. 241:291 (1987).

121. R.Burkhardt, P von Wichert, J.J.Batenburg and L.M.G. van Golde, Fatty acids stimulate phosphatidylcholine synthesis and CTP: choline-phosphate cytidylyltransferase in type II pneumocytes isolated from adult rat lung, Biochem.J. 254:495 (1988).

EFFECT OF MATERNAL DIABETES ON FETAL

LUNG MATURATION

Angeles Zapata, José M. Hernández-García, Cristina Grande,
Isabel Martínez and Jesús Pérez

Hospitales "La Paz"
and "12 de Octubre"
28034 Madrid, Spain

INTRODUCTION

The newborn of diabetic mothers are at a higher risk of hyaline
membrane disease (HMD), the foremost cause of neonate death in
industrialized countries (1). The etiology of HMD is inadequate pulmonary
surfactant synthesis and the major risk factor is pre-term delivery, a
frequent complication in diabetic pregnancy. Determination of fetal lung
maturity in these women by evaluation of the "lung profile" in amniotic
fluid is useful for avoiding iatrogenic lung immaturity. Most researchers
agree that the presence of phosphatidylglycerol (PG) in amniotic fluid is
associated with lung maturity. In normal pregnancy synthesis of this
phospholipid begins at 35-36 weeks of amenorrhea but in pregnant diabetics
fetal lung maturation may deviate from normal (2,3). The purpose of this
study was to report our results on fetal lung maturation in women with
pregnancy complicated by diabetes mellitus as compared to normal pregnant
women.

RESULTS

Amniotic fluid was obtained from 167 pregnant women. Eighty five were
diagnosed as high-risk: 70 class A, B, C diabetics and 15 class D, F and H
diabetics. The control group was 82 normal pregnant women (Table 1). The
echographic characteristics of the placenta and amount of amniotic fluid
coincided with duration of amenorrhea. Amniotic fluid was only obtained
when there was a medical indication. The control group consisted of
repeated cesarean sections, with no other complications.

Phospholipids were determined by thin-layer chromatography (4). Fetal
lung maturity was predicted, under our assay conditions, on the basis of
the lecithin/sphingomyelin ratio (L/S) and presence of PG in amniotic fluid
using the criteria found by us during normal pregnancy (4): L/S < 2.7 and PG
not detected: immature surfactant; L/S $\geq$ 2.7 and PG not detected: incomplete surfactant; and L/S $\geq$ 2.7 and presence of PG, mature surfactant.
The diagnosis of hyaline membrane disease (HMD) and transient tachypnea of
the newborn (TTN) was made independently on the basis of clinical and
radiological criteria by the staff of the Neonatal Intensive Care Service
without knowledge of the biochemical results. Results are given as mean ±

Endocrine and Biochemical Development of the Fetus and Neonate
Edited by J. M. Cuezva *et al.*
Plenum Press, New York, 1990

251

Table 1. Characteristics of diabetic and control groups

Diagnosis	Weeks	No. of subjects	N° delivered within 48 hours of amniocentesis
Diabetes (a):			
A, B, and C	31-40	70	32
D, F, and H	30-35	15	9
Normal			
Pregnancies	29-40	82	23

(a) White's classification

SD. Student's "t" test was used to compare means. To compare two qualitative values Pearson's chi-square was used. Yates' correction was applied to 2 x 2 contingency tables when any of the expected values was smaller than 5.

L/S ratios in amniotic fluid

Figure 1 shows the L/S ratios in the amniotic fluid samples from class A, B, and C diabetes and a control group. In the diabetics the L/S ratios rose from 2.3 ± 0.5 to 3.4 ± 1.3 between weeks 35-36 of pregnancy ($t=2.31$; $p<0.05$). There were no differences between these two groups of subjects.

Mature surfactant in amniotic fluid

At 32-35 weeks, the incidence of L/S ≥2.7 in class D, F and H diabetics, was 100% (11/11) as compared to only 19.3% (6/31) in normal pregnancies and 22.8% (8/35) in class A,B and C diabetics. These differences were significant ($x^2=18.7$,c. Yates; $p<0.0005$ and $x^2=17.49$,c. Yates;

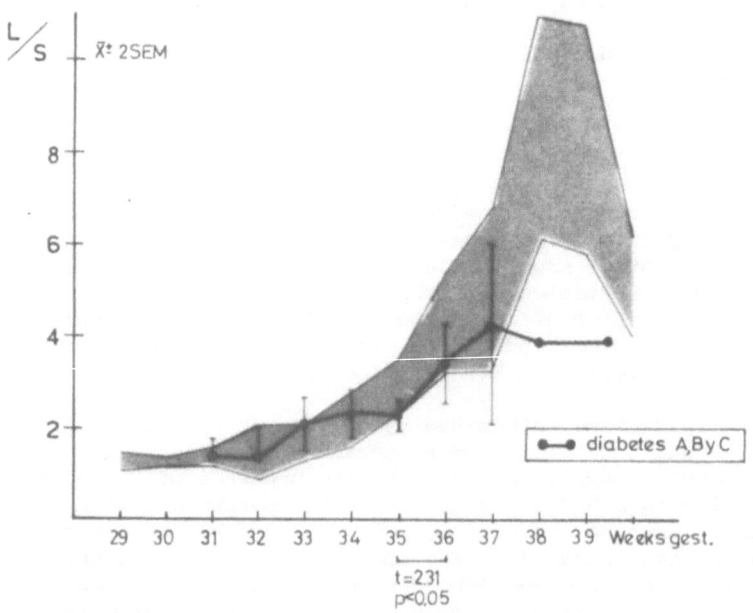

Fig. 1. Evolution of L/S ratio in class A, B and C diabetes. No difference with control could be found.

p<0.0005, respectively). The incidence of mature surfactant from 34 to 35 weeks of amenorrhea was 100% (9/9) in the class D, F and H diabetics, significantly higher than in the controls (5.2%, 1/19). In class A, B and C diabetics the incidence of mature surfactant at 36-38 weeks was lower than in the control group: 35.3% (6/17) and 81% (17/21) respectively (x^2=8.2, p<0.01).

Ratio of phosphatidil-inositol/sphingomyielin (PI/S) in amniotic fluid

In class A, B and C diabetics, the PI/S ratio at 36-38 weeks were significantly greater than in the control group: 1.3±0.2 (n=17) and 1.1±0.2 (n=21) respectively (t=2.98; p<0.01). However, the PI/S ratios were similar in class D,F and H diabetics and normal pregnancies for each gestational period.

C-peptide and cortisol concentration

C-peptide levels were higher in class A, B and C diabetics than in the control group: 4.5±2.2 and 1.4±0.4 ng/ml, respectively (t=12; p<0.001). Cortisol levels were similar in both: 16.6±6.5 and 15.0±4.6 ng/ml, respectively.

Prediction of fetal lung maturity from type of surfactant in amniotic fluid

Of 167 pregnancies studied, in 64 (41 diabetics and 23 normal pregnancies) amniocentesis was performed within 48 hours of delivery (Table 1). Comparing the characteristics of the amnotic surfactant with clinical status of the newborn, we found that in 45 cases with mature surfactant newborns did not have respiratory problems at birth. In 13 cases the surfactant was incomplete and 5 of these newborns had some type of respiratory pathology: 1 HMD (8%) and 4 TTN (31%). Immature surfactant was detected in 6 amniotic samples: 5 developed HMD and the sixth developed TTN.

Of the 11 newborns with respiratory pathology 10 were offspring of class A, B and C diabetics and one was the pre-term infant of a normal mother and was diagnosed as TTN. The L/S ratios in the tracheal aspirate obtained at birth were 1.4±0.4 and 6.1±3 for HMD and TTN, respectively. PG was not detected in any case.

DISCUSSION

In this study, we observed a delay in fetal lung maturation in class A, B and C diabetics. Amniotic C-peptide levels in these women were very high, reflecting fetal hyperglycemia. Since different tissues can synthesize myo-inositol from glucose (5), hyperglycemia in these fetuses accounts for the higher PI/S ratios observed, this elevation being associated with an alteration in PG synthesis. It has been demonstrated (6) that an increase in myo-inositol results in an increase in the synthesis of PI that consume CDP-diacylglycerol which is used as common precursor for both PG and PI synthesis. Furthermore, high myo-inositol levels impair surfactant PG synthesis in HMD (7).

In class D, F and H diabetics, the L/S ratio was higher than in normal pregnancies and PG was present at 34 weeks of amenorrhea. Similar results have been reported by other investigators (8,9). The accelerated pulmonary maturation found in these women could be a response to the chronic stress these fetuses suffer. This metabolic situation may be due to fetal hypo-glycemia by placental vascular affectation.

Of the 13 infants predicted to have incomplete surfactant, 5 developed respiratory pathology. L/S ratio determination alone is thus insufficient for the prediction of fetal lung maturity. In the tracheal aspirate of the newborns with HMD and TTN, selective absence of PG was found at birth.

In summary, we have observed a close relationship between fetal pulmonary maturity and the type of surfactant in amniotic fluid, independent of gestational age or maternal pathology.

REFERENCES

1. M. F. Robert, R. K. Neff, J. P. Hubbell, H. W. Taeusch and M. E. Avery, Association between maternal diabetes and the respiratory-distress sindrome in the newborn, N. Engl. J. Med 294:357 (1976)
2. M. V. Kulovich, M. B. Hallman and L. Gluck, The lung profile I. Normal pregnancy, Am. J. Obstet. Gynecol. 135:57 (1979)
3. M. V. Kulovich and L. Gluck, The lung profile II. Complicated pregnancy, Am. J. Obstet. Gynecol. 135:64 (1979)
4. A. Zapata, J. M. Hernández-García and P. de la Fuente, Amniotic fluid phospholipids and foetal lung maturation, Scand. J. Clin. Lab. Invest. 48:39 (1988)
5. A. I. Winegrad and A. D. Green, Diabetic polyneuropathy: the importance of insulin deficiency, hyperglucemia and alterations in myoinositol metabolism in its pathogenesis, N. Engl. J. Med. 295:1416 (1976)
6. M. B. Hallman and B. L. Epstein, Role of myoinositol in the synthesis of phosphatidyl-glycerol and phosphatidyl-inositol in the lung, Biochem. Biophys. Res. Commun. 92:1151 (1980)
7. M. B. Hallman, O. D. Saugstad, B. L. Epstein and L. Gluck, Acidic surfactant phospholipids in RDS and in adult RDS (ARDS): Role of myoinositol in regulation, Pediatr. Res 15:720 (1981)
8. H. E. Fadel, S. A. Saad and G. Nelson, Effect of maternal-fetal disorders on lung maturation, Am. J. Obstet. Gynecol. 155:544 (1986)
9. E. Amon, J. Lipshitz, B. M. Sibai, T. N. Abdella, D. K. Whybrew and A. El-Naxer, Quantitative analysys of amniotic fluid phospholipids in diabetic pregnant women, Obstet. Gynecol. 68:373 (1986)

LIPOPROTEIN LIPASE IN NEWBORN RAT LIVER: CELLULAR

LOCALIZATION AND SITE OF SYNTHESIS

Julia Peinado, Senen Vilaró, Ferran Burgaya,
Joana Vilanova, Monique Robert, Ignasi Ramírez
and Miquel Llobera

Departamento de Bioquímica y Fisiología
Universidad de Barcelona
08028 Barcelona, Spain

INTRODUCTION

Lipoprotein lipase (LPL) (E.C. 3.1.1.34) is the enzyme responsible for the hydrolysis of triacylglycerols in plasma lipoproteins. The reaction is thought to take place in extrahepatic tissues as adipose tissue, heart, skeletal muscle, lung and lactating mammary gland (1). In recent years we and others described the appearance of LPL activity in the liver of several animal species during the perinatal period as well as in other physiological conditions (2,3).

LPL is synthesized and activated inside parenchymal cell and, by unknown mechanisms, transported to the endothelial cells where it is functional (4). In extrahepatic tissues LPL is found in two major compartments: (i) the luminal side of endothelial cell plasma membranes, and (ii) inside the parenchymal cell (5). The former is commonly referred to as the functional or the heparin-releasable pool, since the enzyme molecules can be released from their binding sites by heparin perfusion. Several lines of evidence indicate that, of all LPL molecules in the tissue, only those in the former compartment have access to the substrate. The second compartment is the so called residual pool since it corresponds to the activity that remains associated to the tissue after heparin perfusion.

Here we present the evidences obtained in our laboratory to indicate that in the neonatal (1 day-old) rat liver, LPL exists in two compartments and that hepatocytes are responsible for the synthesis of the enzyme.

RESULTS AND DISCUSSION

Perfusion of Isolated Livers with Heparin.

When isolated livers from 1-day old rats were perfused in the absence of heparin, no LPL activity was detected in the perfusate. After the addition of heparin to the perfusion fluid (5 U/ml), a sharp peak of LPL activity appeared immediately in the perfusate. Some activity remained in the tissue, but the released activity accounted for about 75% of the total

Endocrine and Biochemical Development of the Fetus and Neonate
Edited by J. M. Cuezva *et al.*
Plenum Press, New York, 1990

255

(released + residual) activity. The hepatic origin of the heparin-releasable LPL activity found in newborn rat livers is evident by the fact that this activity slowly reappears when livers previously perfused with a heparin-containing medium are changed to a heparin-free medium.

The heparin-releasable activity probably represents the functional fraction of the enzyme. In extrahepatic tissues this pool is very sensitive to regulation by the nutritional status of the animal (4), among many other conditions. We previously reported that the LPL activity in the liver of neonatal rats starved for several periods of time, is increased (6). We have recently observed that the size of the heparin-releasable pool is about 2-fold higher in 16-h fasted rats than in fed neonates, whereas the residual activity remains unchanged.

Immunocytochemical Localization

To directly demonstrate that the localization of the heparin-releasable LPL activity we used an immunocytochemical approach. The anti-bovine milk LPL chicken serum used in these experiments, was generously provided by Dr. T. Olivecrona (University of Umeå, Sweden). At the light microscope, we observed a large proportion of LPL bound to endothelial cells. When livers were first perfused with heparin and then used for immunocytochemistry no label was found at the endotelia (7).

Electron microscope showed in great detail the cellular localization of the enzyme. Panel A shows intracellular label inside hepatocytes, pAG particles (arrows) were seen associated to the rough endoplasmic reticulum (rer), little or no specific label was seen in mitochondria (m). Panel B shows the presence of cluster particles (arrows) at the Disse space (DS)

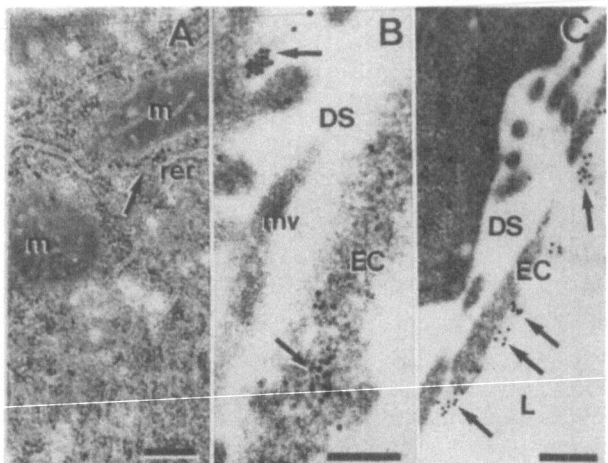

Fig. 1. Immunolocalization of lipoprotein lipase in neonatal rat liver sections. Livers from 1-day old rats were fixed and processed as in 8. Immunolocalization was performed on Lowicryl sections using antibovine LPL IgG from eggs purified by affinity chromatography on LPL-Sepharose columns. Immunoreactive material was visualized with protein A-colloidal gold. Bars = 1 μm.

and inside endothelial cells (EC). Little but no significant label was seen at the microvilli (mv) of hepatocytes. Panel C shows the presence of LPL molecules (arrows) associated to the luminal side (l) of endothelial cells plasma membranes. No pAG particles were seen inside hemopoietic cells (result not shown).

Site of Synthesis

We have also demonstrated that in the adult rat, when LPL activity appears in the liver (as after a fat load in fasted animals), again a substantial amount of the activity is heparin-releasable and associated to the endothelial cells (8). However recent reports indicate that the LPL activity detected in the liver has been synthesized in extrahepatic tissues (9,10). In neonatal rats, as we have shown above, the source of the LPL detected in the liver seems to be the liver itself. The most direct evidence was obtained in isolated livers from neonatal fasted·rats by the incorporation of ^{35}S-methionine into immunoprecipitable LPL (7).

The neonatal liver contains a large proportion of hemopoietic cells (about 70%) which are no longer present in the adult rat liver. Since the ability of the liver to synthesize large amounts of LPL seems a particular feature of the neonate, the possibility that hemopoietic cells were responsible for the synthesis of perinatal rat liver was studied. We have found that isolated hepatocytes from fed neonates contain a considerable amount of LPL activity (97 $\mu U/10^6$ cells), accounting for, at least, 80% of the activity detected in the whole liver after collagenase perfusion (it is known that collagenase, as heparin, releases the LPL bound to endotheliaL cells). On the contrary, isolated hemopoietic cells completely lacked LPL activity (data not shown).

We incubated the isolated neonatal hepatocytes in aminoacid-, vitamin-, and insulin-containing medium for up to 3 hours at 25°C in the presence or the absence of heparin (5 U/ml) and/or cycloheximide (350 μM), a glycosaminoglican that promotes the release to the incubation medium of newly synthesized LPL activity and a protein synthesis inhibitor respectively. In the basal condition (no heparin no cycloheximide), the total activity (cells + medium) was unchanged during the 3 hour incubation (from 237±46 $\mu U/10^6$ cells to 320±47 $\mu U/10^6$ cells). In the presence of cycloheximide the total activity was decreased by 33% (to 159±33 $\mu U/10^6$ cells). In the presence of heparin the total activity was increased 2.4-fold (to 569±52 $\mu U/10^6$ cells). This increase was completely blocked by the presence of cycloheximide (to 259±23 $\mu U/10^6$ cells). These results indicate that neonatal rat hepatocytes synthesize LPL activity.

CONCLUDING REMARKS

From the results presented here, it is concluded that the LPL activity detected in the neonatal rat liver is synthesized inside hepatocytes where it is also activated. By unknown mechanisms most likely involving secretory vesicles, the enzyme is translocated to endothelial cells where, bound to the luminal side of the plasma membrane, it may be functional hydrolyzing tryacylglycerols in tryacylglycerol-rich lipoproteins.

ACKNOWLEDGMENTS

This work was supported by a grant of the Comisión Asesora de Investigación Científica y Técnica of Spain.

REFERENCES

1. "Lipoprotein Lipase", J.Borensztajn ed., Evener Publishers, Chicago (1987)
2. M.LLobera, A.Montes and E.Herrera, Lipoprotein lipase activity in liver of the rat fetus, _Biochem.Biophys.Res.Commun._, 91:272 (1979).
3. S.Vilaró, M.Reina, I.Ramírez and M.Llobera, Intralipid administration induces a lipoprotein lipase-like activity in the livers of starved adult rats, _Biochem.J_, 236:273 (1986).
4. A.Cryer, Tissue lipoprotein lipase activity and its action on lipoprotein metabolism, _Int.J.Biochem._ 13:525 (1981)
5. P.Nilsson-Ehle, A.S.Garfinkel and M.C.Schotz, Lipolytic enzymes and plasma lipoprotein metabolism, _Ann.Rev.Biochem._, 49:667 (1980)
6. D.Grinberg, I.Ramírez, S.Vilaró, M.Reina, M.Llobera and E.Herrera, Starvation enhances lipoprotein lipase activity in the liver of the newborn rat, _Biochim.Biophys.Acta_ 833:217 (1985).
7. S.Vilaró, M.Llobera, G.Bengtsson-Olivecrona and T.Olivecrona, Synthesis of lipoprotein lipase in the liver of newborn rats and localization of the enzyme by immunofluorescence, _Biochem.J_. 249: 549 (1980).
8. S.Vilaró, I.Ramírez, G.Bengtsson-Olivecrona, T.Olivecrona and M. Llobera, Lipoprotein lipase in liver. Release by heparin and immunocytochemical localization, _Biochim.Biophys.Acta_ 959:106 (1988).
9. L.Wallinder, J.Peterson, T.Olivecrona and G.Bengtsson-Olivecrona, Hepatic and extrahepatic uptake of intravenously injected lipoprotein lipase, _Biochim.Biophys.Acta_ 769:513 (1984).
10. J.Peterson, T.Olivecrona and G.Bengtsson-Olivecrona, Distribution of lipoprotein lipase and hepatic lipase between plasma and tissues; effect of hypertriglyceridemia, _Biochim.Biophys.Acta_ 837:262 (1985).

LONG-TERM CONSEQUENCES OF HIGH CARBOHYDRATE INTAKE DURING

THE SUCKLING PERIOD ON LIPID SYNTHESIS IN THE ADULT RAT

Thomas J. Thekkumkara, Peter M. Haney,
Lap Ho and Mulchand S. Patel

Departments of Biochemistry and Nutrition
Pew Center for Molecular Nutrition
Case Western Reserve University School of Medicine
Cleveland, Ohio 44106, U.S.A.

INTRODUCTION

The influence of nutrition during early childhood on the etiology of obesity due to altered lipid metabolism later in life is poorly understood. Alterations in the supply of nutrients due to overnutrition, undernutrition or premature weaning during the first three weeks of life exert long-lasting effects in experimental animals. Most animal studies have relied either on manipulation of litter size to vary the amount of nutrients available during infancy or on alterations in maternal nutrient intake during pregnancy and/or lactation (1-3). Overfeeding of rat pups in early life induces an increased fat storage capacity by increasing differentiation and proliferation rates of adipocyte precursors (3) as well as by increasing hepatic lipogenic enzyme activities (4). However, in these studies overnutrition was achieved by providing substantially increased amounts of mother's milk without attempting to change its composition.

Permanent effects of early dietary modification due to "premature" weaning of rats to a diet high in carbohydrate have also been described (5,6). In most of these studies "premature" weaning was initiated on day 17 or 18 of postnatal life and in no study was it started before day 15 (5), the time limit of this experimental approach.

Messer et al. (7) introduced the artificial rearing procedure for rat pups by implanting intragastric cannulas. Rat milk-substitute formulas (7, 8) with high fat content similar to rat milk have been successfully used to rear pups artificially. Using a modified rat milk-substitute formula (high in carbohydrate and low in fat), we observed striking alterations in liver metabolism in rat pups reared artificially during the first three weeks of life (9,10). Compared to age-matched pups reared naturally by mother and pups reared artificially on the high fat formula, pups reared artificially on the high carbohydrate formula had: (i) significantly higher concentrations of plasma insulin, (ii) precocious induction of both hepatic glucokinase and malic enzyme, and (iii) elevated levels of hepatic glucose-6-phosphate dehydrogenase activity during the first 3 weeks of postnatal life (10). In this communication, we report the the long-term changes in lipid metabolism in the adult rat as a consequence of early and chronic

Endocrine and Biochemical Development of the Fetus and Neonate
Edited by J. M. Cuezva *et al.*
Plenum Press, New York, 1990

intragastric feeding of the high carbohydrate milk formula during the suckling period.

MATERIALS AND METHODS

Pregnant Sprague-Dawley rats were fed Purina Chow and water *ad libitum*. Four day-old naturally born male pups received intragastric cannulas under light ether anaesthesia. Cannulated pups were raised from this point in isolation from their mothers using an artificial rearing system (10). A modified rat milk-substitute formula referred to as the high carbohydrate formula (calorie content of 56% carbohydrate, 20% fat, and 24% protein) (10) was delivered intragastrically every 2 h at a rate of 0.45 kcal/g body weight per day (10). On day 18 of postnatal life the pups were disconnected from the pump but maintained on the same formula provided in liquid diet-feeding tubes and water *ad libitum* until weaning (10). Daily handling and cleaning of these pups were as described previously (10). Naturally reared male rat pups consuming rat milk (caloric content of 8% carbohydrate, 68% fat, and 24% protein) served as controls. All pups, both artificially reared and mother-fed, were weaned to lab chow on day 24 with free access to food and water. Animals were weighed at least once a week. On day 64 both groups of rats were switched to a high-sucrose diet (U.S. Biochem. Corp., #10662A; AIN Semipurified rat diet; calorie content of 48% sucrose; 14% cornstarch, 27% protein and 11% corn oil) to enhance their lipogenic capacity. In two separate experiments, animals were killed on day 100 (Experiment I) or 130 (Experiment II), and their *in vitro* lipogenic capacities in liver and adipose tissues were determined. In Experiment I, liver homogenates prepared in buffered sucrose were centrifuged at 100,000 g for 1 h, and the clear supernatant was assayed for activities of glucose-6-phosphate dehydrogenase (10), malic enzyme (10), fatty acid synthase (11), and glycerol-3-phosphate dehydrogenase (12) at 37°C. In Experiment II, liver slices and pieces of adipose tissues were incubated in Krebs-Ringer bicarbonate buffer containing 10 mM glucose, insulin (0.125 units/3 ml) and 1 mCi of ^3H_2O for 2 h at 37°C (13). Non-saponifiable lipids and fatty acid fractions were isolated and the radioactivity was determined as described previously (14). Statistical differences between control and artificially reared rats were determined by Student's t test.

RESULTS AND DISCUSSION

Artificially reared rats fed the high-carbohydrate formula have been shown to grow at a rate similar to naturally reared pups during the first three weeks of life (10). In Experiment I, although the mean body weight of high-carbohydrate-fed group was slightly lower than that of the mother-fed pups at the time of weaning to lab chow, the high-carbohydrate-fed animals steadily grew heavier on both lab chow and high sucrose diets ($p<0.05$ was noted on day 52 while on lab chow). On day 100 when these rats (Experiment I) were killed, the high-carbohydrate-fed group was approximately 25% heavier than the mother-fed animals (709±47 g vs. 569±13; n=4, $p<0.005$). Similarly, on day 130 in Experiment II the average body weights were heavier in artificially reared rats compared to mother-fed animals (773±40 g vs. 642±60; n=6, $p<0.025$). No significant differences were observed for the weights of liver, heart, kidneys and brain of these two groups of animals in either experiment. Interestingly, increases in epididymal adipose tissue and subcutaneous fat mass were evident in artificially-reared rats compared to mother-fed animals.

In Experiment I, activities of four lipogenic enzymes, namely glucose-6-phosphate dehydrogenase, malic enzyme, fatty acid synthase and glycerol-3-phosphate dehydrogenase were assayed in liver supernatants. The activities

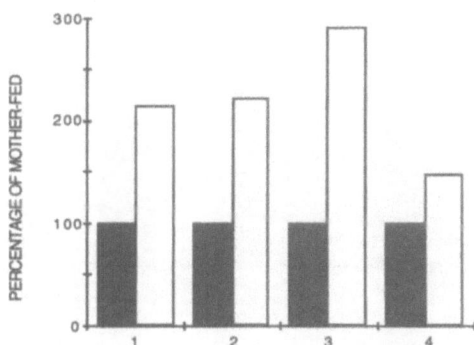

Fig. 1. Long-lasting effects of early exposure to a high
carbohydrate formula on four lipogenic enzyme
activities in rat liver. The results are expressed
as percent of mother-fed adult rats (4 animals in
each group). Enzyme activities (mean±SD) in mother
fed adult rats are shown in parentheses. 1:
Glucose-6-phosphate dehydrogenase (6.1±0.5 units/g
of liver); 2: malic enzyme (0.37±0.2 units/g of
liver), 3: Fatty acid synthase (22.7±6.5 munits/mg
of cytosolic protein); 4: Glycerol-3-phosphate
dehydrogenase (14.5±0.9 munits/mg of cytosol
protein).

of these enzymes were increased 1.5- to 3-fold in livers of artificially-
reared rats compared to mother-fed animals (Figure 1). In Experiment II,
the *in vitro* lipogenic capacities of liver slices and pieces of epididymal
and omental adipose tissues were assessed by measuring the incorporation of
^{3}H into lipids from ^3H_2O. Lipids were then extracted, and the radioactivity
in the non-saponifiable and fatty acid fractions were determined. As seen
in Figure 2A, an approximately 2-fold increase in the incorporation of ^3H_2O
into non-saponifiable lipids was observed in liver slices from artificially
reared animals compared to mother-fed rats. No significant changes were
observed in the ^{3}H incorporation rates into the non-saponifiable lipids by
adipose tissues between two groups of rats (Figure 2A). This is not
surprising because the relative rates of cholesterol synthesis in rat white
adipose tissues are very low. In contrast, the rates of synthesis of fatty
acids measured as the incorporation of ^{3}H from ^3H_2O were increased
approximately 2-fold in the liver and adipose tissues from artificially
reared rats compared to mother-fed animals (Figure 2B).

The mechanism by which early nutritional modifications cause
permanently altered metabolic responses later in life are not well
understood. Proliferation of certain cell types and the resultant increased
capacities for hormonal and metabolic responses may in part be responsible
for permanent effects seen later in life. Evidence shows that over-
nourishment of pups during the suckling period (resulting from reducing
litter size) and premature weaning of rats (on day 16 instead of day 21)
are associated with higher lipogenic capacity later in adult life (1,3-5).
What is evident is that the approaches used for early nutritional
modifications (e.g. overnourished, undernourished, premature weaning, etc.)
are very restrictive in the type of diet provided or in how early the
modification can be instituted. The advent of artificial rearing of rat
pups by intragastric cannulation, and the ease with which a modified milk

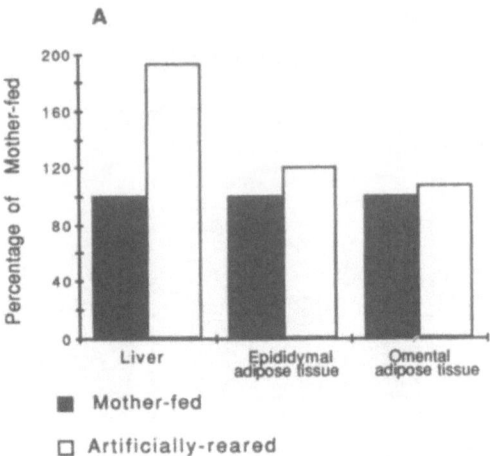

A

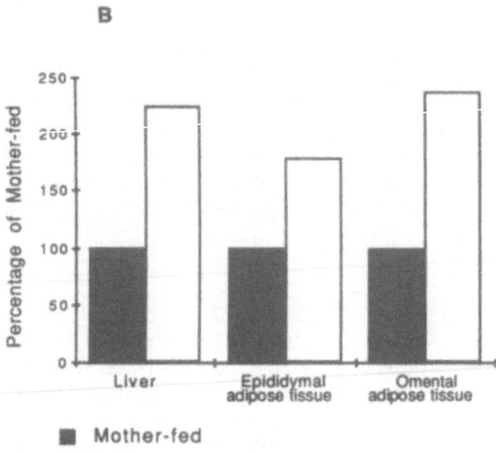

B

Fig. 2. Long-lasting effects of early exposure to a high-
carbohydrate formula on the *in vitro* synthesis of
non-saponifiable lipids (A) and fatty acids (B) in
rat liver and adipose tissues. The results are
expressed as percent of mother-fed adult rats (5
or 6 animals in each group). The incorporation of
^{3}H from ^{3}H$_2$O into the lipid fractions (mean ± SD
cpm/g of tissue/2 h) for mother-fed animals were:
in (A) liver 693±180, epididymal adipose tissue
201±42, and omental adipose tissue 239±65; and in
(B) liver 6406±1337, epididymal adipose tissue
9168±2366 and omental adipose tissue 5285±1720.

formula can be initiated early in the postnatal period, presents an
opportunity to overcome inherent limitations of the other experimental
approaches by initiating "premature weaning" within the first few days of
postnatal life. Among the advantages of the artificial rearing procedure
are the precise control of the amount of milk-substitute formula each pup
receives and the control of the composition of milk formula.

There is little information available on the possible long-term
effects of the consumption of a high carbohydrate diet during the suckling

period in rats on adaptive changes in carbohydrate and lipid metabolism. The results presented here show that artificially-reared rats (fed a high-carbohydrate milk formula during the suckling period) compared to control rats reared naturally but treated identically during the postweaning period accumulated more fat in their adipose tissues and also had higher lipogenic capacities in both liver and adipose tissues (Figures 1 and 2). The mechanisms responsible for these changes in lipid metabolism in these rats remain to be investigated. The artificially-reared rats fed a high carbohydrate formula showed sustained hyperinsulinemia during the first 3 weeks of postnatal life (10). This may have caused an increase in the proliferation and differentiation rates of the precursor cells into adipocytes. A comparable phenomenon has been observed in overnourished suckling rats (3).

In summary, our findings show that alterations in the composition of macronutrients such as fat and carbohydrate in the diet of the rat during the suckling period can exert profound and long-lasting effects on lipid metabolism in adult life. The artificial rearing technique, combined with feeding of modified milk-substitute formulas, provides an opportunity to investigate the mechanism responsible for long-term consequences of nutritional modifications in the early postnatal period on lipid metabolism later in adult life.

ACKNOWLEDGEMENTS

This work was supported by U.S. Public Health Service Grant HD 11089 and by Metabolism Training Grant AM 07319.

REFERENCES

1.   I. M. Faust, P. R. Johnson and J. Hirsch, Long-term effects of early nutritional experience on the development of obesity in the rat, J. Nutr. 110:2027 (1980).
2.   P. R. Johnson, J. S. Stern, M. R. C. Greenwood, L. M. Zucker and J. Hirsch, Effect of early nutrition on adipose cellularity and pancreatic insulin release in the Zucker rats, J. Nutr. 103:738 (1978).
3.   I. Dugail, A. Quignard-Boulange and F. Dupuy, Role of adipocyte precursors in the onset of obesity induced by overfeeding in suckling rats, J. Nutr. 116:524 (1986).
4.   D. A. Duff and K. Snell, Effect of altered neonatal nutrition on the development of enzymes of lipid and carbohydrate metabolism in the rat, J. Nutr. 112:1057 (1982).
5.   P. Hahn, Nutrition and metabolic development in mammals, in: "Nutrition: Pre- and post-natal development", (Human Nutrition: A comprehensive Treatise, vol. 1), M. Winick, ed., Plenum Press, New York, pp. 1, (1979).
6.   D. W. Back and J. F. Angel, Effects of premature weaning on the metabolic responses to dietary sucrose in adult rats, J. Nutr. 112:978 (1982).
7.   M. Messer, E. B. Thoman, A. G. Terrasa and P. R. Dallman, Artificial feeding of infant rats by continuous gastric infusion, J. Nutr. 98: 440 (1969).
8.   K.-Y. Yeh, Small intestine of artificially reared rat pups: Effect of caloric intake and dietary composition on growth and disaccharidase activities, J. Nutr. 113:1496 (1983).
9.   P. M. Haney and M. S. Patel, Regulation of succinyl-CoA: 3-oxoacid CoA-transferase in developing rat brain. Responsiveness associated with prenatal but not postnatal hyperketonemia, Arch. Biochem. Biophys. 240:426 (1985).

10. P. M. Haney, C. Raefsky-Estrin, A. Caliendo and M. S. Patel, Precocious induction of hepatic glucokinase and malic enzyme in artificially reared rat pups fed a high-carbohydrate diet, <u>Arch. Biochem. Biophys.</u> 244:787 (1986).

11. D. B. Martin, G. G. Horning and P. R. Vagelos, Fatty acid synthesis in adipose tissue. I. Purification and properties of a long chain fatty acid-synthesizing system, <u>J. Biol. Chem.</u> 236:663 (1961).

12. L. S. Wise and H. Green, Participation of one isozyme of cytosolic glycero-phosphate dehydrogenase in the adipose conversion of 3T3 cells, <u>J. Biol. Chem.</u> 254:273 (1979).

13. M. S. Patel, M. Jomain-Baum, F. J. Ballard and R. W. Hanson, Pathway of carbon flow during fatty acid synthesis from lactate and pyruvate in rat adipose tissue, <u>J. Lipid Res.</u> 12:179 (1971).

14. M. S. Patel and O. E. Owen, Lipogenesis from ketone bodies in rat brain. Evidence for conversion of acetoacetate into acetyl-coenzyme A in the cytosol, <u>Biochem. J.</u> 156:603 (1976).

FETAL-PLACENTAL INTERACTIONS IN THE

CONTROL OF PRENATAL GROWTH

Colin T.Jones

Laboratory of Cellular and Developmental Physiology
and Nuffield Department of Clinical Medicine
University of Oxford
John Radcliffe Hospital
Oxford OX3 9DU, UK

INTRODUCTION

During the course of prenatal development there is usually a very
close relationship between the growth of the placenta and that of the fetus
(1-4). This relationship does not necessarily persist throughout the whole
of gestation and may not be particularly apparent early in fetal life.
However as the fetus enlarges the relationship appears and dominates
prenatal growth (3). Hence if placental size is restricted the growth of
the fetus suffers in an almost proportionate fashion (2-4). The close
relationship between the growth of these two structures is also reflected
in a clear correlation between the extent of the perfusion of the maternal
placenta and fetal growth (1,2,4,5). Such observations imply that one of
the determining mechanisms underlying prenatal growth rate may be the
substrate supply across the placenta. Although there is some evidence in
favor of this mechanism it does not adequately explain control of prenatal
growth and particularly maturation (6-9). In addition to its transport
functions the placenta has a wide range of endocrine roles, one of which
could be the control of prenatal growth (6). Although there are few clear
indications of the mechanisms that could integrate substrate supply across
the placenta with its endocrine functions there are indications that this
is an important pathway regulating prenatal development (6-9). Some of the
pertinants observations will be reviewed here.

NUTRIENT SUPPLY AND UTILIZATION AT DIFFERENT GROWTH RATES

One fairly common feature of high growth rates characteristic of
prenatal life is substantial substrate cycling such that only 10-20% of
substrate utilization is directed to growth (10). This means that
substantial reserve exists in the proportion of nutrient turnover that
is directed to heat production as compared to accretion. The function of
the high prenatal rate of substrate turnover is remains unclear.

Coupling between substrate supply and fetal growth is poorly
understood. It is often assumed from data on rates of unidirectional flux
and glucose turnover that supply exceeds demand (11-13). However, such
analysis is considerably complicated by the fact that placental utilisation
of substrate may exceed that of the fetus and much of such uptake is

Endocrine and Biochemical Development of the Fetus and Neonate
Edited by J. M. Cuezva *et al.*
Plenum Press, New York, 1990

265

Table 1. Substrates available to the normal and growth retarded 45-50 days fetal guinea pig

Substrates	Plasma concentration	
	Normal	Growth retarded
Glucose	4.4 ±0.5 (8)	1.4 ±0.6 (8)*
Lactate	2.7 ±0.6 (8)	4.2 ±1.5 (8)*
Citrate	0.16±0.05(8)	0.29±0.08(8)*
Alanine	0.10±0.03(9)	0.45±0.13(9)*
Serine	0.12±0.03(8)	0.33±0.10(8)*
Glutamine	0.43±0.12(9)	0.29±0.09(9)*
Leucine	0.08±0.03(9)	0.23±0.07(9)*
α-aminonitrogen	12.9 ±2.6 (9)	6.3 ±2.11(9)*
Ammonia	<0.04	0.39±0.04(9)*
Acetate	0.31±0.07(9)	0.37±0.12(9)

The data are for umbilical venous plasma collected from fetal guinea pigs of normal and 32-35% of normal size, the latter were growth-retarded because of uterine artery ligation at day 30 of pregnancy. P value for comparison of normal and small fetuses is *, <0.01.

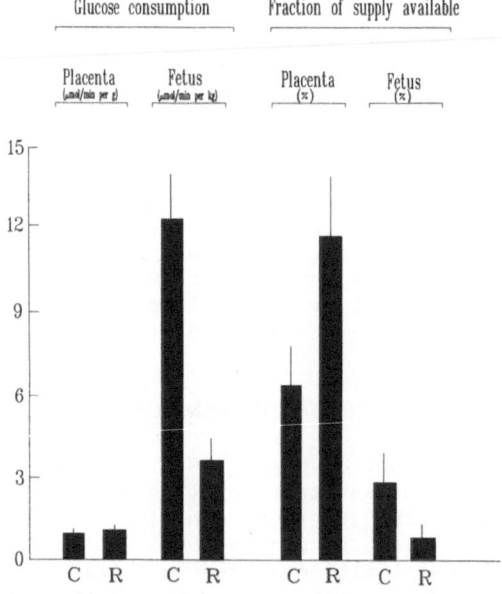

Fig. 1. Glucose uptake *in vivo* by placenta and fetus of 58-61 days normal (C) and growth-retarded fetal guinea pigs. The fraction of uterine glucose that is estimated to be available to the fetus is also plotted. The growth-retarded fetuses were < 40% of normal size.

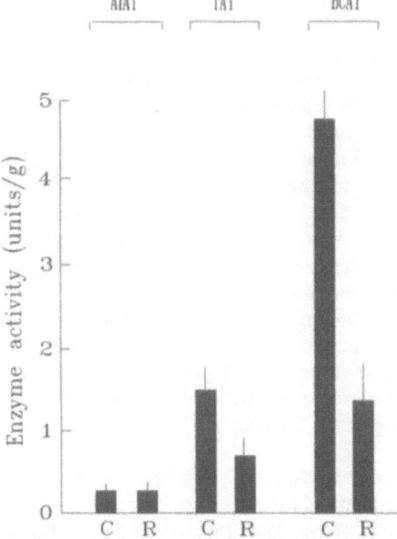

Fig. 2. Hepatic aminotransferase activities
in normal (C) and growth-retarded (R)
fetal guinea pigs of 45-50 days.
Alanine aminotransferase (AlAT),
Tyrosine aminotransferase (TAT),
branched-chain aminotransferase
(BCAT). Other details in Figure 1.

derived from the fetal circulation (11-13). Consequently in assessing the
relationship between substrate supply and fetal growth the interaction
between placenta and fetus must be quantified; except possibly for glucose
such information is not yet available.

Qualitative evaluation is therefore predominantly used. The simplest
and most clearcut of these is to follow changes in the plasma
concentrations of major substrates that occurs at different rates of
prenatal growth. Data for changes induced by intrauterine growth
retardation are summarized in Table 1.

The finding that glucose and fetal growth rate show a strong positive
correlation is very common, whether caused by restricted supply as in
diabetic pregnancy (3,4,8,13-16). Such observations indicate that glucose
supply appears to be a potentially important determinant of fetal growth
(3,4,6,8,11). Hence a common finding in growth retardation is fetal hypoglycaemia and reduced glucose consumption (3,4,14,15) (Figure 1). The
availability of other substrates, especially aminoacids, is also affected
(3,11,17) (Table 1). In most instance, unlike glucose, plasma amino acid
levels are elevated. The reason for this is the utilization falls
dramatically (Figures 2 and 3). Placental use of amino acids in growth
retardation probably increases and this could account for the elevation of
plasma ammonia (18). Fatty acid supply is depressed (Table 1) as is the
capacity of fetal tissues to oxidize them.

These observations taken together indicate that supply and use of the
major fuels available to the fetus falls when fetal growth rate is reduced.
However it appears that this decline in substrate availability and use is
much greater than the reduction in fetal growth. Hence it seems likely that
the efficiency of substrate use increases. Therein is one explanation for
the high rate of susbtrate turnover in growing tissues.

The close relationship between placental and fetal growth described above imply a matching of placental performance to fetal development. Previous descriptive studies have lead to the conclusion that placental size does not limit fetal growth until a particular time late at gestation (3). However as placental growth is restricted in growth retardation placental performance must limit fetal growth (1,2,4). The simplest manner in which placental performance could restrict fetal growth would result from a decline in placental perfusion limiting substrate availability to the placenta. Clearly normal fluctuations in uterine blood flow are able to achieve this (13). However it is interesting to observe that whilst substrate availability to the placenta and fetal growth show a fairly close linear relationship over much of the normal range, that relationship changes abruptly when fetal size is less than 40% of normal (4). Below this fetal size there is a dramatic effect also upon the maturation of fetal tissues with their development being delayed substantially (6,8,9,14,20).

The reason for the relationship between placental perfusion, and particularly glucose availability to the fetus requires closer inspection as it exposes an important role and pathway for the placenta. It is unusual that growth retarded fetuses are hypoglycaemic as might be expected. Clearly fetal glucose consumption is depressed per unit fetal mass (Figure 1) and hence an impairment of placental transport might be expected. Indeed there is some evidence in the fetal sheep that this might be so (3). However studies with isolated perfused placentas from growth retarded fetal guinea pigs show that transport capacity is normal per unit mass placenta (9).

The explanation for the apparently impaired transport of glucose to the fetus and the altered relationship between placental perfusion and fetal glucose concentration is to be found in the handling of substrates by the placenta. The placenta extracts a large proportion of its substrate from the fetal circulation and this fraction rises are uterine blood flow and hence maternal glucose supply to the placenta falls so the proportion extracted by the placenta rises so that fetal needs cannot be sufficiently

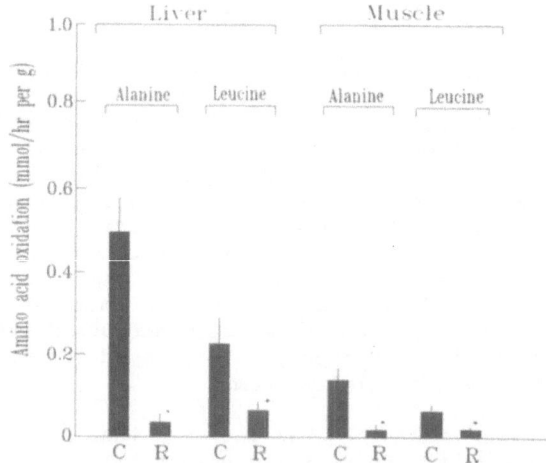

Fig. 3. Oxidation of alanine and leucine by perfused
liver or hindlimb muscle of normal (C) and
growth-retarded (R) fetal guinea pigs of
45-50 days. Other details as for Figure 1.

sustained and fetal hypoglycaemia and reduced uptake ensue (9,11,13) (Figure 1).

Such organization means that fetal needs are secondary to maintaining the placenta as far as glucose, and possibly amino acids, is concerned. Hence the control of nutrient supply probably resides with the placenta and not just because of its transport functions. Moreover the placenta is likely to be the first point of sensing changes in maternal nutritional state, particularly as caused by altered uterine perfusion, and therefore ideally placed to signal changes to and responses from the fetus.

FETAL-PLACENTAL SIGNALING AND THE CONTROL OF GROWTH

The role that the placenta plays in providing signals, either through its metabolism or output of hormones has been reviewed elsewhere (6-9,11, 14). It appears likely that fluctuations in uterine blood flow and hence substrate supply to the placenta cause subtle or large changes in hormone output across the placenta to the fetus. Clearly some of these placental responses contribute to the dramatic changes in fetal endocrine state that occurs when prenatal growth is restricted (3,6-9,14,17,20,21) (Table 2). Of the major changes in endocrine state those in fetal plasma insulin, presumably driven by changes in glucose (Figure 4) and fetal plasma IGF-I and sulphation promoting activity (Figure 5) show the closest correlation with fetal growth.

To what extend do the changes in fetal endocrine state account for the alterations in fetal growth in growth retardation caused by substrate restriction? In attempting to answer this question the dramatic effects that growth restriction has on fetal organ maturation cannot be ignored. Hence the normal development of bone, liver, voluntary muscle and even

Table 2. Plasma hormone and growth factor concentration in normal and growth retarded 45-50 days fetal guinea pigs

Hormone	Plasma concentration (ng/ml)	
	Normal	Growth retarded
ACTH	0.21± 0.09(8)	0.10± 0.06(9)
Cortisol	38.4 ± 11.3 (9)	2.7 ± 6.1 (9)*
Androstendione	1.67± 0.42(5)	3.11± 0.39(5)*
Insulin	12.9 ± 3.6 (9)	4.54± 0.93(9)*
IGF-I	157 ± 29.8 (8)	69.8 ± 10.3 (8)*
IGF-II	437 ±132 (8)	2309 ±872 (8)*
Sulphation activity	0.48± 0.07(9)	0.07± 0.02(9)*
Glucagon	0.16± 0.07(9)	0.42± 0.14(9)*
Growth hormone	128 ± 37 (9)	93.2 ± 35.1 (9)*
T_3	48.3 ± 13.6 (9)	21.4 ± 7.35(9)*
T_4	1.53± 0.31(9)	0.61± 0.21(9)*
rT_3	3.62± 0.85(9)	1.57± 0.48(9)*

The values are for umbilical venous plasma. Fetal guinea pigs were either of normal size of 32-35% of normal because of uterine-artery ligation at day 30 of pregnancy. P value for comparison of normal and growth-retarded fetuses is *,<0.01.

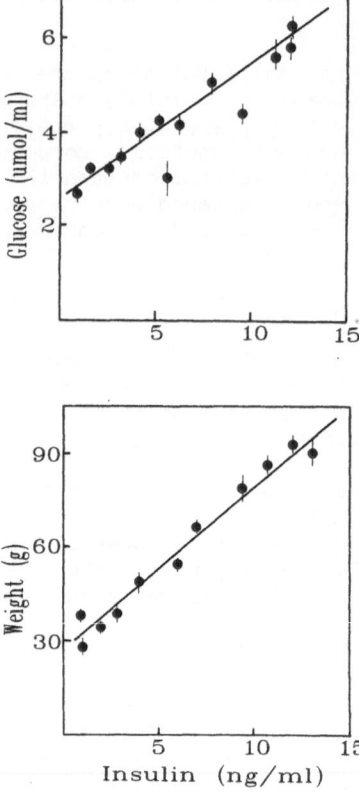

Fig. 4. The relationships between glucose,
insulin and weight in guinea pigs
of 60-63 days.

brain are substantially slowed (6-9,11,14,19,21). These gross developmental
delays are amply illustrated by cell proteins, such as the delayed
appearance of the actinomyosin complex in muscle or of the enzymes of
glucose and fat metabolism in the liver.

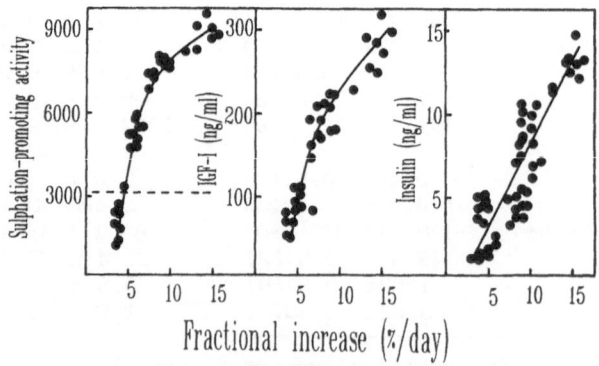

Fig. 5. The relationship between fractional increase
in weight and plasma insulin and growth factor
concentrations in fetal guinea pigs between
days 45-55 of gestation.

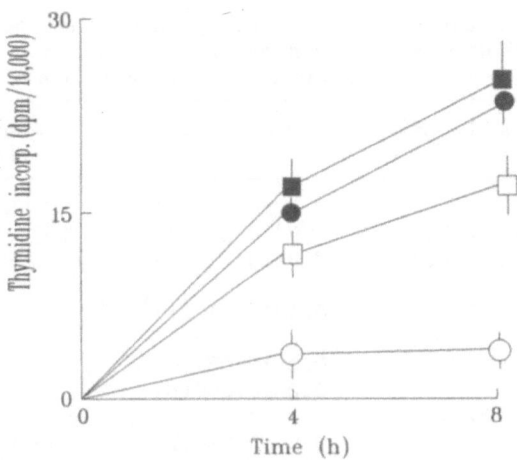

Fig. 6. [3H]Thymidine incorporation into DNA by
hepatocytes from 40 days fetal guinea
pigs. Hepatocytes were incubated in
medium containing insulin, thyrosine,
cortisol and EGF together with plasma
at 20% (v/v) from the umbilical vein
of normal fetal sheep of 125 days (■)
or from those subjected to a 55-65%
reduction in uterine blood flow for 60
min (□), or form normal (●) or growth-
retarded (○) 45 fetal guinea pigs.

It is possible to account for some of the maturational delays in terms
of changes in plasma concentrations of hormones such as insulin, glucagon
and thyroid hormones, although it is increasingly clear that other factors
are involved (19). Hence plasma taken from growth-retarded fetuses inhibits
sulphate incorporation into cartilage and thymidine incorporation into
hepatocytes (19,21-23). The origin of the inhibitory factors is at present
unclear, although one site of production appears to be the placenta (6-9,

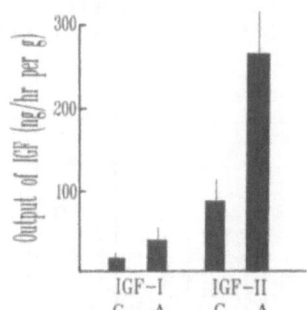

Fig. 7. Output of IGF-I and >IGF-II from
dually perfused placenta of 55-60
days pregnant guinea pig. The
perfusion for 3 hours recirculating
with Krebs' bicarbonate buffer
containing 10 mM glucose was either
control (C) or during administration
of adrenaline (A) at 10 ng/ml.

271

Table 3. The induction of PEP-Carboxykinase in hepatocytes of
40-45 days fetal guinea-pigs in response to exposure
to glucagon and growth-inhibiting factors

Incubation conditions	Enzyme activity	Immunoreactive protein
Control	0.27±0.06 (6)	20.8± 4.6 (6)
Glucagon (100 ng/ml)	1.94±0.49 (6)	157.2±48.3 (6)
Glucagon (1000 mg/ml)	2.84±0.73 (5)	243.9±68.2 (5)*
Glucagon (100 ng/ml) ± TGF-β (20 µg/ml)	0.49±0.11 (7)*	35.8± 9.3 (7)*,*
Control	0.24±0.052(6)	18.9± 3.8 (6)
GR-plasma	0.20±0.050(5)	13.7± 4.3 (5)
Glucagon (100 ng/ml) ± GR-plasma	0.88±0.25 (7)**	73.6±20.3 (7)**
Glucagon (100 ng/ml) ± GRFI-35K (10 µg/ml)	0.59±0.09 (7)**	42.4±11.3 (7)**

Isolated hepatocytes were incubated for 24 hours in Krebs' bicarbonate buffer with 10 µg/ml cortisol and 10 ng/ml IGF-I
together with the additions outlined above. GR-plasma refers
to plasma obtained from growth retarded fetal guinea pigs that
were 30-40% of normal size, it was added at a final dilution
of 10%. GRFI-35K is a 35K growth-inhibiting peptide isolated
from plasma of growth-retarded fetal guinea pigs. P value for
comparison of glucagon treated against controls is: *,<0.01,
and for comparison of the effects of inhibitors on induction
by glucagon is: **,<0.05.

19,20) (Figure 6). Reduction of uterine blood flow stimulates the placenta
to produce growth-inhibitory (Figure 6) as well as growth-stimulatory
substances (Figure 7). The production of which is closely related to the
extent of reduction in uterine blood flow (Figure 8).

The existence of growth inhibiting substances in plasma of fetuses
subjected to reduction in uterine blood flow, and particularly in response
to severe growth retardation can be shown by a variety of different
methods. In addition to effects upon sulphate and thymidine incorporation
factors can be isolated from fetal plasma that will inhibit enzyme

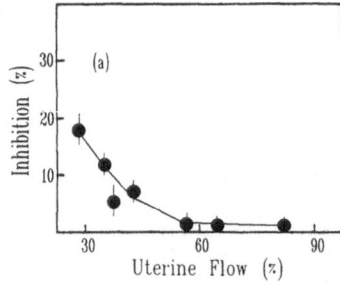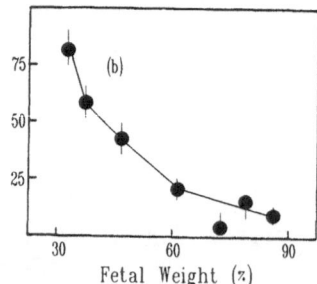

Fig. 8. The relationship between the ability of plasma from
from growth retarded fetal guinea pigs (a) or from
fetal sheep subjected to reduced uterine blood flow
(b) to inhibit [³H]thymidine incorporation into DNA
in isolated fetal hepatocytes.

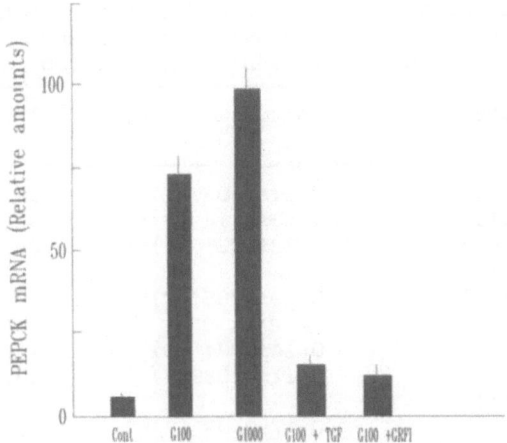

Fig. 9. PEP-carboxykinase mRNA induction in hepato-
        cytes from 40 days fetal guinea pigs
        incubated for 48 hours either in the absence
        (Cont) or presence of glucagon at 100 ng/ml
        (G100) or 1000 ng/ml (G1000) to which may be
        added 10 μg/ml TGF-B (+TGF) or 10 μg/ml of
        35K growth-inhibiting peptide (+GRFI).

induction. Hence plasma isolated from growth retarded fetal guinea pigs
will block the ability of glucagon to induce PEP-carboxykinase (Table 3,
Figure 9) and of insulin to induce enzymes of fatty acid synthesis (Table
4, Figure 10) in fetal hepatocytes. This effect can be stimulated by
transforming growth factor 2-β, an inhibitory well known to suppress enzyme
induction in and growth of hepatocytes. However this is not the primary

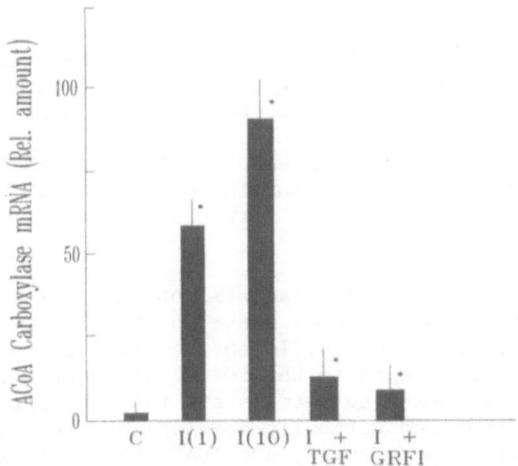

Fig. 10. Acetyl CoA carboxylase mRNA induction in
         hepatocytes from 40 days fetal guinea pigs
         incubated for 48 hours either in the absence
         (C) or presence of insulin at 1 μg/ml (I(1))
         or 10 μg/ml (I(10)) to which may be added 10
         μg/ml TGF-B (+TGF) or 10 μg/ml of 35K growth-
         inhibiting peptide (+GRFI).

Table 4. The induction of Fatty Acid Synthetase in hepatocytes
of 40-45 days fetal guinea pigs in response to exposure
to insulin and growth-inhibiting factors

Incubation Conditions	Enzyme activity	Immunoreactive protein
Control	0.11±0.037(7)	10.9± 2.7 (7)
Insulin (1 µg/ml)	0.43±0.14 (6)*	42.9± 8.9 (6)*
Insulin (10 µg/ml)	0.72±0.21 (5)**	65.8±14.9 (5)**
Insulin (1 µg/ml) ± TGf-β (20 µg/ml)	0.19±0.05 (7)*	17.2± 6.3 (7)*
Control	0.14±0.04 (6)	13.9± 3.4 (6)
GR-plasma	0.11±0.039(5)	13.1± 4.0 (5)
Insulin (1 µg/ml) ± GR-plasma	0.17±0.03 (7)**	20.1± 4.3 (7)**
Insulin (1 µg/ml) ± GRFI-35K (10 µg/ml)	0.15±0.05 (7)**	17.2± 4.9 (7)**

Isolated hepatocytes were incubated for 24 hours in Krebs'bicarbonate buffer with 10 ng/ml T_3, 1 µg/ml cortisol and 10 ng/ml
IGF-I together with the additions outlined above. GR-plasma
refers to plasma obtained from growth-retarded fetal guinea pigs
that were 30-40% of normal size, it was added to a final dilution
of 10%. GRFI-35K is a 35K growth-inhibiting peptide isolated from
plasma of growth-retarded fetal guinea pigs. P value for
comparison of insulin treated against controls is: *, <0.01, and
for comparison of the effects of inhibitors on induction by
glucagon is:**, <0.05.

inhibitory factor suppressing hepatocyte development and growth in growth-
retarded fetuses. An inhibitory glycoprotein of about 35 Kdaltons has been
isolated from plasma of growth-retarded fetuses and this has the ability to
suppress hepatocyte growth and enzyme induction (Tables 3 and 4).

Such observations imply that the placenta has the capacity to regulate
fetal tissue maturation by controlling the release of factors inhibitory to
development. Such capacity is likely to be counteracted by the release of
stimulatory agents such as IGF-I and -II (Figure 7). IGF-II for instance,
appears to have the capacity to increase the placental extraction of
glucose from the fetus (9) and to maintain the deposition of glycogen in
the fetal liver against even when the fetus is hypoglycaemic and
insulinaemic (9).

It therefore appears that the placenta not only has the capacity to
sense changes in maternal nutritional state, particularly if caused by
alterations in uterine perfusion sufficient to alter placental metabolism,
but also to signal these changes to the fetus. Although some of the
signaling is directed at changing fetal state (6) many are involved in
adapting fetal maturation and metabolic development. This implies that the
placenta functions not only to supply substrate to the fetus for growth,
but also conducts a coordinating role sensing environmental changes and
modifying fetal development accordingly.

ACKNOWLEDGMENTS

The work reported in this study was supported by the Wellcome Trust,
Birthright, Action Research, and the Medical Research Council.

REFERENCES

1. N.W.Bruce and R.S.Abdul-Karim, Relationship between fetal weight, placental weight and maternal circulation in th rabbit at different stages of gestation, J.Reprod.Fertil. 32:14 (1937).
2. M.Gilbert and A.Leturque, Fetal weight and its relationship to placental blood flow and placental weight in experimental growth retardation in the rat, J.Develop.Physiol. 4:237 (1982).
3. J.E.Harding, C.T.Jones and J.S.Robinson, Studies on experimetal growth retardation in sheep. The effects of a small placenta in restricting transport to and growth of the fetus, J.Develop.Physiol. 7:427 (1985).
4. C.T.Jones and J.T.Parer, The effect of alterations in placental blood flow on the growth of and nutrient supply to the fetal guinea pig, J.Physiol 343:525 (1983).
5. S.L.B.Duncan and B.V.Lewis, Maternal placental and myometrila blood flow in pregnant rabbit, J.Physiol. 202:471 (1969).
6. C.T.Jones, Pathways of communication between the placenta and fetus, potentially important in the regulation of growth and state, in: "Fetal and Neonatal Development", C.T.Jones ed. Perinatology Press, Ithaca, pp. 68 (1988).
7. C.T.Jones, Endocrine mechanisms of communication between palcenta and fetus, in: "Proceedings of the XI European Congress of Perinatal Medicine", Harewood, London (1989).
8. C.T.Jones, Observations on the control of prenatal growth in the guinea pig, in: "Proceedings of the Liggins Symposium", Perinatology Press, Ithaca (1989).
9. C.T.Jones, J.E.Harding, W.Gu and H.N.Lafeber, Placental metabolism and endocrine effects in relation to the control of fetal and placental growth, in: "The Endocrine Control of the Fetus. Physiologic and Pathophysiologic Aspects", A.Jensen and W.Kunzel, eds. Springer-Verlag, Berlin, pp. 213 (1988).
10. C.T.Jones and T.P.Rolph, Metabolism during fetal life: a functional assessment of metabolic development, Physiol.Rev. 65:357 (1985).
11. C.T.Jones, Relationship between alteratios in uterine blood flow and the handling of glucose by fetus and placenta, in: "The Endocrine Control of the Fetus. Physiologic and Pathophysiologic Aspects", A.Jensen and W.Kunzel, eds. Springer-Verlag, Berlin pp. 333 (1988).
12. W.W.Hay, J.W.Sparks, F.C.Battaglia and G.Meschia, Maternal-fetal glucose exchange: necessity of a 3-pool model, Amer.J.Physiol. 246: E258 (1984).
13. C.T.Jones, W.Gu and J.E.Harding, Metabolism of glucose by fetus and placenta in sheep. The effects of normal fluctuations in uterine blood flow, J.Devel.Physiol. 9:369 (1987).
14. C.T.Jones, H.N.Lafeber and M.M.Roebuck, Studies on the growth of the fetal guinea pig. Changes in plasma hormone concentration during normal and abnormal growth, J.Devel.Physiol. 6:461 (1984).
15. J.S.Robinson, I.C.Hart, E.J.Kingston, C.T.Jones and C.D.Thorburn, Studies on experimetal growth retardation in sheep. The effects of removal of endometrial caruncles on fetal sites and metabolism, J.Devel.Physiol. 1:379 (1979).
16. J.B.Susa, K.L.McCormick, J.A.Widness, D.B.Singer, W.Oh, K.Adamsons and R.Scwartz, Effects of hyperinsulinaemia in the rhesus monkey: effects of physiological hyperinsulinaemia on fetal growth and composition, Diabetes 28:1058 (1979).
17. J.S.Robinson, I.C.Hart, E.J.Kingston, C.T.Jones and G.D.Thorburn, Studies on the growth of the fetal sheep. The effects of reduction of placental size on hormone concentrations in fetal plasma, J.Devel.Physol. 2:239 (1980).
18. J.S.Robinson and J.A.Owens (personal communication).

19. C.T.Jones, Reprogramming of metabolc development by rstriction of fetal growth, <u>Biochem.Soc.Trans.</u> 13:89 (1985).
20. C.T.Jones, T.P.Rolph, H.N.Lafeber, W.Gu, J.E.Harding and J.T.Parer, Experimental studies on the control of fetal growth, <u>in</u>: "The Physiological Development of the Fetus and Newborn", C.T.Jones and P.W.Nathanieslz eds., Academic Press, London pp. 11 (1985).
21. H.N.Lafeber, T.P.Rolph and C.T.Jones, some consecuences of intrauterine growth retardation, <u>in</u>: "Nutrition and Metabolism of the Fetus and Infant", H.K.A.Visser, ed. Martinus Nijhoff Publishers, The Hague. pp. 43 (1979).
22. C.T.Jones, W.Gu, J.E.Harding, D.A.Price and J.T.Parer, Effects of surgical reduction in placental size or experimental manipulation on uterine blood flow on plasma sulphation-promoting activity and on the concentration of insulin-like growths factors I and II, <u>J.Devel. Physiol.</u> 10:179 (1988).
23. C.T.Jones, H.N.Lafeber, D.A.Price and J.T.Parer, Studies on the growth of the fetal guinea pig. Effects of reduction in uterine blood flow on the concentration of insulin-like growth factors I and II, <u>J.Devel.Physiol.</u> 9:181 (1987).

FETAL/MATERNAL PLASMA AMINO ACID RELATIONSHIPS

IN THE STREPTOZOTOCIN DIABETIC RAT

Antonia Martín, Manuel Palacín, Miguel A. Lasunción
and Emilio Herrera

Departamento de Bioquímica e Investigación
Hospital Ramón y Cajal and
Universidad Alcalá de Henares
2034 Madrid, Spain

INTRODUCTION

Gestational diabetes mellitus (GDM) affects placental composition and intrinsic metabolism both in humans (1) and rats (2), and causes abnormalities in maternal-fetal metabolite tranfer (3). These changes may actively contribute to the altered fetal growth that occurs in GDM, which directly depends on the quality and quantity of metabolic fuels crossing the placenta. These fuels may be used as building blocks for fetal accretion and also modulate fetal beta-cell insulin release which is recognized as an important growth factor for the fetus (4).

A major factor that modulates the placental transfer of any metabolite is its concentration in maternal circulation. We previously found that whereas plasma concentration of total amino acids in streptozotocin-diabetic pregnant rats was unaffected, the level of the same parameter was intensely reduced in their fetuses indicating either an impaired placental transfer or an enhanced utilization in the fetal side.

The aim of the present study was to establish whether maternal diabetes during late pregnancy in the rat impairs placental amino acid transfer, and to elucidate the mechanism through which this effect is produced. For this purpose, on day 7 of gestation, rats were made diabetic by a single intravenous injection of 45 mg streptozotocin/Kg dissolved in citrate buffer pH 4.5. Other rats were injected with only buffer and used as controls. All the animals were studied on day 20 of gestation. Placental amino acid transfer was measured using an in situ placental preparation (5), and placental blood flow was estimated by infusion of 99mTc labelled microspheres to other animals under the same experimental condition (6). The in $vitro$ uptake and utilization of $(U-^{14}C)$-L-alanine and $(U-^{14}C)-\alpha$-aminoisobutiric acid was studied in placental slices from diabetic and control animals.

RESULTS

As reported previously (3), on the 20th day of gestation, streptozotocin diabetic pregnant rats show reduced maternal body weight

Endocrine and Biochemical Development of the Fetus and Neonate
Edited by J. M. Cuezva *et al.*
Plenum Press, New York, 1990

277

free of conceptus structures, conceptus and fetus weights as well as plasma levels of radioimmunoassayable insulin both in mothers and fetuses. Litter size and placental weight was not modified in diabetics as compared with controls. Maternal plasma glucose concentration was approximately 6 times higher in diabetics than in controls. Plasma levels of glucose in fetuses were below values in their respective mothers. Glycemia in fetuses of diabetic mothers was about 11 times greater than in those of controls. Consequently, the fetal/maternal plasma glucose ratio was significantly enhanced in diabetic rats.

In spite of the intense diabetic condition, maternal plasma amino acid levels were modified very slightly, with significant reductions in glutamine and serine, and increments in glycine and ornithine, giving an unmodified value for total plasma amino acid levels. Different to the mothers, in fetuses from diabetic rats there was a reduction in the plasma level of most individual amino acids as compared with controls. Consequently, as shown in Figure 1, the fetal/maternal plasma amino acids ratio was significantly reduced in diabetic rat as compared to controls, with the exception of alanine, glutamine and glutamate which did not differ between the two groups. As expected, fetal/maternal plasma ratio was above 1 for most amino acids (Figure 1).

Due to the stable fetal/maternal plasma alanine ratio in diabetic rats and the fact that placental alanine transfer seems to take place by both the ASC and A transport systems proposed by Christensen, which are the same as those used by most amino acids, we decided to study how the transfer of L-alanine is affected in the diabetic rat. For comparison, the study was also extended to determine the placental transfer of glucose in animals kept under similar conditions.

Placental alanine and glucose transfer was determined by infusion of either $(U-^{14}C)$-L-alanine or $(U-^{14}C)$-D-glucose through the maternal left uterine artery, following our already validated technique (5). Animals were anesthesized with sodium pentobarbital (33 mg/Kg, intravenously) and they were infused for 20 min with medium containing the ^{14}C-labelled tracer through a cannula placed counter-current into the left external iliac artery to the level of the left uterine artery with the colateral vessels clamped. In this way the infusion medium becomes diluted with maternal blood reaching the left uterine artery, and the left uterine horn receives the tracer directly before it becomes diluted in the mother's general circulation. Results of radioactivity were always corrected by considering $1x10^6$ dpm as the total infused radioactivity per rat. Comparison of radio-activity present in fetal plasma from the left and right uterine horns, specific radioactivity of the tracer in the left uterine artery and uterine blood flow were used to calculate the actual placental transfer (equation shown in Figure 2). Left uterine blood flow was determined by infusing other experimental rats with 99mTc labelled albumin microspheres, (6) through a cannula placed in the carotid artery and determining the amount of radioactivity appearing in the left uterine horn as compared to that present in blood withdrawn from the femoral artery. As show in Table 1, 20 min after constant infusion, maternal plasma ^{14}C-glucose values were similar in diabetic and control rats, whereas plasma ^{14}C-glucose specific radioactivity was significantly lower in diabetics, and the difference of total radioactivity between left minus right fetuses was significantly higher in these animals. However, when L-alanine was the infused tracer neither of these parameters differed between the two groups (Table 1). Left uterine blood flow was significantly reduced in diabetic rats (Table 1). When these parameters are used to calculate maternal-fetal transfer (Figure 2), it appears that, as expected, the transfer of glucose is intensely enhanced, whereas that of L-alanine is significantly decreased in diabetic rats as compared to controls. When the infused tracer was alanine,

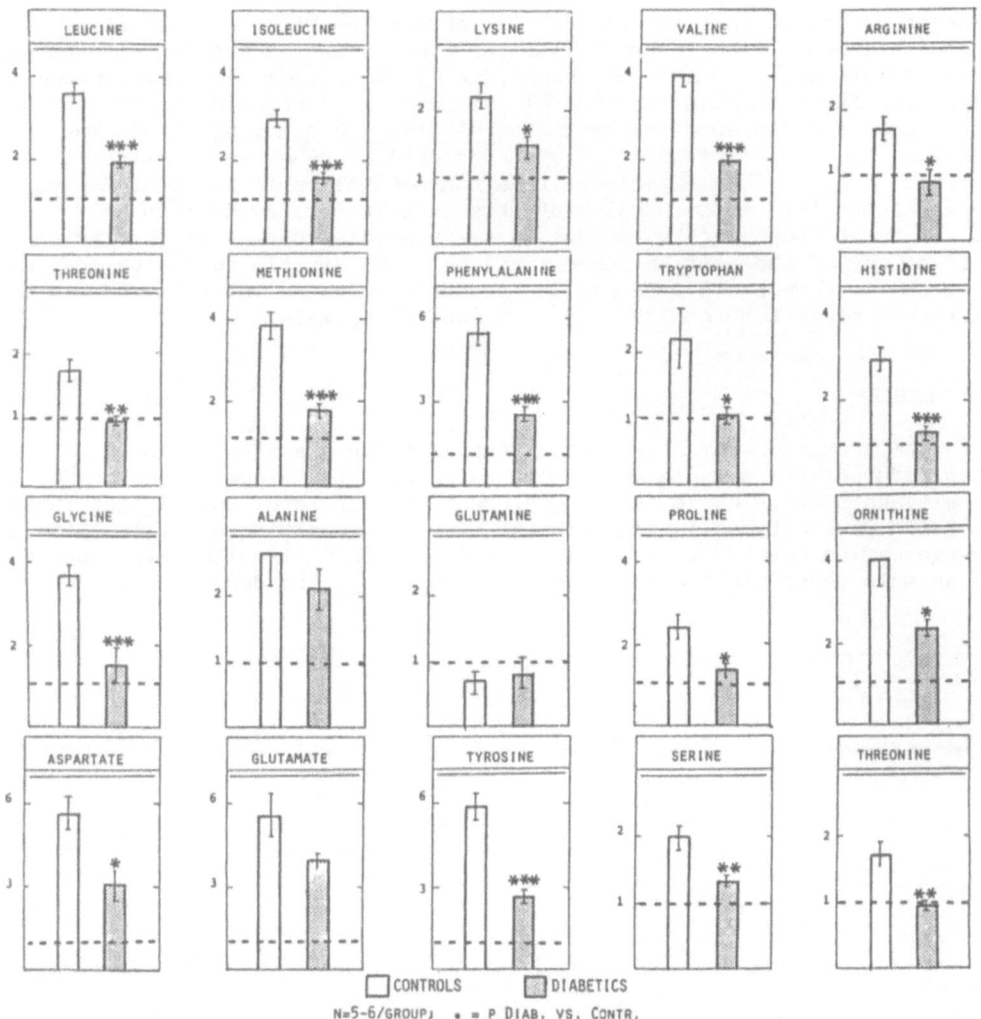

Fig. 1. Fetal/maternal plasma amino acids ratio in control (open bars) and diabetic (shadowed bars) 20 day pregnant rats. Means ± S.E.M. of 5-6 rats/group. Significance of the difference between controls and diabetics is shown by: *, p< 0.05; **, p< 0.01 and ***, p < 0.001. Dotted line corresponds to the amino acid ratio value of 1.

the only parameter used to calculate the transfer that differed between the two groups was the uterine blood flow, with no difference in either maternal alanine specific radioactivity or the radioactivity present in the left minus the right fetuses (Table 1).

To determine whether an altered intrinsic placental metabolism together with the reduced uterine blood flow could be contributing to this alanine decreased placental transfer, we studied the *in vitro* utilization of $(U-^{14}C)$-alanine by placental slices from diabetic and control animals. Placental slices were incubated for 50 min in Krebs Ringer bicarbonate supplemented with 0.4 mM of $(U-^{14}C)$-L-alanine. It was observed that neither alanine uptake or its conversion into CO_2 or lactate differed between the two groups. The same experiment was carried out but using different L-alanine concentrations in the media to obtain the kinetic constant of these parameters. The inverse plots for alanine uptake and its conversion into lactate and CO_2 show that both Vmax and Km values were similar for placentas from diabetic and control animals. In order to determine whether this lack of difference in the placenta metabolic handling of alanine between the two groups could be also the case of other amino acids, it was decided to extend the experiment to a non-metabolizable amino acid such as α-aminoisobutiric acid. Results of this experiment showed again no difference between the two groups in the placental uptake of this amino acid at any of the studied concentrations. Used kinetic constants derived from the inverse plots were similar to those found for alanine and did not differ in placentas from diabetic and control animals.

DISCUSSION

The present study shows that streptozotocin diabetic pregnant rats have reduced maternal and fetal body weights, fetal plasma amino acids levels and blood flow to the uterus. Reduction in fetal body weight in the diabetic pregnant rat is in agreement with previous findings (7). This change may be related to reductions in fetal plasma insulin levels present in animals under similar conditions (3), although the decreased uterine

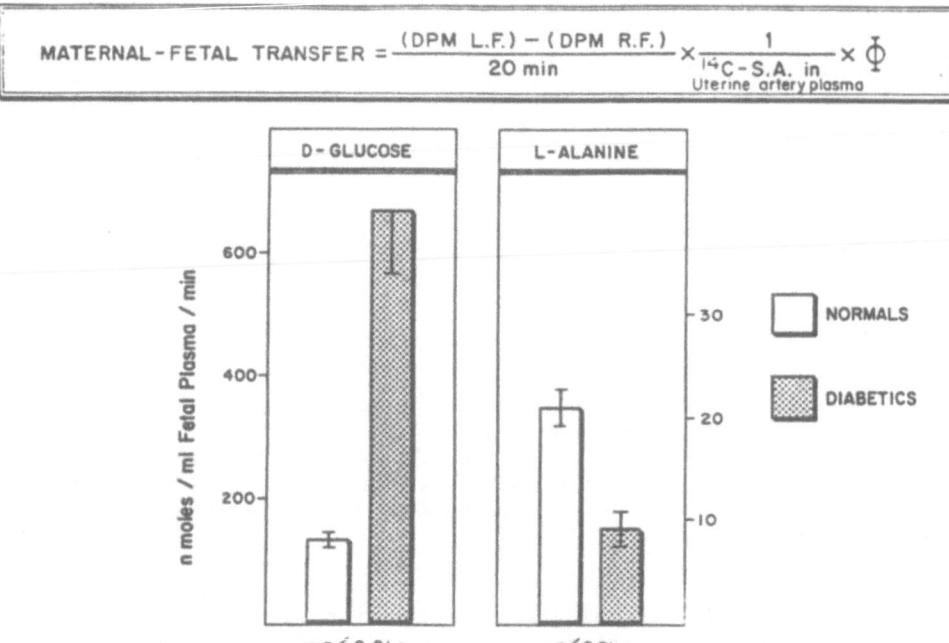

Fig. 2. Placental transfer of glucose and alanine in normal (open bars) and diabetic (shadowed bars) 20 day pregnant rats. Means ± S.E.M. of 5-7 rats/group. Placental transfer was estimated as the difference in total plasma radioactivity between the left (L.F) and right fetuses (R.F) divided by the time of infusion with $(U-^{14}C)$-D-glucose or $(U-^{14}C)$-L-alanine through the maternal left uterine artery (20 min) and by the ^{14}C specific radioactivity (SA) in the maternal artery, and the blood flow (Φ) to the left uterine horn (equation shown in the upper part of the Figure).

Table 1. Parameters used to calculate placental glucose and alanine transfer in STZ-diabetic 20 day pregnant rats

	CONTROLS	DIABETICS	p
Left uterine blood flow (ml/min)	5.31±0.8	2.44±0.23	*
$(U-^{14}C)$-D-Glucose Maternal glucose (mM)	5.1±0.3	24.40±2.30	x
Maternal glucose S.A (dpm/μmol)	1242±236	310±52	*
(DPM L.F.- DPM R.F) (dpm/ml)	2868±529	7497±1092	*
$(U-^{14}C)$-L-Alanine Maternal alanine (μmoles/L)	390±35	402±29	N.S.
Maternal ALA S.A. (dpm/μmol)	2215±206	2336±601	N.S.
(DPM L.F.-DPM R.F.) (dpm/ml)	11173±887	8916±1557	N.S.

Means ± S.E.M. of 7-9 rats/group. Statistical comparations vs. controls are shown: *, $p < 0.01$ and x, $p < 0.001$.

blood flow and plasma amino acid levels, could be also responsible for the limited growth of the fetuses. From studies with pregnant guinea pigs under steady-state conditions, it has been proposed that a direct relationship exists between placental blood flow and fetal body weight (8).

Placental D-glucose transfer values were previously found to be linearly correlated to maternal glycemia in spite of reduced uterine blood flow and augmented placental glucose transfer is a direct consequence of maternal hyperglycemia (3). Reduced placental transfer of L-alanine in the diabetic rat may be however related to the decreased uterine blood flow. This parameter was significantly reduced in diabetic rats whereas neither of the other parameters used to calculate materno-fetal L-alanine transfer differ between the two experimental groups. Variations in uterine blood flow have been reported to affect the transfer to the fetus of molecules with high placental extraction coefficients in other experimental conditions (9). Thus, the reduced uterine blood flow in the diabetic rat may participate in the decreased placental L-alanine transfer observed in these animals.

Placental slices produce a considerable amount of lactate and CO_2 from alanine. These findings demonstrate that rat placenta actively synthesizes lactate from alanine supporting the hypothesis that placenta modifies the quality of maternal metabolites reaching the fetus, lactate being the major product of this effect. Lactate production from other metabolites besides alanine crossing the placenta was previously shown by us (6) and these findings are of special interest under the current belief that lactate is an important energy fuel for the fetus (10). The present results show that neither alanine uptake or its conversion into lactate by placental slices differ between diabetic and control rats. These findings demonstrate that intrinsic placental metabolism in diabetic rat was not altered.

If L-alanine may be considered representative of most amino acids, we may therefore conclude from present results that reduced circulating amino acids in fetuses from streptozotocin diabetic rats is due to an impairment in mother to fetus amino acid transfer secondary to reduced uterine blood flow and not to an altered placental metabolic capacity to handle amino acids coming from maternal circulation.

ACKNOWLEDGMENTS

This study was supported by a grant from the Fondo de Investigaciones Sanitarias de la Seguridad Social, Spain.

REFERENCES

1. Y. Z. Diamant, B. E. Metzger, N. Freinkel and E. Shafrir, Placental lipid and glycogen content in human and experimental diabetes mellitus, Am. J. Obstet. Gynecol. 144:5 (1982).
2. V. Barash, A. Gutman and E. Shafrir, Fetal diabetes in rats and its effect on placental glycogen, Diabetologia 28:244 (1985).
3. E. Herrera, M. Palacín, A. Martín and M. A. Lasunción, Relationship between maternal and fetal fuels and placental glucose transfer in rats with maternal diabetes of varying severity, Diabetes 34 (suppl. 2):42 (1985).
4. P. S. Cooke and C. S. Nicoll, Role of insulin in the growth of fetal rat tissues, Endocrinology 114:638 (1984).
5. M. A. Lasunción, X. Testar, M. Palacín, R. Chieri and E. Herrera, Method for the study of metabolite transfer from rat mother to the fetus, Biol. Neonate 44:83 (1983).
6. M. Palacín, M. A. Lasunción, A. Martín and E. Herrera, Decreased uterine blood flow in the diabetic pregnant rat does not modify the augmented glucose tranfer to the fetus, Biol. Neonate 48:197 (1985).
7. J. M. Cuezva, E. S. Burkett, D. S. Kerr, H. M. Rodman and M. S. Patel, The newborn of diabetic rat. I. Hormonal and metabolic changes in the postnatal period, Pediat. Res. 16:632 (1982).
8. L. L. H. Peeters, J. W. Sparks, G. Grutters, J. Girard and F. C. Battaglia, Utero-placental blood flow during pregnancy in chronically catheterized guinea pigs, Pediat. Res. 16:716 (1982).
9. P. Rosso and R. Kava, Effects of food restriction on cardiac output and blood flow to the uterus and placenta in the pregnant rat, J. Nutr. 110:2350 (1980).
10. R. A. Rhoades and D. A. Ryder, Fetal lung metabolism, response to maternal fasting, Biochim. Biophys. Acta 663:621 (1981).

EFFECT OF PHYSIOLOGICAL CHANGES IN CYSTATHIONASE ACTIVITY

ON AMINO ACID AND GLUTATHIONE METABOLISM IN RAT FETUS

José Viña, Esperanza Gasco, Federico V. Pallardo,
Juan Sastre, José M. Estrela and Francisco Rodrigo

Departamento de Fisiología
Facultad de Medicina
46010 Valencia, Spain

INTRODUCTION

The trans-sulphuration pathway serves to synthesize cysteine from methionine. This pathway is important to ensure an adequate supply of cysteine and, indeed, if the flux through the pathway is decreased, cysteine becomes an essential amino acid. A key enzyme of the pathway is cystathionase, which catalyzes the cleavage of cystathionine to yield cysteine.

Previous work by Sturman et al. (1) showed that cystathionase activity is absent in fetal liver from premature infants and suggested that cysteine might be an essential amino acid in the prematures. However, Zlotkin and Anderson (2) challenged this view by measuring the development of cystathionase activity in extrahepatic tissues and opened the question that this activity might be sufficient to ensure an adequate supply of cysteine from methionine.

We decided to investigate this problem in the rat by measuring not only cystathionase activity but also the possible physiological relevance of changes in this activity, studying the amino acid profile in fetal blood and the rate of trans-sulphuration in isolated hepatocytes from fetal livers.

RESULTS AND DISCUSION

Cystathionase Activity in Adult and Fetal Rat Liver

Cystathionase activity was measured as described by Sturman et al. (1). In liver from adult rats we found an activity of 3.19 ± 0.93 nmol of cysteine formed x mg protein^{-1} x min^{-1}. However, in livers from 20 day-old fetuses we found an activity of 0.86 ± 0.76 (10) nmol x mg^{-1} x min^{-1}, i.e. significantly lower than in adult rats.

Amino Acid Concentrations in Blood from Fetuses and from Adult Rats

Amino acids were measured as previously described (3). Table 1 shows the concentration of amino acids in whole arterial blood from adult rats and also in whole blood from 20-day old fetuses. The main steps of the

Endocrine and Biochemical Development of the Fetus and Neonate
Edited by J. M. Cuezva *et al.*
Plenum Press, New York, 1990

283

Table 1. Amino acid concentration (μM) in whole blood
from fetuses and adult rats

Amino acids	Adult rats (4 month-old)	Fetuses (20 day-old)
Taurine	201 ± 37 (3)	283 ± 65 (15)*
Threonine	164 ± 35 (3)	459 ±137 (15)**
Serine	114 ± 10 (3)	204 ± 72 (15)**
Glutamic acid	102 ± 8 (3)	302 ± 97 (15)**
Glutamine	448 ±147 (3)	1978 ±782 (14)**
Proline	153 ± 47 (3)	141 ± 64 (15)
Glycine	113 ± 21 (3)	373 ± 96 (15)**
Alanine	351 ± 53 (3)	643 ±285 (15)**
Valine	126 ± 10 (3)	235 ± 78 (15)**
Cysteine	53 ± 14 (3)	8 ± 5 (14)*
Methionine	57 ± 29 (3)	26 ± 29 (11)
Isoleucine	90 ± 36 (3)	93 ± 33 (15)
Leucine	84 ± 16 (3)	167 ± 53 (15)**
Tyrosine	68 ± 15 (3)	152 ±119 (15)*
Phenylalanine	48 ± 23 (3)	86 ± 25 (15)
Ornithine	43 ± 7 (3)	118 ± 56 (15)**
Lysine	275 ± 34 (3)	663 ±213 (15)**
Histidine	202 ±124 (3)	168 ± 29 (15)
Arginine	135 ± 24 (3)	161 ± 88 (15)

Data are means ± S.D. for the number of observations
in parentheses. Values different from adults are
shown: * $p < 0.05$, **$p < 0.01$.

trans-sulphuration pathway are methionine ---> homocysteine; homocysteine + serine ---> cystathionine, and cystathionine ---> cysteine + homoserine. Cystathionine and homocysteine are undetectable in blood from both fetuses and adult rats. This is not the case in human beings in which cystathionine levels are detectable (see below).

Thus, in rats the relevant amino acids (apart from homocysteine and cystathionine) are methionine, serine and cysteine. If the low levels of cystathionase have physiological importance to determine the amino acid profile in blood, we would expect the ratio methionine + serine/ cysteine to be higher in fetal than in adult rats. Our measurements show that indeed this is the case (the ratio for adults rats is 3.22 and for fetuses it is 28.7). This is in agreement with the postulate that the low cystathionase activity in fetal liver has physiological importance, at least to determine the amino acid profiles in blood.

Cysteine and Glutathione (GSH) Synthesis from Methionine in Fetal Hepatocytes

Hepatocytes from 20-day old fetuses were isolated as described (4). Isolated hepatocytes are a good model to study the trans-sulphuration pathway. Indeed we showed that, in isolated hepatocytes from adult rats methionine actively supports GSH synthesis (5). Thus, we measured the rate of synthesis of cysteine and of GSH in cells incubated with methionine or with N-acetyl cysteine as sulphur donor.

Table 2. Glutathionine and cysteine synthesis in fetal isolated hepatocytes pre-incubated with 0.5 mM dietyl maleate

Additions	GSH	Cysteine	ATP
T0	2.0±1.5(6)	1.2±1.0(4)	6.0±3.7(5)
T60	2.1±0.9(7)	1.1±0.7(4)	6.1±1.7(7)
AA mix$_1$	3.2±0.7(5)	4.6±3.8(4)	6.0±2.7(6)
AA mix$_2$	8.4±0.6(4)**	60.0±36.4(4)*	5.3±2.6(4)

Values are μmol/g dry wt. Data are means ± S.D. for the number of observations in parentheses. Values for fetuses that are different from controls are shown by asterisks *p <0.05, **p <0.01.

ACKNOWLEDGEMENTS

This work was supported by a grant from FISS No. 88/2124 to J.V.

Since physiological levels of GSH inhibit GSH synthesis, we pretreated isolated hepatocytes with low concentrations of diethyl maleate (0.5 mM) which is known to decrease hepatic GSH content (6). We have previously shown that this concentration of dietyl maleate does not cause non-specific cell damage to hepatocytes from adult rats (7). Table 2 shows that low concentrations of diethyl maleate do not cause cell damage to fetal hepatocytes, as evident by the maintenance of normal ATP levels.

When we incubated fetal hepatocytes for 30 min with a mixture of amino acids containing glutamine (5 mM), glycine (2 mM), serine (1 mM) and methionine (0.2 mM) (see AA mix$_1$ in Table 2) or glutamine (5 Mm), glycine (2 Mm) and N-acetyl cysteine (1 mM) (see AA mix$_2$ in Table 2) as cysteine precursors, we observed that methionine supported neither cysteine nor GSH synthesis in these hepatocytes. However, N-acetyl cysteine did promote a very active cysteine and GSH synthesis in these cells (Table 2). These results are in marked contrast with the previous findings using isolated hepatocytes from adult rats which showed that methionine was very effective to maintain high rates of cysteine and GSH synthesis (5). The failure of methionine to support GSH synthesis is a consequence of the low rate of cysteine synthesis from methionine due to lack of cystathionase activity. However, since deacylation of N-acetyl cysteine does not depend on the activity of cystathionase, cysteine and glutathione synthesis from N-acetil cysteine would not be impaired in fetal liver. Indeed, fetal hepatocytes show high rates of cysteine and GSH synthesis from N-acetyl cysteine (Table 2).

The main findings reported in this paper are: i) Cystathionase activity in livers from fetal rats at day 20 of gestation is much lower than in livers from adult rats. ii) These changes are reflected in the amino acid profile in blood from fetuses when compared with that of adult rats, and iii) Isolated fetal hepatocytes with low cystathionase activity, do not show a significant rate of cysteine and GSH synthesis from methionine, but do show a high rate from N-acetyl cysteine, in marked contrast with hepatocytes from adult rats.

REFERENCES

1. J. A. Sturman, G. Gaull and N. C. R. Raiha, Absence of cystathionase in human fetal liver: Is cysteine essential?, <u>Science</u> 169:74 (1970).
2. S. H. Zlotkin and H. Anderson, The development of cystathionase activity during the first year of life, <u>Pediatr. Res.</u> 16:65 (1982)
3. J. Viña, I. R. Puertes and J. R. Viña, Effect of premature weaning on amino acid uptake by the mammary gland of the lactating rat, <u>Biochem. J.</u> 200:705 (1981).
4. M. Lorenzo, C. Roncero and M. Benito, The role of prolactine and progesterone in the regulation of lipogenesis in maternal and fetal rat liver *in vivo* in isolated hepatocytes during the last day of gestation, <u>Biochem. J.</u> 239:135 (1986).
5. J. Viña, R. Hems and H. A. Krebs, Maintenance of glutathione content of isolated hepatocytes, <u>Biochem. J.</u> 170:627 (1978).
6. E. Boylang and L. F. Chasseaud, Enzyme-catalysed conjugations of glutathione with unsaturated compounds, <u>Biochem. J.</u> 104:95 (1967).
7. G. T. Saez, F. J. Romero and J. Viña, Effect of glutathione depletion on gluconeogenesis in isolated hepatocytes, <u>Arch. Biochem. Biophys.</u> 241:75 (1985).

AMINO ACID UPTAKE BY LIVER

IN PREGNANT AND LACTATING RATS

Marçal Pastor-Anglada, Antonio Felipe,
Javier Casado and Xavier Remesar

Departamento de Bioquímica y Fisiología
Universidad de Barcelona
08028 Barcelona, Spain

INTRODUCTION

Two new sites of active amino acid metabolism develop during the reproductive cycle of mammals, the fetoplacental unit and the mammary gland. Amino acid requirements are fulfilled by different means before and after parturition; thus, pregnancy is associated with a nitrogen sparing effect which is mediated by a low ureagenic flux (1), whereas lactation is characterized by a marked hyperphagia (2). It has been claimed that a low substrate availability and/or a low liver capacity to take up amino acids should be the major point(s) responsible for the low ureagenic flux during pregnancy (1) and that the amino acid excess during lactation should be counterbalanced by a high liver uptake (2). It is well stablished now that the gestational hypoalaninemia which develops at mid-pregnancy in the rat is related to an increase in alanine turnover and alanine metabolism in liver (3,4), but the role of this organ in amino acid homeostasis before parturition and during lactation has been scarcely studied. The next lines are addressed to discuss this specific point, especially the availability and net uptake of amino acids by the liver in these physiological situations, as well as the mechanisms by which the observed changes are being mediated.

RESULTS

Table 1 shows the hepatic availability and uptake of amino acids in anaesthetized pregnant and lactating rats. Both parameters were calculated by combining hepatic blood flow measurements with the estimation of blood amino acid content in afferent and efferent vessels of the liver. Total amino acid availabilities were unchanged although some individual differences were found: decreased taurine and citrulline availabilties in both pregnant and lactating rats, low availability of aspartate, glutamate, glycine and ornitine in pregnant and of tryptophan, lysine, glutamine and arginine in lactating rats; asparagine and glycine availabilities were increased in the latter group of animals. Amino acid uptake by liver was significantly enhanced in either pregnant or lactating rats. This was mainly accounted for an increase in transport of essential and gluconeogenic amino acids. In both physiological conditions alanine was responsible for a 40% of the total amino acid uptake by liver.

Endocrine and Biochemical Development of the Fetus and Neonate
Edited by J. M. Cuezva *et al.*
Plenum Press, New York, 1990

Table 1. Amino acid availability and uptake (nmol/min. g of liver) by liver of late pregnant and mid-lactating rats

	Availability			Uptake		
	Virgin	Pregnant	Lactating	Virgin	Pregnant	Lactating
THR	805±62	632±64	736±28	26±21	135±19**	133±12**
MET	66±14	95±16	33±5	-57±31	19±12	2±5
VAL	487±33	370±59	416±26	-43±31	119±17**	71±16*
ILE	252±26	197±45	223±24	0±12	19±12	40±4*
LEU	370±33	321±76	349±31	17±12	93±26*	76±11**
PHE	178±17	165±25	150±13	14±5	47±10**	48±7**
TRP	100±14	72±10	46±7**	-14±7	0±5	-4±4
LYS	1883±89	1605±150	1246±35***	19±57	95±85	216±70*
EAA	4014±178	3225±330*	3105±75***	-100±128	535±125**	647±116***
Tau	1316±114	812±110**	683±66***	93±59	-32±6	194±62
SER	636±57	585±102	617±29	33±14	75±35	117±13***
ASP	176±21	95±20*	161±13	17±19	17±10	31±7
ASN	197±14	267±22*	286±9***	57±9	70±7	79±7
GLU	506±26	402±32*	498±15	-17±31	0±35	48±13
GLN	1928±140	2550±287	1517±64*	-59±88	-212±105	-165±95
GLY	799±33	562±45***	984±70*	119±38	255±40*	348±33***
ALA	1782±178	2180±152	2083±92	385±135	752±25*	951±77**
Cit	425±31	305±10**	266±15***	-31±17	30±17*	57±7***
TYR	131±17	97±15	112±7	24±9	40±7	44±2
HIS	273±28	242±27	211±11	17±9	32±10	51±7*
Orn	261±14	210±17*	253±11	-24±19	22±10	0±9
ARG	838±60	652±70	522±26***	69±24	40±40	37±44
TAA	13846±119	12775±1107	11870±440	31±264	1767±522**	2532±308***

Statistical comparisons vs. virgin *p<0.05, **p<0.01, ***p<0.001 (n=7-8)
Cit: Citrulline, Orn: Ornithine, Tau: Taurine, EAA: Essential Amino Acids,
TAA: Total Amino Acids.

Influx and efflux components of transport determine the net balance of any substrate across any tissue. L-alanine transport into liver parenchymal cells is the limiting step of its metabolism (5), and thus, it is likely that changes in kinetic properties of alanine carriers at the plasma membrane level shound underlie on the mechanisms contributing to the enhanced alanine uptake by liver seen *in vivo*. Thus, we studied L-alanine transport into plasma membrane vesicles from rat liver. These preparation retain the enhancement of Na^+-dependent alanine transport induced in intact cells by glucagon (6) and diabetes (7), and they can be considered as a good tool to ascertain if changes in substrate uptake are mediated by stable changes at the plasma membrane level (i.e. insertion of more carriers or intrinsic higher activity of preexisting carriers).

Table 2 shows the Na^+-dependent L-alanine transport rates into plasma membrane vesicles from livers of virgin, pregnant and lactating rats. Purification of plasma membrane vesicles was carried out by means of a Percoll gradient centrifugation method (3). Vesicles have been proved to be a useful tool to study uptake of natural substrates, such as amino acids (3,6,7) or lactate (8). Preparations were tested according to their contamination by other subcellular membranes, which was low and similar in all the experimental groups (data not shown). Mean enrichment of 5'-nucleotidase activity (a plasma membrane marker) was over eight fold. Vesicular volumes and response to hyperosmotic conditions were similar also in all the studied preparations (data not shown). Vesicles were incubated with physiological concentrations of L-alanine (from 0.1 to 2 mM). For all

Table 2. L-Alanine uptake rates into plasma membrane vesicles
from rat liver

	L-alanine concentration (mM)				
	0.1	0.25	0.5	1	2
Control	72±1	203±9	291±21	555±13	706±61
Pregnant	120±6**	264±13*	390±12*	665±28*	1067±10**
Lactating	103±3***	210±13	329±26	479±22*	746±48

Uptake rates are expressed as pmol/10s per μg of protein.
Statiscal comparisons vs. virgin *p<0.05, **p<0.01, ***p<0.001
(n=3).

the concentrations assayed transport rates were higher in those
preparations from pregnant than in those from virgin rats. This was not the
case for the preparations purified from lactating animals, which showed
little variations in transport capacity.

DISCUSSION

The present results show that late pregnancy is characterized by an
enhanced amino acid uptake by liver and thus, it seems unlikely that a low
substrate availability could explain the low ureagenic flux reported to
occur during pregnancy. Even after a 24 hour fast, when amino acid
availability decreases to a greater extent in pregnant than in virgin rats,
the net hepatic uptake was similar in both experimental groups (data not
shown), thus confirming the view that intracellular events might be fully
responsible of the impairement of the ureagenic flux in gravid mammals.
These events have been recently summarized and controversy has been raised
on whether or not they apply to both, experimental animals and women (9).

During lactation, the amino acid drainage by the mammary gland is
extremely dependent on the nutritional status of the animal (2). In our
experimental conditions the uptake by this tissue has been proved to be
enhanced, at a time when neither the hepatic amino acid availability nor
the uptake are decreased, but they are even markedly enhanced. In this case
it should be correlated with the marked hyperphagia which is greater than
that of pregnancy, but the possibility that our data reflect an activation
of the endogenous amino acid metabolism should not be discarded.

In both physiological situations it is likely that the generalized
amino acid uptake by liver reflects a need for nitrogen for anabolic goals
in the splachnic bed, hypothesis which is supported by the fact that the
liver of pregnant and lactating rats has developed hypertrophya and
hyperplasia.

Our results also show that the enhanced amino acid uptake by liver at
late-pregnancy and mid-lactation does not correlate with amino acid
availability, which is slightly modified or even decreased. Only during
lactation glycine availability increased by 30% over the control values,
but it hardly explains the enhancement of its hepatic uptake which was
three-fold higher in lactating than in virgin animals. It obviously means
that the hepatic fractional extraction of amino acids should be
significantly enhanced (13% and 20% for pregnant and lactating rats,
respectively), and thus, factors other than availability itself are
modulating liver amino acid uptake during the reproductive cycle of the
rat. In an attempt to determine the mechanisms contributing to the

enhancement of liver amino acid uptake, we purified plasma membrane vesicles retaining Na^+-dependent amino acid transport activity. Stable changes at the plasma membrane level were found in those preparations purified from pregnant rats and it is likely that a higher capacity to take up L-alanine mediates the enhancement of L-alanine net uptake observed in *in vivo*. This is not the case for mid-lactating animals. An interpretation to this apparent controversy may be easily found. One could assume that an enhanced amino acid transport capacity, mediated by the insertion of new carriers at the plasma membrane level, could be partially lost during the purification procedure. This is unlikely to occur, because no differences were found in the yield of transport capacity in those preparations from rats fed a high protein diet purified by the same method, either in the presence or in the absence of proteases inhibitors (Bourdel, Forestier, Gouhot, personal communication). A more likely explanation might be the direct hormonal effects acting *in vivo*, which are lost after plasma membrane vesicles purification. Thus, it has been shown that glucagon may exert short-term mediated actions on amino acid transport in isolated hepatocytes, by a protein synthesis-independent mechanism which implies membrane hyperpolarization (10). The altered hormonal status of both pregnant and lactating rats should contribute to modulate liver amino acid uptake *in vivo*.

A major conclusion of the present work is that, neither at late-pregnancy nor at mid-lactation, evidence for nitrogen sparing at the inter-organ relationship level exists. On the contrary, amino acid metabolism at the hepatic level is very active and at least at late-pregnancy, a higher capacity of some Na^+-dependent amino acid transport systems underlies in this metabolic adaptation.

REFERENCES

1. N. Freinkel, B. E Metzger, M. Nitzan, J. W. Hare, G. E. Shambaugh, R. T. Marshall, B. Z. Surmaczynska and T. C. Nagel, Accelerated starvation and mechanisms for the conservation of maternal nitrogen during pregnancy, Isr. J. Med. Sci. 8:426 (1972).
2. D. H. Williamson, Regulation of metabolism during lactation in the rat, Reprod. Nutr. Develop. 26:597 (1986).
3. M. Pastor-Anglada, X. Remesar and G. Bourdel, Alanine uptake by liver at midpregnancy in rats, Am. J. Physiol. 252:E408 (1987).
4. M. Pastor-Anglada, O. Champigny, P. Ferré, X. Remesar and J. Girard, Alanine turnover rate and its hepatic metabolism are increased in the midpregnant rat, Biol. Neonate 54:126 (1988).
5. A. K. Groen, H. J. Sips, R. C. Vervoorn and J. M. Tager, Intracellular compartmentation and control of alanine metabolism in rat liver parenchymal cells, Eur. J. Biochem. 122:87 (1982).
6. M. A. Schenerman and M. S. Kilberg, Maintenance of glucagon-stimulated system A amino acid transport activity in rat liver plasma membrane vesicles, Biochim. Biophys. Acta 856:428 (1986).
7. N. R. Rosenthal, R. Jacob and E. Barret, Diabetes enhances activity of alanine transport in liver plasma membrane vesicles, Am. J. Physiol. 48:E581 (1985).
8. I. Quintana, A. Felipe, X. Remesar and M. Pastor-Anglada, Carrier-mediated uptake of L-(+)-lactate in plasma membrane vesicles from rat liver, FEBS Lett. 235:224 (1988).
9. M. Pastor-Anglada, Letter to the Editor, Metabolism 38:290 (1989).
10. S. K. Moule, N. M. Bradford and J. D. McGivan, Short-term stimulation of Na^+-dependent amino acid transport by dibutyryl cyclic AMP in hepatocytes, Biochem. J. 241:737 (1987).

REGULATION OF THE UREA CYCLE DURING

LACTATION

Teresa Barber, María A. Betrán, Javier Martí,
Concepción García and Juan R. Viña

Departamento de Bioquímica y Biología Molecular
Universidad de Valencia
46010 Valencia, Spain

INTRODUCTION

We have recently shown that the uptake of amino acids by rat lactating mammary gland is 15 mmoles/day (1). This effect is achieved by an increase in food intake during lactation and probably by changes in amino acid metabolism in other tissues. Rat liver removes 75% of the total amino acids derived from a protein meal (2); therefore plays a key role in the regulation of plasma amino acid concentration.

Others have reported the absence of any nitrogen-sparing mechanism at the level of inter-organ flux of amino acids during lactation; however, the intake of protein in the non-lactating control rats was not kept equal to the lactating rats (3). It is worthwhile to emphasize that the amount of protein in the diet is an important factor in the rate of metabolism/ degradation of amino acids and in the rate of urea synthesis by the liver.

The aim of this study was to determine whether lactating rats spared amino acids from degradation as compared to virgin rats consuming the same amount of protein. We used virgin rats and rats as 14th day of lactation. All the rats were fed with a liquid diet (4) for two weeks. This diet contained 23% protein supplied by casein and was formulated to meet The National Research Council-National Academy of Sciences (USA) 1978 recommendations.

The lactating rats were fed *ad libitum*. The virgin rats were divided in two groups: the first group was fed *ad libitum*; the second received the same volume of liquid and a similar amount of carbohydrates and lipids as the first group, but the same amount of protein as the lactating group.

RESULTS

Physiological Parameters

The protein intake in the lactating group and in the virgin group fed a high-protein diet was kept equal. The virgin rats fed *ad libitum* had a protein intake significantly lower than the other two groups. The protein intake (g/day) was: 7.4±0.6 (n=7) for the lactating rats, 6.8±0.5 (n=7) for

Endocrine and Biochemical Development of the Fetus and Neonate
Edited by J. M. Cuezva *et al.*
Plenum Press, New York, 1990

the high-protein virgin rats and 3.6±0.1 (n=7) for the virgin rats. The serum urea (μmoles/ml) was significantly higher in the high-protein virgin group than in the other two groups. The values were: 10.5±0.4 (n=7) for the lactating group, 13.2±1.8 (n=5) for the high-protein virgin rats and 7.8±0.7 (n=7) for the virgin group. The liver weight was significantly higher in the lactating group than in the others two groups. The values were as follows: 16.9±1.0 (n=7) for the lactating rats, 9.7±0.8 (n=7) for the virgin rats and 9.7±0.8 (n=7) for the virgin rats fed a high-protein diet.

Activities of the Urea Cycle Enzymes in Virgin, High-protein Virgin Rats and Lactating Rats

Table 1 shows that the activities of the urea cycle enzymes were decreased in the livers of lactating mothers as compared with the other two groups, except the argininosuccinate-lyase that did not change in comparison to the virgin group and arginase activity that increased slightly. When differences in liver weight were considered, the total enzyme activities in lactating rats were higher than those in virgin rats, but similar to those in virgin rats fed the same amount of protein (high-protein virgin group). The only exception was the arginase, which was higher in the liver of lactating rats than in the other two groups (data not shown).

Acetylglutamate Content in Liver

Table 1 shows the concentration of N-acetylglutamate (NAGA) in the three experimental groups, expressed as nmol/g liver. This metabolite was lower in the lactating group than in the other two groups. The high-protein virgin rats had an impressive increase in the hepatic concentration of NAGA. When the total liver is considered the content of NAGA was slightly higher in the lactating than in the virgin group. However, the

Table 1. Urea cycle enzyme activities (U/g liver) and
Acetylglutamate content (nmol/g liver) in virgin,
lactating and high-protein virgin rats

	Virgin	Lactating	High-protein virgin
Carbamoylphosphate synthetase	322±47	227±37**	423±44*
Ornithine carbamoyltransferase	8390±1120	7142±651*	11328±2170*
Argininosuccinate synthetase	142±20	118±15*	202±29*
Argininosuccinate lyase	284±35	291±28	459±48*
Arginase	58177±5294	66661±7648*	61814±7485
Acetylglutamate	30.4±6.1	18.0±6.3*	51.0±9*

Results are means ± SD for n= 5-7. Lactating vs. virgin:
*P<0.05; **P<0.01. High-protein virgin vs. lactating:
*P<0.001.

concentration of this metabolite in the high-protein virgin rats remained significantly higher.

Urea Synthesis in Isolated Hepatocytes

The rate of urea synthesis by isolated hepatocytes was a 28% higher in the high-protein virgin rats than in the lactating group when NH_4^+, ornithine and lactate were used as precursors. This difference is statistically significant (data not shown).

Amino Acid Concentrations in the Liver

The steady-state concentration of L-taurine, L-aspartate, L-proline, L-valine, L-methionine, L-isoleucine, L-leucine, L-phenylalanine and L-ornithine in liver was lower in the lactating group than in the high-protein virgin group; the concentration of glycine was higher.

Amino Acid Uptake by Liver

The uptake values were calculated as a percentage of total supply of each amino acid. We have studied the *in vivo* uptake of the amino acids that are transported by the amino acid transport system A: L-alanine, glycine, L-proline and L-serine (5). In the lactating group, the uptake of these amino acids was lower than in the high-protein virgin group (results not shown).

Table 2. Concentration of amino acids (nmol/g liver) in virgin, lactating and high-protein virgin rats

	Virgin	Lactating	High-protein virgin
L-Aspartic acid	3006±222	2019±948*	3109±382*
L-Serine	894±236	765±363	695±298
L-Glutamic acid	2614±575	3712±1812	2700±665
L-Glutamine	3994±958	3434±515	3396±845
L-Proline	231±110	175±71	391±131*
Glycine	1863±333	1875±292	1404±336*
L-Alanine	2918±194	3008±1196	2399±50
L-Valine	240±19	177±38*	397±41**
L-Cystine	42±20	40±19	48±11
L-Methionine	39±12	33±8	57±20*
L-Isoleucine	121±21	97±13	185±9**
L-Leucine	219±25	170±24*	310±27**
L-Tyrosine	124±14	103±11	129±27
L-Phenylalanine	105±16	96±16	117±19*
L-Ornithine	150±26	120±45	204±23**
L-Lysine	762±80	915±332	1107±247
L-Histidine	640±94	636±107	630±115
L-Arginine	36±15	23±6	28±11

Results are means ± SD for n=5-6. Lactating vs. virgin: *P<0.05. High-protein virgin vs. lactating: *P<0.05; **P<0.005.

DISCUSSION

The lactating rats should be compared with virgin rats fed with equal amount of protein because the intake of nitrogen is a key factor in the degradation of amino acids and in the rate of urea synthesis. Our results show that, except for arginase, the activities of the urea cycle enzymes were equal in the lactating rats and in the virgin rats fed a high protein diet when the values were expressed in units/total liver weight. It has been reported that the high intake of protein raises the glucagon/insulin ratio (6) as well as lactation (7). Glucagon stimulates the synthesis of the urea cycle enzymes by increasing their transcription rates (8).

N-acetylglutamate, the physiological allosteric activator of carbamylphosphate synthetase I (CPS), was higher in the virgin rats fed a high-protein diet than in the lactating rats. This was also the case when the values were expressed relative to total liver weight. Variations of NAGA concentrations correlate with the ureogenic capacity of the animals. Although the virgin rats fed a high protein diet had the same CPS activity (units/total liver) as the lactating rats, the higher values of NAGA (nmoles/total liver) suggest that the rate of urea synthesis should be higher in the former group. This is in agreement with the higher values of uremia found in the virgin rats fed a high protein diet when they are compared with the lactating rats.

It is also interesting that the percentage of uptake of the individual amino acids studied was higher in the virgin rats fed a high protein diet than in the lactating rats. All these results support the hypothesis that during lactation there is a nitrogen-sparing mechanism in the liver.

ACKNOWLEDGMENT

Supported by a grant from CICYT (PB 86-0289).

REFERENCES

1.   J.R.Viña, I.R.Puertes, A.Rodríguez, G.T.Saez and J.Viña, Effect of
          Fasting on Amino Acid Metabolism by Lactating Mammary Gland: Studies
          in Women and Rats, J.Nutr. 117:533 (1987).
2.   P.Lund and D.H.Williamson, Inter-Tissue Nitrogen Fluxes, British Med.
          Bull. 41:251 (1985).
3.   J.Casado, M.Pastor-Anglada and X.Remesar, Hepatic uptake of amino
          acids at mid-lactation in de rat, Biochem.J. 245:297 (1987).
4.   C.Lee-Chuan and F.L.Cerklewski, Formulation of a Liquid Diet for
          Ethanol Studies Involving Gestation and Lactation in the Rat,
          J.Nutr. 114:634 (1984).
5.   M.S.Kilberg, System A-mediated amino acid transport: metabolic control
          at the plasma membrane, Trends Biochem.Sci. 11:183 (1986).
6.   R.H.Unger, Glucagon and the insulin: glucagon ratio in diabetes and
          other catabolic illnesses, Diabetes 20:834 (1971).
7.   A.F.Burnol, A.Leturque, P.Ferré and J.Girard, Glucose metabolism
          during lactation in the rat: quantitative and regulatory aspects,
          Am.J.Physiol. 245 (Endocrinol.Metab.8):E351 (1983).
8.   J.C.Ryall, M.A.Quantz and G.C.Shore, Rat liver and intestinal mucosa
          differ in the development pattern and hormonal regulation of
          carbamoyl-phosphate syntethase I and ornithine-carbamoyl transferase
          gene expression, Eur.J.Biochem. 156:453 (1986).

ONTOGENESIS OF COPPER AND ZINC

CARRIERS IN RAT PLASMA

Alberto Mas, Jaume Solé, Emilia Martínez-Lista
and Luis Arola

Departamento de Ingeniería Química y Bioquímica
Universidad de Barcelona-VII
43005 Tarragona, Spain

INTRODUCTION

In recent years, interest in copper and zinc as micronutrients has grown, especially in the awareness of marginal intakes. The fact that more than 200 enzymes are dependent on these metals has focused the attention of many researchers.

The major carriers of the two metals have been described more than 20 years ago (see 1 for review), yet new developments have questioned the exclusive role of previously reported proteins. Although there are variations among different species, in rodents and humans, ceruloplasmin is the main copper-containing protein in plasma with an estimate of 90% of the total copper (1,2). However, this estimate has been recently challenged by the description of a new high molecular weight copper binding protein, which has been tentatively named transcuprein (3). According to that report, transcuprein accounts for 30% of the copper previously considered as bound to ceruloplasmin. The 10% of the copper left is considered to be loosely bound to albumin in equilibrium with amino acids and other low molecular weight compounds. Due to the difficulty of removal of copper from ceruloplasmin (1), the cooper bound to albumin and low molecular weight components have been considered the "exchangeable pool", which would be increased by the amount of copper bound to transcuprein. Owing to the strong binding and to the oxidase and ferroxidase activities of ceruloplasmin, the function of this protein as a copper carrier can be questioned. Furthermore, the low levels of ceruloplasmin in diseases such as Wilson's disease without apparent copper deficiency is an additional argument against a role for ceruloplasmin as the major copper carrier.

In regard to the transport of zinc, most of the zinc in blood is associated with erythrocytes (1), but plasma zinc is considered to be the metabolically active compartment, although some exchange may take place between erythrocytes and plasma. Plasmatic zinc transport is also species-related and is considered to be mostly bound to albumin (approximately 65% (4)) in equilibrium with amino acids and other low molecular weight components. The portion left is considered to be bound to $\alpha2$-macroglobulin (4) with very high association constant (1), which thus limits the possibility as a carrier. The function of this strong binding zinc to $\alpha2$-macroglobulin is still unknown.

Endocrine and Biochemical Development of the Fetus and Neonate
Edited by J. M. Cuezva *et al.*
Plenum Press, New York, 1990

The presence of these two metals is essential during the growth
period, mainly due to the high activity of the numerous enzymes containing
copper or zinc. The postnatal period exemplifies most of the deleterious
effects that genetic disorders or marginal intake can produce during this
particular period. Some of the above mentioned carriers are low (5), and to
the date, complete study of the transport mechanisms and distribution of
these metals have been difficult. Our initial approach to the problem
consisted in using gel filtration (Sephadex G-150) to fractionate the
plasma and subsequent determination of copper and zinc levels in the eluent
by atomic absorption spectrophotometry. In the present work, we report
carefully investigate changes in copper and zinc plasma distribution in the
rat during the postnatal period.

RESULTS AND DISCUSSION

After drawing blood from adults rats and pups of different ages,
erythrocytes were sedimented by centrifugation, and plasma was immediately
applied to the columns to avoid re-distribution of copper and zinc. The
elution profile of plasma copper is shown in Figure 1. The fractionation of
plasma through Sephadex G-150 allows a good separation between proteins
higher than 200,000 (which elute in the void volume, peak 1), Ceruloplasmin
(detected by the copper content and the oxidase activity, peak 2) albumin,
along with other proteins such as transferrin, histidine rich glycoprotein,
α-fetoprotein (peak 3) and low molecular weight components (less than 5,000
daltons, which elute in the bed volume of the column, peak 4). The results
illustrated in Figure 1 represent the changes at different periods of
development and it can be easily seen that the continuous increase in the
amounts of copper bound to ceruloplasmin (from 0.6 at birth to 2 μkat/l in
the adults) is in accord with the data reported by Terao and Owen (5). It
is apparent from Fig 1 that there is a change in the elution pattern from
the fetal life, where three peaks are evident (with similar amounts of
copper in them), compared to the development in 20 day-old pups, where most
of the copper is associated with high molecular weight proteins and
ceruloplasmin. The amount bound to albumin is negligible at mid
development. In adult, the main copper peak is found to be associated to

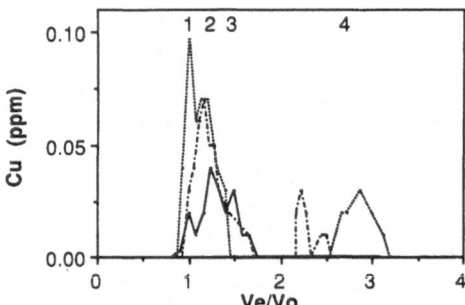

Fig. 1. Sephadex G-150 elution profiles of copper
in rat plasma at different ages: Fetuses 21
d(——), 20 day old (···) and adult rat
(·—·-). Copper levels are expressed in ppm
(μg/l). Numbers indicate peaks of known
standards: 1, void volume (as for blue
dextran): 2, Ceruloplasmin (as for oxidase
activity); 3, Bovine serum albumin (BSA);
4, bed volume.

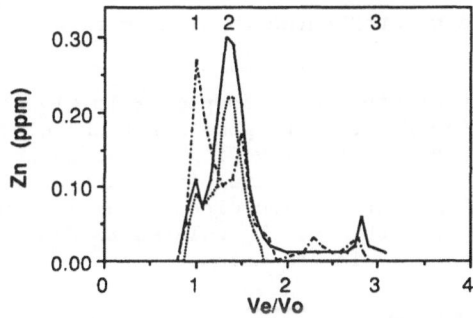

Fig. 2. Sephadex G-150 elution profiles of zinc
in rat plasma at different ages: Fetuses
21 d(——), 10 day old(· · ·) and adult rat
(·—·-). Zinc levels expressed in ppm.
Standards: 1, void volume; 2, BSA; 3, bed
volume.

high molecular weight components and albumin peaks appear as only shoulders
of the ceruloplasmin peak.

It must be emphasized that the albumin peak overlaps with other
proteins such as α-fetoprotein and histidine-rich glycoprotein (6), which
could also bind copper. It is known that α-fetoprotein is present in the
fetal blood in appreciable quantities and could account for a larger amount
of copper observed in peak 3. Interestingly enough, the changes in the
amounts of copper associated with high molecular weight proteins could be
related to the proposed role of transcuprein, which may be a protein
functioning as an effective carrier of copper in situations when this
element is especially needed (such as in the case the postnatal period).

The presence of low molecular weight copper binding components must
also be considered in terms of copper transport and availability. To date,
involvement of copper-histidine complexes or small peptides have been
proposed to act as carriers for copper transport. In the present
fractionation system, although the levels were very low, two different
peaks were observed, which are suggested to be age-related. The two peaks
eluted in the low molecular weight region and are very similar to those
peaks observed with [67]Cu in cord or adult human sera after *in vitro*
equilibration with this isotope (7). The role of these two different
components may also be related to the tissue specific distribution of
copper within the body. Those found in the human cord blood or in the
developing rat are probably more effective in terms of transport than the
ones present in adults.

In the case of zinc distribution the major changes are shown in the
late development of the rat. The profile is similar at the end of the fetal
life to the adult developing animal. There are no significant changes
observed as a results of the zinc bound to albumin or related proteins (in
this particular case a more significant role for histidine-rich
glycoprotein as a zinc carrier protein has been proposed (6); previous
reports have shown than transferrin may also bind zinc (1)), hence changing
the pattern towards a larger amount of zinc bound to high molecular weight
proteins (presumably α2-macroglobulin). In the present study, in which most
of the zinc in adult rat plasma in bound to high molecular weight protein,
the zinc distribution is in contrast to previous reports for other species
(1,4). The amount of zinc associated with peak 2 in Figure 2 in adult rat
plasma was slightly but significantly shifted from the main absorbance peak

at 280 nm, which could be attributed to binding to other components rather than albumin.

As for the amount of zinc bound to peak 2 in Figure 2 during rat development, albumin (or for this purpose α-fetoprotein) has a low association constant for zinc (1) favoring higher amounts of zinc to be associated during the development state. In Figure 2 it becomes more evident that the total amount of zinc circulating in plasma decreases during the development and this may be related to depletion of the main storage of zinc (and copper) in the animal.

It can be seen in Figure 2 that the low molecular weight zinc binding components are observed in the adults and also present in the late fetus, which follows a similar pattern as that described in the equilibration of ^{65}Zn in cord and adult human sera (7). The above statements for the low molecular weight components which bind copper can be applied for zinc as well.

As for the origin of the plasma carriers, the metal incorporation can follow two different patterns. The first involves the binding to circulating ligands (i.e. albumin, α-fetoprotein, possibly α2-macroglobulin and most likely the low molecular weigh components), their relative amounts present and their association constants. All the proteins following this pattern are known to exist in the late fetus and the neonate, in proportions similar to the adults, except in the case of α-fetoprotein, which has been found to be in amounts much higher in the fetuses or neonates (8). However, the zinc binding to α2-macroglobulin is not completely understood and the possibility of different binding sites or mechanistic properties for zinc incorporation has to be considered.

The second pattern involves the synthesis *de novo*. This synthesis *de novo* is known to take place in the liver for ceruloplasmin. Ceruloplasmin synthesis is active in the late fetus, probably due to the induction by maternal estrogens (8), but fails dramatically in the newborn, related to the lack of induction. However, at the end of the fetal life, both copper and zinc accumulate in the form of liver metallothionein (9) and the presence of copper in liver may be an inducer of ceruloplasmin synthesis. The levels of these metals sharply decrease during the neonatal period and this might be the main source of these two metals (and also cysteine) for the development (10). The large amounts of metallothionein present in the neonatal liver could reduce the levels of available copper for cerulo-plasmin induction and only when the levels of metallothionein drop, the synthesis of ceruloplasmin reach the adult expression.

Further studies involving different species and the mechanisms of induction of the above discussed plasma carriers are currently underway in our laboratory.

REFERENCES.

1. R.J.Cousins, Absorption, transport, and hepatic metabolism of copper and zinc: Special reference to metallothionein and ceruloplasmin, Physiol.Rev 65:238 (1985).
2. N.Marceau, N.Aspin and A.Sass-Kortsak, Absortion of copper 64 from grastointestinal tract of the rat, Am.J.Physiol. 218:373 (1970).
3. K.C.Weiss and M.C.Linder, Copper transport in rats involving a new plasma protein, Am.J.Physiol. 249 (Endocrinol. Metab, 12):E77 (1985).
4. E.L.Giroux, Determination of zinc distribution between albumin and alfa-2-macroglobulin in human serum, Biochem.Med 12:258 (1975).

5. T. Terao and C. A. Owen, Copper metabolism in pregnant and post partum rat and pups, _Am. J. Physiol._ 232:E172 (1977)

6. W. T. Morgan, Iteractions of the histidine-rich glycoprotein of serum with metals, _Biochemistry_ 20:1054 (1981)

7. S. J. Lau and B. Sarkar, Comparative studies of manganese (II)-, nickel (II)-, zinc (II)-, copper (II)-, cadmium (II)-, and iron (III)- binding components in human cord and adult sera, _Can. J. Biochem. Cell. Biol._ 62(6):449 (1984).

8. D. Gitlin and J. D. Gitlin, Fetal and neonatal development of human plasma proteins, _in_: "The Plasma Proteins", Vol. 2. F. W. Putnam (ed.), Academic Press, New York. pp. 264, (1975).

9. H. K. Ohtake, K. Hasegawa and M. Koga, Zinc binding protein in the liver of neonatal, normal and partially hepatectomized rats, _Biochem. J._ 174:999 (1978).

10. S. H. Zlotkin and G. M. Cherian, Hepatic metallothionein as a source of zinc and cysteine during the first year of life, _Pediatr. Res._ 24(3):326 (1988).

Aláez, C.
Instituto de Bioquímica
Centro Mixto C.S.I.C.-U.C.M
Facultad de Farmacia
Universidad Complutense
28040 Madrid
Spain

Alcázar, A.
Servicio de Bioquímica
Departamento de Investigación
Hospital Ramón y Cajal
28034 Madrid
Spain

Alconada, A.
Departamento de Biología Molecular
Centro de Biología Molecular
U.A.M.-C.S.I.C.
Universidad Autónoma de Madrid
28049 Madrid
Spain

Alvárez, C.
Instituto de Bioquímica
Centro Mixto C.S.I.C.-U.C.M.
Facultad de Farmacia
Universidad Complutense
28040 Madrid
Spain

Andrés, A.
Departamento de Biología Molecular
Centro de Biología Molecular
Universidad Autónoma de Madrid
28049 Madrid
Spain

Aprille, J.R.[*]
Mitochondrial Physiology Unit
Department of Biology
Tufts University
Medford, MA 02155
U.S.A.

Arinze, I.J.[*]
Department of Biochemistry
Meharry Medical College
Nashville, TN 37208
U.S.A.

Arizmendi. C.[*]
Departamento de Bioquímica y
    Biología Molecular
Facultad de Farmacia
Universidad de Salamanca
Salamanca
Spain

Arola, L.
Departamento de Ingeniería Química
    y Bioquímica
Universidad de Barcelona-VII
43005 Tarragona
Spain

Azuara, M.
Servicio de Bioquímica
Departamento de Investigación
Hospital Ramón y Cajal
28034 Madrid
Spain

Barber, T.
Departamento de Bioquímica y
    Biología Molecular
Universidad de Valencia
Avda. Blasco Ibáñez, 13
46010 Valencia
Spain

Batenburg, J.J.
Laboratory of Veterinary
    Biochemistry and Small Animal
    Clinic
University of Utrecht
Utrecht
The Netherlands

Benito, M.
Departamento de Bioquímica y
    Biología Molecular
Centro Mixto C.S.I.C./U.C.M.
Facultad de Farmacia
Universidad Complutense
28040 Madrid
Spain

Betrán, M.A.
Departamento de Bioquímica y
    Biología Molecular
Universidad de Valencia
Avda. Blasco Ibáñez, 13
46010 Valencia
Spain

Bolaños, J.P.
Departamento de Bioquímica y
    Biología Molecular
Facultad de Farmacia
Universidad de Salamanca
Salamanca
Spain

Bonet, B.
Departamento de Bioquímica e
    Investigación
Universidad de Alcalá de Henares y
    Hospital Ramón y Cajal
Ctra. Colmenar Km 9
28034 Madrid
Spain

Burgaya, F.
Departamento de Bioquímica y
    Fisiología
Universidad de Barcelona
Avda. Diagonal, 645
08028 Barcelona
Spain

Camps, M.
Unidad de Bioquímica y Biología
    Molecular
Departamento de Bioquímica y
    Fisiología
Facultad de Biología
Universidad de Barcelona
08028 Barcelona
Spain

Carothers, D.J.
Departments of Biochemistry and
    Nutrition and Pew Center for
    Molecular Nutrition
Case Western Reserve University
    School of Medicine
Cleveland, 44106 Ohio
U.S.A.

Carrascosa, J.M.[*]
Departamento de Biología Molecular
Centro de Biología Molecular-
    C.S.I.C.
Universidad Autónoma de Madrid
28049 Madrid
Spain

Casado, J.
Unidad de Bioquímica y Biología
    Molecular B
Departamento de Bioquímica y
    Fisiología
Universidad de Barcelona
08028 Barcelona
Spain

Casteilla, L.
Laboratoire de Production de
    Viande, INRA
Theix, and Centre de Recherches
    sur la Nutricion, CNRS
Meudon-Bellevue
France

Chamorro, M.
Departamento de Biología Molecular
Centro de Biología Molecular
(U.A.M.-C.S.I.C.)
Universidad Autónoma de Madrid
28049 Madrid
Spain

Clercx, C.
Laboratory of Veterinary
    Biochemistry and Small Animal
    Clinic
University of Utrecht
Utrecht
The Netherlands

Christine Coupé
Centre de Recherches sur la
    Nutrition, C.N.R.S.
92190 Meudon-Belleuve
France

Cuezva, J.M.[*]
Departamento de Biología Molecular
Centro de Biología Molecular
(U.A.M.-C.S.I.C.)
Universidad Autónoma de Madrid
28049 Madrid
Spain

Escobar del Rey, F.
Instituto de Investigaciones
    Biomédicas, C.S.I.C.
Facultad de Medicina
28029 Madrid
Spain

Escrivá, F. [*]
Instituto de Bioquímica
Centro Mixto C.S.I.C.-U.C.M.
Facultad de Farmacia
Universidad Complutense
28040 Madrid
Spain

Estrela, J.M.
Departamento de Fisiología
Facultad de Medicina
Avda. Blasco Ibáñez, 17
46010 Valencia
Spain

Felipe, A.
Unidad de Bioquímica y Biología
    Molecular B
Departamento de Bioquímica y
    Fisiología
Universidad de Barcelona
08028 Barcelona
Spain

Fernández, E.
Departamento de Bioquímica y
    Biología Molecular
Facultad de Farmacia
Universidad de Salamanca
Salamanca
Spain

Fernández, M.
Departamento de Bioquímica y
    Biología Molecular
Centro Mixto C.S.I.C./U.C.M.
Facultad de Farmacia
Universidad Complutense
28040 Madrid
Spain

Ferré, P.
Centre de Recherches sur la
    Nutrition, C.N.R.S.
92190 Meudon-Belleuve
France

García, C.
Departamento de Bioquímica y
    Biología Molecular
Universidad de Valencia
Avda. Blasco Ibáñez, 13
46010 Valencia
Spain

García-Higuera, I.
Departamento de Biología Molecular
Centro de Biología Molecular
Universidad Autonóma de Madrid
28049 Madrid
Spain

Gasco, E.
Departamento de Fisiología
Facultad de Medicina
Avda. Blasco Ibáñez, 17
46010 Valencia
Spain

Geelen, M.J.H.
Laboratory of Veterinary
    Biochemistry and Small Animal
    Clinic
University of Utrecht
Utrecht
The Netherlands

Giralt, M. [*]
Unidad de Bioquímica y Biología
    Molecular B
Departamento de Bioquímica y
    Fisiología
Universidad de Barcelona
08028 Barcelona
Spain

Girard, J.
Centre de Recherches sur la
    Nutrition, C.N.R.S.
92190 Meudon-Belleuve
France

Gómez-Coronado, D.
Departamento de Bioquímica e
    Investigación
Universidad de Alcalá de Henares y
    Hospital Ramón y Cajal
Ctra. Colmenar Km 9
28034 Madrid
Spain

Goya, L. [*]
Instituto de Bioquímica
Centro Mixto C.S.I.C.-U.C.M.
Facultad de Farmacia
Universidad Complutense
28040 Madrid
Spain

Grande, C.
Hospitales "La Paz"
y "12 de Octubre"
28034 Madrid
Spain

Guerri, C. [*]
Instituto de Investigaciones
    Citológicas
(Centro Asociado al CSIC)
46010 Valencia
Spain

Haagsman, H.P.
Laboratory of Veterinary
    Biochemistry and Small Animal
    Clinic
University of Utrecht
Utrecht
The Netherlands

Haney, P.M.
Deparments of Biochemistry and
    Nutrition
Pew Center for Molecular Nutrition
Case Western Reserve University
    School of Medicine
Cleveland, Ohio 44106
U.S.A.

Hernández-García, J.M.
Hospitales "La Paz"
y "12 de Octubre"
28034 Madrid
Spain

Herrera, E.[*]
Departamento de Bioquímica e
    Investigación
Universidad de Alcalá de Henares y
    Hospital Ramón y Cajal
Ctra. Colmenar Km 9
28034 Madrid
Spain

Ho, L.
Departments of Biochemistry and
    Nutrition
Pew Center for Molecular Nutrition
Case Western Reserve University
    School of Medicine
Cleveland, 44106 Ohio
U.S.A.

Hou, Q.-C.
Department of Pharmacology
Duke University Medical Center
Durham, North Carolina 27710
U.S.A.

Iglesias, R.
Unidad de Bioquímica y Biología
    Molecular B
Departamento de Bioquímica y
    Fisiología
Universidad de Barcelona
08028 Barcelona
Spain

Issad, T.
Centre de Recherches sur la
    Nutrition, C.N.R.S.
92190 Meudon-Belleuve
France

Izquierdo, J.M.
Departamento de Biología Molecular
Centro de Biología Molecular
(U.A.M.-C.S.I.C.)
Universidad Autónoma de Madrid
28049 Madrid
Spain

Jiménez, E.[*]
Departamento de Bioquímica y
    Biología Molecular
Universidad de Málaga
Málaga
Spain

Jones, C.T.[*]
Laboratory of Cellular and
    Developmental Physiology and
    Nuffield Department of Clinical
    Medicine
University of Oxford
John Radcliffe Hospital
Oxford OX3 9DU
U.K.

Jost, A.[*]
Collège de France
75231 PARIS Cedex 05
France

Kawai, Y.
Department of Biochemistry
Meharry Medical College
Nashville, TN 37208
U.S.A.

Kudlacz, E.M.
Department of Pharmacology
Duke University Medical Center
Durham, North Carolina 27710
U.S.A.

Lasunción, M.A.
Departamento de Bioquímica e
    Investigación
Universidad de Alcalá de Henares y
    Hospital Ramón y Cajal
Ctra. Colmenar Km 9
28034 Madrid
Spain

López-Fando, J.
Servicio de Bioquímica
Departamento de Investigación
Hospital Ramón y Cajal
28034 Madrid
Spain

Luis, A.M.
Departamento de Biología Molecular
Centro de Biología Molecular
(U.A.M.-C.S.I.C.)
Universidad Autónoma de Madrid
28049 Madrid
Spain

Llobera, M.*
Departamento de Bioquímica y
    Fisiología
Universidad de Barcelona
Avda. Diagonal, 645
08028 Barcelona
Spain

Solange Magre
Collège de France
75231 PARIS Cedex 05
France

Mampel, T.
Unidad de Bioquímica y Biología
    Molecular B
Departamento de Bioquímica y
    Fisiología
Universidad de Barcelona
08028 Barcelona
Spain

Martí, J.
Departamento de Bioquímica y
    Biología Molecular
Universidad de Valencia
Avda. Blasco Ibáñez, 13
46010 Valencia
Spain

Martín, A.*
Departamento de Bioquímica e
    Investigación
Universidad de Alcalá de Henares y
    Hospital Ramón y Cajal
Ctra. Colmenar Km 9
28034 Madrid
Spain

Martin, I.
Unidad de Bioquímica y Biología
    Molecular B
Departamento de Bioquímica y
    Fisiología
Universidad de Barcelona
08028 Barcelona
Spain

Martín, M.E.
Servicio de Bioquímica
Departamento de Investigación
Hospital Ramón y Cajal
28034 Madrid
Spain

Martínez, C.
Departamento de Biología Molecular
Centro de Biología Molecular
C.S.I.C.
Universidad Autónoma de Madrid
28049 Madrid
Spain

Martínez, I.
Hospitales "La Paz"
y "12 de Octubre"
28034 Madrid
Spain

Martínez-Lista, E.
Departamento de Ingeniería Química
    y Bioquímica
Universidad de Barcelona-VII
Plza. Imperial Terraco, 1
43005 Tarragona
Spain

Mas, A.*
Departamento de Ingeniería Química
    y Bioquímica
Universidad de Barcelona-VII
Plza. Imperial Terraco, 1
43005 Tarragona
Spain

Mayor jr., F.*
Departamento de Biología Molecular
Centro de Biología Molecular
Universidad Autonóma de Madrid
28049 Madrid
Spain

Medina, J.M.*
Departamento de Bioquímica y
    Biología Molecular
Facultad de Farmacia
Universidad de Salamanca
Salamanca
Spain

Montiel, M.
Departamento de Bioquímica y
    Biología Molecular
Universidad de Málaga
Málaga
Spain

Morreale de Escobar, G. [*]
Instituto de Investigaciones
    Biomédicas, C.S.I.C.
Facultad de Medicina
28029 Madrid
Spain

Obregón, M.J.
Instituto de Investigaciones
    Biomédicas, C.S.I.C.
Facultad de Medicina
28029 Madrid
Spain

Palacín, M.
Unidad de Bioquímica y Biología
    Molecular
Departamento de Bioquímica y
    Fisiología
Facultad de Biología
Universidad de Barcelona
08028 Barcelona
Spain

Pallardó, F.V.
Departamento de Fisiología
Facultad de Medicina
Avda. Blasco Ibáñez, 17
46010 Valencia
Spain

Pascual-Leone, A.M. [*]
Instituto de Bioquímica
Centro Mixto C.S.I.C.-U.C.M.
Facultad de Farmacia
Universidad Complutense
28040 Madrid
Spain

Pastor-Anglada, M. [*]
Unidad de Bioquímica y Biología
    Molecular B
Departamento de Bioquímica y
    Fisiología
Universidad de Barcelona
08028 Barcelona
Spain

Patel, M.S. [*]
Departments of Biochemistry and
    Nutrition
Pew Center for Molecular Nutrition
Case Western Reserve University
    School of Medicine
Cleveland, 44106 Ohio
U.S.A.

Peinado, J.
Departamento de Bioquímica y
    Fisiología
Universidad de Barcelona
Avda. Diagonal, 645
08028 Barcelona
Spain

Peñas, M.
Departamento de Bioquímica y
    Biología Molecular
Centro Mixto C.S.I.C./U.C.M.
Facultad de Farmacia
Universidad Complutense
28040 Madrid
Spain

Pérez, J.
Hospitales "La Paz"
y "12 de Octubre"
28034 Madrid
Spain

Porras, A. [*]
Departamento de Bioquímica y
    Biología Molecular
Centro Mixto C.S.I.C./U.C.M.
Facultad de Farmacia
Universidad Complutense
28040 Madrid
Spain

Portha, B. [*]
Laboratoire de Physiologie du
    Développement
Université Paris 7, Tour 33
75252 Paris Cedex 05
France

Ramírez, I.
Departamento de Bioquímica y
    Fisiología
Universidad de Barcelona
Avda. Diagonal, 645
08028 Barcelona
Spain

Remesar, X. [*]
Unidad de Bioquímica y Biología
    Molecular B
Departamento de Bioquímica y
    Fisiología
Universidad de Barcelona
08028 Barcelona
Spain

Renau-Piquera, J.
Instituto de Investigaciones
    Citológicas
Centro Asociado al CSIC
C/ Amadeo de Saboya, 4
46010 Valencia
Spain

Rivero, F.
Instituto de Bioquímica
Centro Mixto C.S.I.C.-U.C.M.
Facultad de Farmacia
Universidad Complutense
28040 Madrid
Spain

Robert, M.
Departamento de Bioquímica y
    Fisiología
Universidad de Barcelona
Avda. Diagonal, 645
08028 Barcelona
Spain

Rodrigo, F.
Departamento de Fisiología
Facultad de Medicina
Avda. Blasco Ibáñez, 17
46010 Valencia
Spain

Rodríguez, C.
Instituto de Bioquímica
Centro Mixto C.S.I.C.-U.C.M.
Facultad de Farmacia
Universidad Complutense
28040 Madrid
Spain

Ruiz, M.
Departamento de Bioquímica y
    Biología Molecular
Universidad de Málaga
Málaga
Spain

Ruiz, P.
Departamento de Biología Molecular
Centro de Biología Molecular,
    C.S.I.C.
Universidad Autónoma de Madrid
28049 Madrid
Spain

Salinas, M. [*]
Servicio de Bioquímica
Departamento de Investigación
Hospital Ramón y Cajal
28034 Madrid
Spain

Sancho-Tello, M.
Instituto de Investigaciones
    Citológicas
(Centro Asociado al CSIC)
C/ Amadeo de Saboya, 4
46010 Valencia
Spain

Sastre, J.
Departamento de Fisiología
Facultad de Medicina
Avda. Blasco Ibáñez, 17
46010 Valencia
Spain

Seidler, F.J.
Department of Pharmacology
Duke University Medical Center
Durham, North Carolina 27710
U.S.A.

Slotkin, T.A. [*]
Department of Pharmacology
Duke University Medical Center
Durham, North Carolina 27710
U.S.A.

Solé, J.
Departamento de Ingeniería Química
    y Bioquímica
Universidad de Barcelona-VII
Plza. Imperial Terraco, 1
43005 Tarragona
Spain

Testar, X.
Unidad de Bioquímica y Biología
    Molecular
Departamento de Bioquímica y
    Fisiología
Facultad de Biología
Universidad de Barcelona
08028 Barcelona
Spain

Thekkumkara, T.J.
Deparments of Biochemistry and
    Nutrition
Pew Center for Molecular Nutrition
Case Western Reserve University
    School of Medicine
Cleveland, Ohio 44106
U.S.A.

Valcarce, C. [*]
Departamento de Biología Molecular
Centro de Biología Molecular
(U.A.M.-C.S.I.C.)
Universidad Autónoma de Madrid
28049 Madrid
Spain

van Golde, L.M.G. [*]
Laboratory of Veterinary
    Biochemistry and Small Animal
    Clinic
University of Utrech
Utrech
The Netherlands

Vicario, C.
Departamento de Bioquímica y
    Biología Molecular
Facultad de Farmacia
Universidad de Salamanca
Salamanca
Spain

Vilanova, J.
Departamento de Bioquímica y
    Fisiología
Universidad de Barcelona
Avda. Diagonal, 645
08028 Barcelona
Spain

Vilaró, S.
Departamento de Bioquímica y
    Fisiología
Universidad de Barcelona
Avda. Diagonal, 645
08028 Barcelona
Spain

Villarroya, F. [*]
Unidad de Bioquímica y Biología
    Molecular B
Departamento de Bioquímica y
    Fisiología
Universidad de Barcelona
08028 Barcelona
Spain

Viña, J. [*]
Departamento de Fisiología
Facultad de Medicina
Avda. Blasco Ibáñez, 17
46010 Valencia
Spain

Viña, J.R. [*]
Departamento de Bioquímica y
    Biología Molecular
Universidad de Valencia
Avda. Blasco Ibáñez, 13
46010 Valencia
Spain

Viñas, O.
Unidad de Bioquímica y Biología
    Molecular B
Departamento de Bioquímica y
    Fisiología
Universidad de Barcelona
08028 Barcelona
Spain

Zapata, A. [*]
Hospital "La Paz"
28034 Madrid
Spain

Zaragoza, R.
Instituto de Investigaciones
    Citológicas
Centro Asociado al CSIC
C/ Amadeo de Saboya, 4
46010 Valencia
Spain

Zorzano, A. [*]
Unidad de Bioquímica y Biología
    Molecular B
Departamento de Bioquímica y
    Fisiología
Facultad de Biología
Universidad de Barcelona
08028 Barcelona
Spain

312